AF342490

266
Topics in Current Chemistry

Topics in Current Chemistry

Recently Published and Forthcoming Volumes

Microwave Methods in Organic Synthesis

Volume Editors: Mats Larhed · Kristofer Olofsson

With contributions by

P. Appukkuttan · K. Ersmark · E. Van der Eycken · C. O. Kappe
J. M. Kremsner · M. Larhed · I. Mutule · P. Nilsson · K. Olofsson
A. Stadler · C. R. Strauss · E. Suna · R. S. Varma · J. Wannberg
W. Zhang

 Springer

The series *Topics in Current Chemistry* presents critical reviews of the present and future trends in modern chemical research. The scope of coverage includes all areas of chemical science including the interfaces with related disciplines such as biology, medicine and materials science. The goal of each thematic volume is to give the nonspecialist reader, whether at the university or in industry, a comprehensive overview of an area where new insights are emerging that are of interest to a larger scientific audience.

As a rule, contributions are specially commissioned. The editors and publishers will, however, always be pleased to receive suggestions and supplementary information. Papers are accepted for *Topics in Current Chemistry* in English.

In references *Topics in Current Chemistry* is abbreviated Top Curr Chem and is cited as a journal.

Visit the TCC content at springerlink.com

ISSN 0340-1022
ISBN-10 3-540-36757-8 Springer Berlin Heidelberg New York
ISBN-13 978-3-540-36757-4 Springer Berlin Heidelberg New York
DOI 10.1007/11535799

Springer is a part of Springer Science+Business Media

springer.com

© Springer-Verlag Berlin Heidelberg 2006

Cover design: WMXDesign GmbH, Heidelberg
Typesetting and Production: LE-TEX Jelonek, Schmidt & Vöckler GbR, Leipzig

Printed on acid-free paper   02/3100 YL – 5 4 3 2 1 0

Preface

We are delighted to present this volume with contributions from some of the most renowned and experienced microwave chemists today.

The delivery and introduction of energy has been closely connected with the discovery and investigation of new chemistry. It is with pleasure that we have seen an increased use of microwave irradiation over the years and we hope that this volume will reflect the current interest in expanding the scope of microwave applications in both organic and medicinal chemistry. One important explanation behind the growth of microwave-enhanced chemistry has been the introduction of dedicated microwave reactors.

As a result of this development we are proud to present a diverse set of reviews. Apart from chapters spanning the scope that is usually associated with microwave methods, such as heterocyclic chemistry – an intriguing, but frustratingly diverse field that is excellently presented in one of the reviews – and transition metal-catalyzed reactions, we also present a review on microwave-assisted natural product chemistry, a topic that is of high interest and neither often nor widely covered. A contribution on microwave-accelerated synthesis of protease inhibitors underlines the usefulness of microwave heating in medicinal chemistry and a review of fluorous microwave chemistry highlights the importance of the combination of high-speed reactions and quick separations. Two separate chapters on scaled-up microwave reactions and green and sustainable chemistry give an overview of aspects of microwave chemistry that might be of great use in both industrial and small-scale applications.

We would like to take this opportunity to express our sincere gratitude to the contributors of this volume for their valuable time and efforts. We believe that the presented work will further promote the use of controlled microwave heating in both academia and industry.

Uppsala, September 2006   Mats Larhed and Kristofer Olofsson

Contents

Contents *Topics Heterocyclic Chemistry Vol. 1*

Microwave-Assisted Synthesis of Heterocycles

Volume Editors: Erik Van der Eycken · Oliver Kappe
ISBN: 3-540-30983-7

Top Curr Chem (2006) 266: 1–47
DOI 10.1007/128_051
© Springer-Verlag Berlin Heidelberg 2006
Published online: 8 July 2006

Microwave-Assisted Natural Product Chemistry

Prasad Appukkuttan[1] · Erik Van der Eycken[2] (✉)

[1]Department of Medicinal Chemistry, Organic Pharmaceutical Chemistry, BMC,
Uppsala University, P.O. Box. 574, SE 751 23 Uppsala, Sweden

[2]Department of Chemistry, University of Leuven, Celestijnenlaan 200F, B-3001 Leuven,
Belgium
erik.vandereycken@chem.kuleuven.be

Abstract An overview of the application of microwave irradiation in natural product synthesis is presented, focusing on the developments in the last 5–10 years. This contribution covers the literature concerning the total synthesis of natural products and their analogues, the synthesis of alkaloids and the construction of building blocks of interest for natural product synthesis. As microwave irradiation appeared on the scene only recently, we are at an early stage of its application in natural product chemistry, even though some nice examples have been communicated recently. The application of dedicated microwave instruments as well as domestic microwave ovens is discussed, giving emphasis to the microwave-enhanced transformations.

Keywords Alkaloids · Microwave irradiation · Natural products · Steroids · Total synthesis

Abbreviations

Boc	*tert*-butoxycarbonyl
CNS	central nervous system
m-CPBA	*m*-chloroperbenzoic acid
CTH	catalytic transfer hydrogenation
o-DCB	*o*-dichlorobenzene

DDQ   2,3-dichloro-5,6-dicyano-1,4-benzoquinone
DMF   *N,N*-dimethylformamide
DMSO   dimethyl sulfoxide
HIV   human immunodeficiency virus
IEDDA   inverse electron-demand Diels–Alder
PEA   phenethylamine
PMB   *p*-methoxybenzyl
PTSA   *p*-toluene sulfonic acid
RCAM   ring-closing alkyne metathesis
RCM   ring-closing metathesis
TBAF   *N,N,N,N*-tetrabutylammonium fluoride
TBDMS   *tert*-butyl dimethyl silyl
THF   tetrahydrofuran

1
Introduction

For many years the (total) synthesis of natural products has inspired many chemists to develop synthetic approaches that many times are conceptually real beauties. Nature is an inexhaustible source of diverse chemical compounds and many of them possess interesting biological activities. That is exactly what makes natural products so important for mankind: they represent a nearly unlimited reservoir of precious starting compounds for the development of new medicines by combinatorial chemistry, high throughput screening and medicinal chemistry. Taxol and penicillin are two of the striking examples of what nature is offering us.

In contrast with the enormous effort directed at natural product synthesis, the application of microwave irradiation in this field is rather scarcely investigated, and a systematic use of this technique for most of the conversions in a (total) synthesis sequence is still a challenging target. We have just reached dawn in the development of microwave-assisted natural product synthesis, although unquestionably, some beautiful examples have already been described.

We attempted to give an overview of the last 5–10 years and via different searches we tried to retrieve the relevant literature, which was not an obvious task. Nevertheless, we hope that we have covered as much as possible of the published research, although obviously some work will be missing. For the sake of clarity we divided the collected literature into five subsections: (1) Total synthesis of various classes of natural products; (2) Synthesis of alkaloids, as this is a clear and well-defined subcategory of natural products; (3) Modifications of natural products (i.e. synthesis starting from natural products); (4) Synthesis of unnatural analogues of natural products (i.e. synthesis not starting from natural products) and finally; (5) Synthesis of interesting building blocks for natural product synthesis. We decided not to incorporate peptide chemistry and related topics in this work, as we judged that this belongs to a separate field from natural product chemistry. Moreover, the literature dealing with the application of microwave irradiation in peptide synthesis is, to date, relatively scarce.

As dedicated microwave instruments appeared rather recently on the market and several interesting applications of microwave irradiation for natural product synthesis were described applying domestic microwave ovens, we decided to include also research performed with domestic ovens, although one could argue that some of these experiments lack reproducibility. On the other hand, the development of safe and reproducible synthetic routes for domestic instruments, which are cheap and at the disposal of every research lab all over the world, is a challenge worth the task as this should tremendously speed up the introduction of microwave irradiation in organic synthesis in general.

2
Total Synthesis of Various Classes of Natural Products

Natural products are undoubtedly the most challenging class of compounds for total synthesis, due to their structural diversity and complexity as well as the interesting biological activity inherent in many of them. During the last decades many examples of beautifully designed synthetic approaches have been published, using the plethora of available reagents and methods of the time. However, as microwave irradiation appeared on the scene only recently, its application has not been used in full strength for natural product synthesis, although some attractive examples have appeared in the recent literature.

Turrianes [1] are naturally occurring cyclophane derivatives (Fig. 1). These compounds are of particular interest as they are proven to be potent DNA cleaving agents under oxidative conditions [2–4] when administered in the presence of copper ions. An ingenious total synthesis has been elaborated by A. Fürstner et al. [5], applying a Ring Closing Metathesis (RCM) [6–10] or a Ring Closing Alkyne Metathesis (RCAM) [11–14].

The sequence starts with the synthesis of the sterically hindered biaryl entity, formed by the Grignard reaction, followed by further conversions

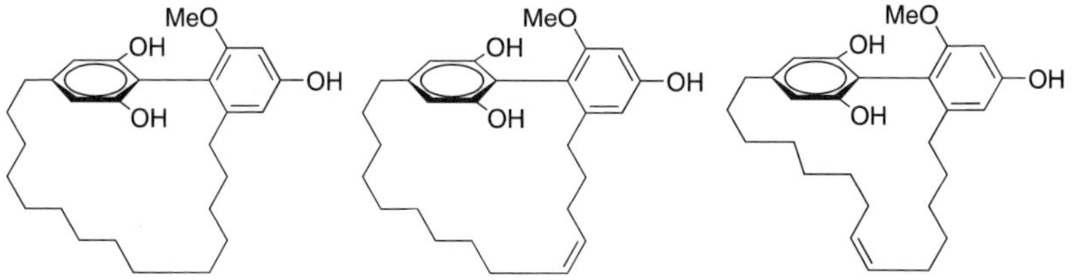

Fig. 1 Naturally occurring cyclophane derivatives belonging to the Turriane family

allowing the introduction of two different unsaturated tethers for macrocyclization (Scheme 1). PMB-ethers (PMB = p-methoxybenzyl) were found to be compatible protective groups with the diverse reaction conditions *en route* to the final natural compounds. Macrocyclization was investigated via RCM starting from the alkene-tethered substrates applying the Grubbs catalyst or a phenylindenylidene analogue in refluxing CH_2Cl_2. As expected, these compounds cyclize smoothly to the corresponding 20-membered rings, although unfortunately in a mixture of both stereoisomers with the undesired (E)-alkene prevailing (Fig. 2).

This problem could be circumvented when RCAM was applied starting from the acetylene-tethered compounds, as the initially formed cycloalkynes could be stereoselectively reduced to the Z-alkenes using Lindlar hydrogenation (Fig. 2). For the RCAM two different catalyst systems were evaluated: t-$(BuO)_3W \equiv CCMe_3$ in toluene at 80 °C for 16 h and $Mo(CO)_6/F_3CC_6H_4OH$ in chlorobenzene at 135 °C for 4–6 h, resulting in good yields of the cyclized products. However, the reaction times were rather long and, as in the

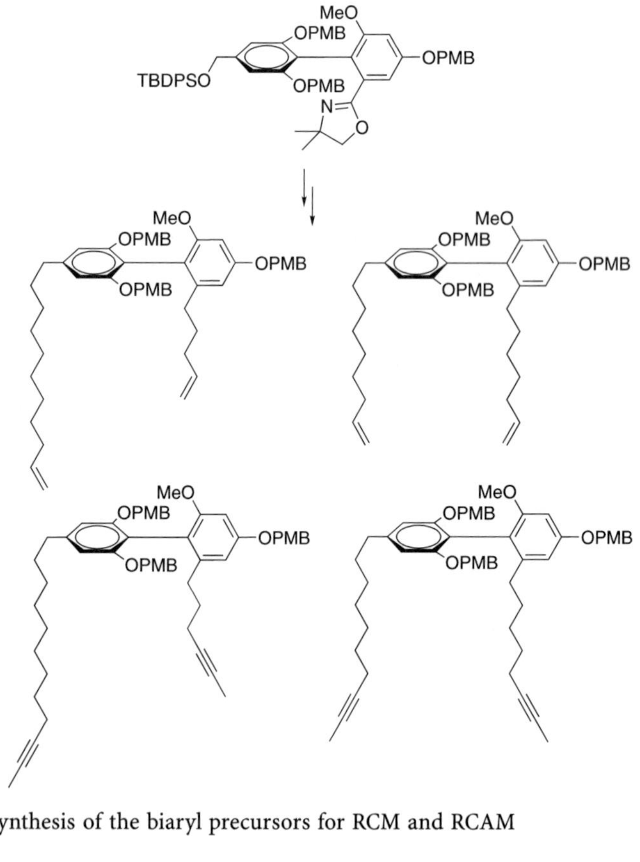

Scheme 1 Synthesis of the biaryl precursors for RCM and RCAM

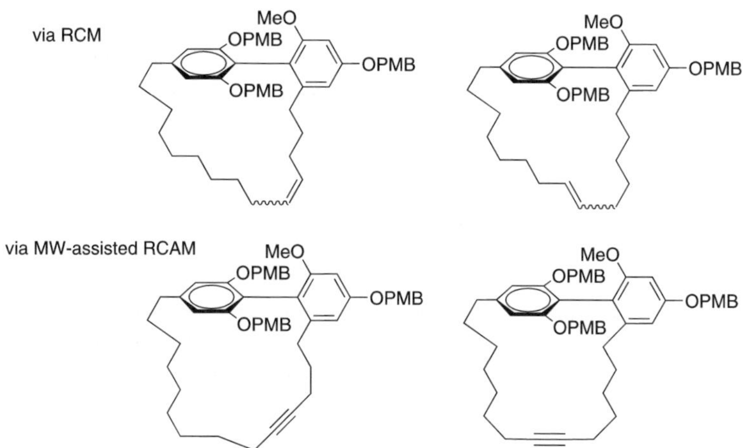

Fig. 2 Macrocycles formed via RCM or microwave-assisted RCAM

latter case a high temperature was required, the authors tried out focused microwave irradiation in chlorobenzene at 150 °C.

This resulted in a dramatic decrease of the reaction time to a mere 5 min with comparable yields. The sequence was accomplished upon full (H_2/Pd – C) or partial and stereoselective reduction (Lindlar catalyst) of the double bond and cleavage of the PMB-ethers.

Another interesting class of natural products is the Serinol marine compounds isolated from Tunicate *Didemnum* sp. (Scheme 2). These structures shows promising biological activity as for example HIV-1 integrase inhibitors [15]. All these Serinolipids possess a unique serinol component and a 6,8-dioxabicyclo[3.2.1]octane core structure, which make them attractive targets for synthesis. S.V. Ley et al. [16] developed a multistep sequence starting from D-(or L-) serinol and a known butanediacetal (BDA)-protected chiral building block [17] to prove by synthesis the absolute and relative stereochemistry of (+)-Didemniserinolipid B [18]. After the accomplishment of the sequence, comparison of the ^{1}H NMR spectra of the synthesized compound with that of the natural product revealed that the peaks associated with the Serinol unit of the molecule were shifted significantly up field. In view of the fact that related natural products were sulfated on the Serinol unit, the authors believed that the structure should be reassigned as the monosulfated one. This was supported by re-measurements of the HR-FAB-MS spectrum. Therefore, the authors started to synthesize the monosulfate derivative (Scheme 2). To achieve monosulfation, the amino group was protected as the fluorenylmethoxycarbonyl (Fmoc) group. However, standard sulfation conditions ($SO_3 \times$ Py or $SO_3 \times NMe_3$ in DMF, 1,4-dioxane and/or pyridine at 20–100 °C) failed to derivatize the diol and only starting material was obtained. On the contrary, when the reaction was performed with the aid

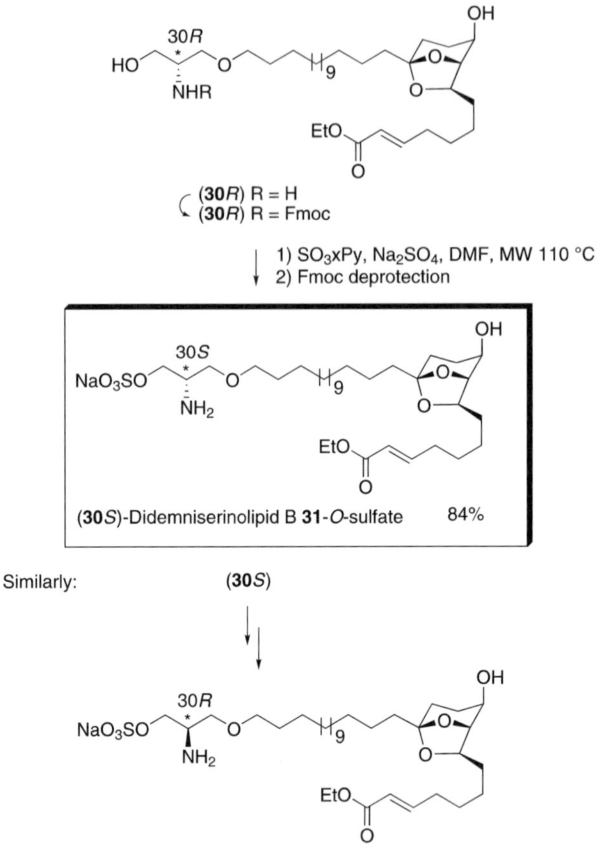

(**30R**) R = H
(**30R**) R = Fmoc

1) SO$_3$xPy, Na$_2$SO$_4$, DMF, MW 110 °C
2) Fmoc deprotection

(**30S**)-Didemniserinolipid B **31**-*O*-sulfate   84%

Similarly: (**30S**)

Scheme 2 Microwave-assisted synthesis of (30R)- and (30S)-Didemniserinolipid B 31-*O*-sulfate

of a dedicated microwave instrument, treatment of the diol (30R) with 1 equiv SO$_3$ × Py at 110 °C for 1 h provided the desired 31-*O*-sulfate as the major product. Deprotection of the Fmoc group under mild conditions (piperidine in DMF, rt) furnished 31-*O*-sulfate (30S). Similarly the (30R) diastereomer was synthesized for spectral comparison. It was shown that the spectroscopic data as well as the specific rotation of the (30S) diastereomer were in full agreement with those of the natural product, establishing a final proof for the structure of the natural product.

The two novel chromane derivatives rhododaurichromanic acids A and B [19] as well as the known Daurichromenic acid [20] are members of a new interesting class of anti-HIV agents and are therefore attractive syn-thetic targets (Scheme 3). These compounds were successfully synthesized by Z. Jin et al. [21] in good overall yields, applying a microwave-assisted tandem condensation and intramolecular S$_N$2$'$-type cyclization to form the

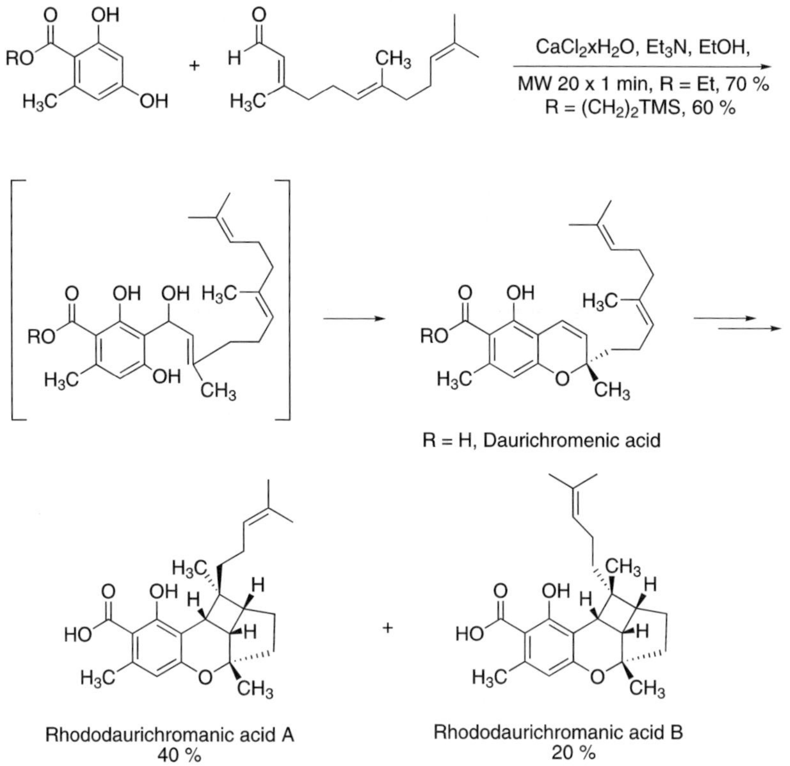

CaCl$_2$xH$_2$O, Et$_3$N, EtOH,

MW 20 x 1 min, R = Et, 70 %
R = (CH$_2$)$_2$TMS, 60 %

R = H, Daurichromenic acid

Rhododaurichromanic acid A
40 %

Rhododaurichromanic acid B
20 %

Scheme 3 Microwave-assisted synthesis of Daurichromenic Acid and Rhododaurichromanic Acids A and B

2H-benzopyran core structure. This approach for the construction of the 2H-benzopyran scaffold was developed by Shigemasa [22]. However, the authors found that the reaction of the ethyl ester with the aldehyde was extremely slow under Shigemasa's conditions.

The mixture gave only a 15% yield of the desired product after refluxing for 4 days. The yield was improved to 32% when the mixture was heated at 90 °C in a sealed tube for 1 day. As the intramolecular S_N2'-type cyclization is believed to be fast, the overall slow reaction is presumably due to the high activation energy in the condensation step. Therefore, the authors tried microwave irradiation, applying a domestic microwave oven. This resulted in a dramatic increase of the rate and the yield (70%) of the reaction. Optimized conditions were found to be microwave irradiation of a mixture of the ethyl ester (1 equiv) with the aldehyde (2.0 equiv) for 20 × 1 min, after which time an additional 1.0 equiv of the aldehyde was added and the mixture was irradiated again for 20 min. As the hydrolysis of the ethyl ester moiety to Daurichromenic acid proved to be extremely difficult, the authors switched to the application of a β-trimethylsilyl ethyl ester, which could be easily converted to the carboxylic acid upon treatment with TBAF. The microwave-assisted condensation reaction using the optimized conditions, yielded the β-trimethylsilyl ethyl ester of Daurichromenic acid in 60% yield upon irradiation in a Teflon pressure vessel for 20 × 1 min. After deprotection of the carboxylic acid functionality, resulting in the formation of Daurichromenic acid, the compound was converted into the rhododaurichromanic acids A and B upon irradiation with a low-pressure mercury lamp for about 5 days, to afford a mixture of the desired compounds in 40 and 20% respectively, calculated on recovered starting material. The physical data of the synthesized material were identical with those reported for the natural products.

Two other natural products, Neorautane and Neorautanin [23], belonging to the class of pterocarpans [24], were also synthesized via the formation of intermediate chromenes (Scheme 4). G.K. Trivedi et al. reported an approach applying the microwave-assisted cyclization of propargyl phenyl ethers [25]. Thus, heating of these starting materials to reflux (170 °C) in dry xylene resulted in the formation of the angular chromene as the sole product in good yield (76–83%). Microwave irradiation (domestic microwave, sealed tube) sped up these reactions tremendously (25–30 min) and a mixture of the desired linear chromene was formed, together with the angular chromene, albeit in low yield (22–32%).

These chromenes were further converted into the natural products Neorautane (37%) and Neorautanin (51%), upon Heck reaction with the appropriate phenol derivative in the presence of LiCl and $PdCl_2$ in dry acetone. The 7,15-methano-15H-dibenzo[d,g][1,3]dioxocins were isolated as undesired side compounds (R = OMe, 15%; R = H, 26%).

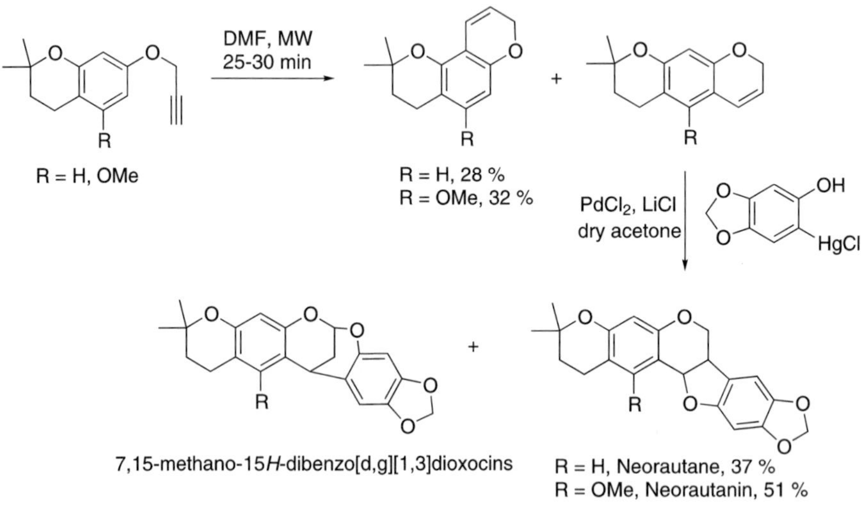

Scheme 4 Microwave-assisted synthesis of Neorautane, Neorautanin

An interesting application of microwave irradiation, although with a domestic oven, is found in the total synthesis of (+)-Longifolene [26, 27], a bridged-ring sesquiterpene (Scheme 5). Its unusual topology inspired many organic chemists and several superb total syntheses have been published. The approach of A.G. Fallis et al. [28] starts from an epoxyfulvene that was converted into an unsaturated lactone via a four-step sequence including a resolution with (–)-menthol chloroformate and a condensation with the anion derived from methyl 3-methylcrotonate. The cyclopropane ring was regioselectively cleaved applying $BF_3 \times Et_2O$ in methanol, affording a rapidly equilibrating mixture of substituted cyclopentadiene lactone isomers, due to the facile 1,5-sigmatropic rearrangement. Surprisingly, one single adduct was formed in 97% yield upon microwave irradiation (domestic oven) of a toluene solution of this triene mixture in a sealed tube for 2.5 h. This compound was further transformed into the desired natural product.

As part of the proof of the identity of three newly isolated chalcones, B.M. Abegaz et al. developed a synthetic route which involves a microwave irradiation promoted Ullmann coupling [29]. A mixture of the dimethylated phenolic chalcone and the bromochalcone in the presence of potassium carbonate and copper(I)chloride was microwave irradiated (domestic oven) at 180 °C, giving 19% yield of the resulting bichalcone (Scheme 6). This bichalcone was converted into a mixture of the desired Rhuschalcones, upon treatment with BBr_3, followed by chromatographic separation.

A one-step synthesis of the antifungal and larvicidal natural product methyl 3-(2,4,5-trimethoxyphenyl)propionate was described by J. Tamariz et al. [30] (Scheme 7). Reaction of methyl acrylate with 1,2,4-trimethoxybenzene catalyzed by $AlCl_3$ in 1,1,2,2-tetrachloroethane took 168 h (7 days) at 80 °C, yielding the compound in the low yield of 37%. However, when the reaction was run under microwave irradiation (domestic oven) in a Teflon

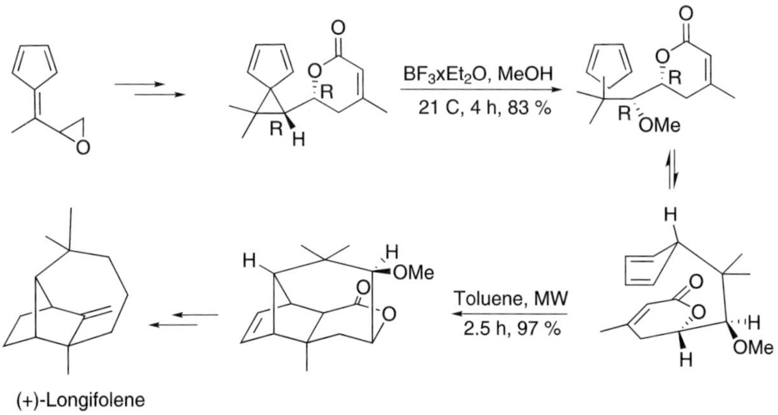

Scheme 5 Total synthesis of (+)-Longifolene

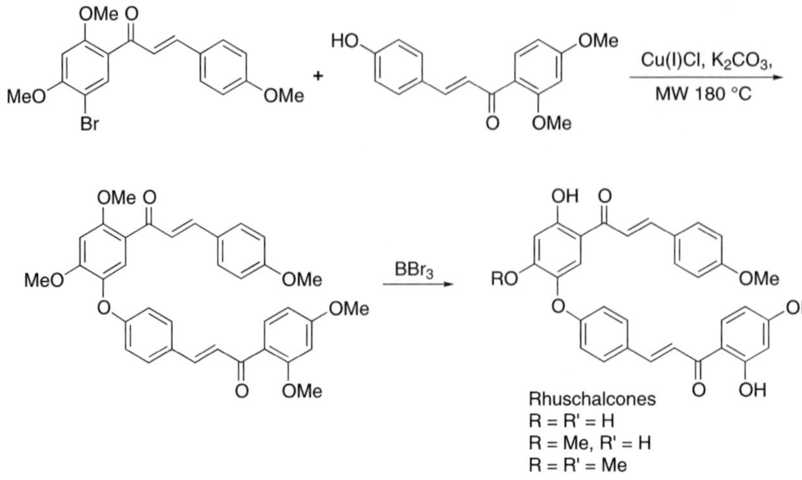

Scheme 6 Microwave-assisted synthesis of Rhuschalcones

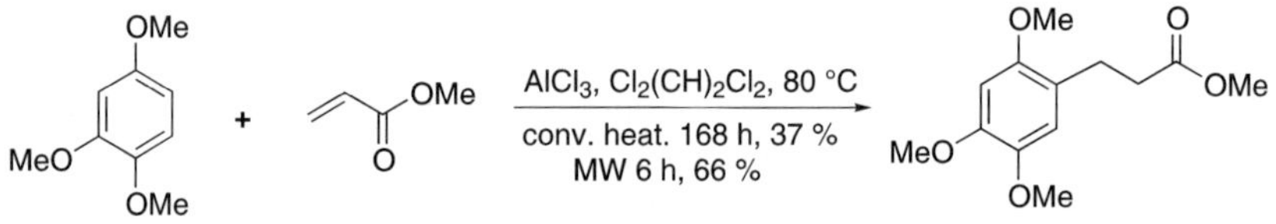

Scheme 7 Microwave-assisted synthesis of methyl 3-(2,4,5-trimethoxyphenyl)propionate

screw-capped glass tube at 80 °C, the reaction time was drastically shortened to 8 h and the yield was increased to 66%.

A short total synthesis of Phycopsisenone, a new phenolic secondary metabolite from the sponge *Phycopsis* sp, was performed by G.L. Kad et al. [31]. A microwave irradiation-induced (domestic oven) aldol condensation of 4-hydroxy-benzaldehyde and acetone in aqueous NaOH solution afforded the α,β-unsaturated ketone in 65% yield (Scheme 8). This was further converted in Phycopsisenone applying a two-step sequence.

Several members of the Cuparene class of sesquiterpenoids show interesting biological activity, including potent antifungal, neurotrophic and lipid peroxidation inhibition activity. Moreover, the presence of a hindered, quaternary

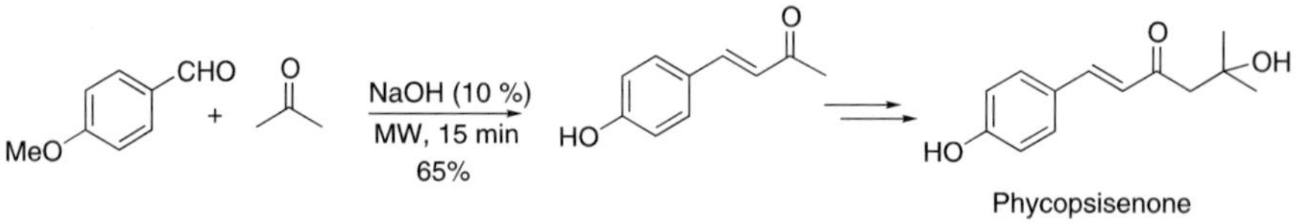

Scheme 8 Microwave-assisted synthesis of Phycopsisenone

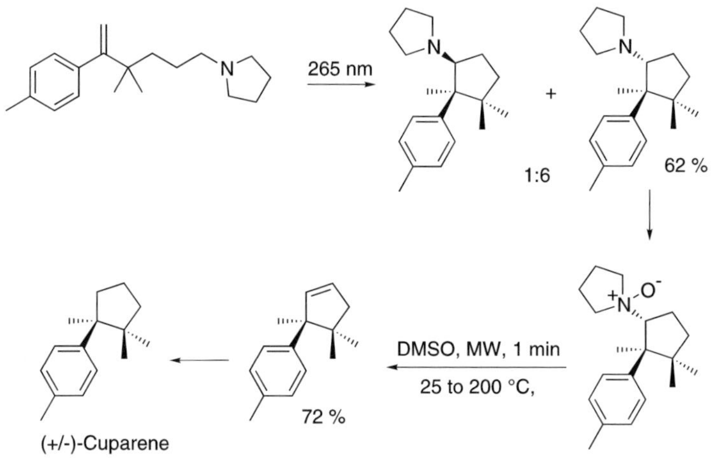

Scheme 9 Photomediated and microwave-assisted synthesis of Cuparene

stereocenter on a five-membered ring has made these sesquiterpenoids popular synthetic targets. A photomediated asymmetric synthesis of (–)-Cuparene is described by R.S. Grainger et al. [32] (Scheme 9). The key step of the sequence is the photomediated ring closure of an aminostyrene to cyclopentane according to the procedure of Lewis et al. [33, 34]. Irradiation of this aminostyrene at 265 nm under high dilution conditions (0.005 M) resulted in the quantitative formation of a mixture of two cycloadducts in a 6 : 1 ratio. The major diastereomer could be isolated in 62% yield. For the further conversion to Cuparene the authors used the thermally induced Cope elimination (THF, 60 °C, 8 h), after oxidation of the cycloadduct with m-CPBA. This proved to be troublesome and yielded the desired alkene in only 40% yield. However, applying focused microwave irradiation on a solution of the N-oxide in DMSO for 1 min going from 25–200 °C in temperature, afforded the product in 72% yield. Finally, hydrogenation of the double bond completed a racemic synthesis of Cuparene. An asymmetric variant was elaborated applying (S)-(+)-2-(methoxymethyl)pyrrolidine, rendering (S)-Cuparene with an optical rotation identical to that reported for the natural product.

An improved synthesis of the C_{11}-terpenic lactone Dihydroactinidiolide was described by A.K. Bose et al. [35]. The synthesis started from a commercially available aldehyde that was subjected to treatment with m-CPBA and a catalytic amount of PTSA (Scheme 10). The intermediate epoxy acid underwent cyclization resulting in the formation of Aeginetolide. Dehydration has been reported earlier by heating of this compound with aqueous NaOH, at 60 °C for 24 h, or with $SOCl_2$ and pyridine at room temperature for 5 h. The authors observed expeditious and convenient dehydration of this compound supported on silica gel, under microwave irradiation for 5–10 min (domestic oven), yielding Dihydroactinidiolide in 80% yield.

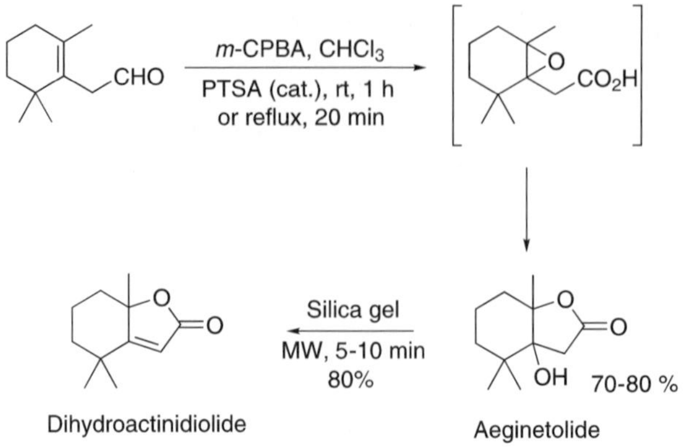

Scheme 10 Improved synthesis of Dihydroactinidiolide

The natural cyclic tripeptide Biphenomycin B [36–39], structurally belonging to the still expanding family of macrocyclic natural products with an endo aryl-aryl bond, displays potent activities against Gram-positive, β-lactam-resistant bacteria. The combination of structural novelty and biological activity motivated J. Zhu et al. to develop a concise asymmetric total synthesis for this compound [40] (Scheme 11). The key step is a microwave-assisted intramolecular Suzuki–Miyaura reaction of a linear tripeptide, for the formation of a 15-membered *meta,meta*-cyclophane. This tripeptide could be prepared from three non-proteinogenic amino acids, by standard peptide chemistry. However, even after extensive optimization of the reaction parameters for macrocyclization via the Suzuki–Miyaura coupling, the

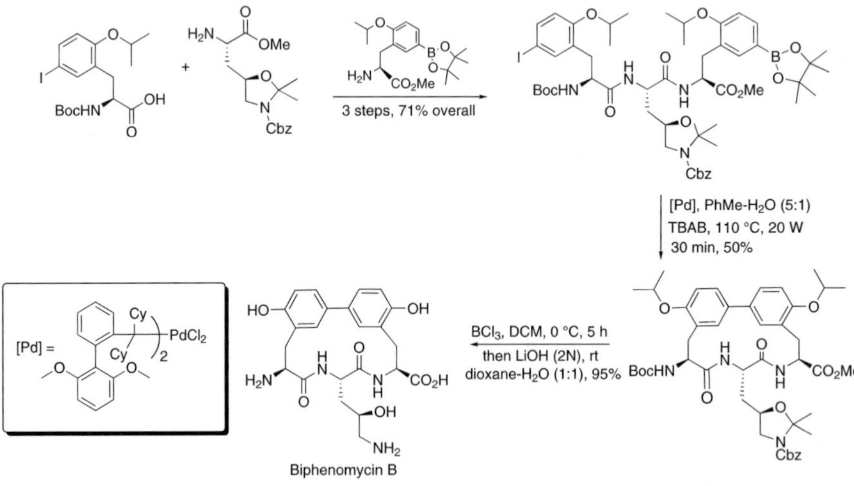

Scheme 11 Synthesis of Biphenomycin B

yield of the desired compound did not exceed 20%. On the contrary, upon microwave irradiation, the yield was increased and the reaction rate was dramatically raised. Under optimized conditions the intramolecular Suzuki–Miyaura coupling afforded the desired macrocycle in 50% yield. After final deprotection of all functional groups, the total synthesis of Biphenomycin B was accomplished.

3
Synthesis of Alkaloids

Among the various classes of natural products that have stimulated extensive research in the fields of both synthetic and pharmaceutical chemistry, alkaloids can arguably be described as the most important class. While their structural complexity has invited the attention of synthetic organic chemistry, their diverse biological activities and somewhat limited abundance has demanded the interest of pharmaceutical and medicinal research. With the ever-increasing demand for novel lead- and drug-like substrates for the purpose of biological screening, a plethora of literature has been published in recent years related to the synthesis, structural modification and biological studies of alkaloids and alkaloid-like structures. Consequently, research related to the total and partial syntheses of alkaloids has stimulated extensive research in the fields of methodology development. Automated synthesis and related high throughput technologies have been investigated to achieve fast and efficient generation of natural product related libraries. The use of microwave-assisted methods in the synthesis of alkaloids and alkaloid-related structures has therefore attracted an enormous current interest in both academia and industry [41–46], owing to the potential of microwave-assisted methods in rapid, easy and efficient synthesis of target libraries.

Because of the large number of biologically interesting natural products bearing highly functionalized indole skeletons [47–49], Fischer indole synthesis has received considerable synthetic interest. As a consequence, microwave-assisted Fischer indole synthesis has recently been investigated as a key step in the synthesis of a large number of indole alkaloids and their structural analogues. As a typical example, the microwave-assisted synthesis of iso-meridianins was explored by J.A. Palermo et al. [50], using a $ZnCl_2$ promoted microwave-assisted Fischer indole synthesis as the key step (Scheme 12). Iso-Meridianins are close structural analogues of the naturally occurring indole alkaloids Meridianins and Psammopemmins which exhibit high antitumor activity [51–53]. Iso-meridianins bear a pyrimidine ring at the C-2 position of the indole skeleton, in comparison with the parent compounds containing a pyrimidine ring at the C-3 position. As the corresponding 2-amino pyrimidines and phenylhydrazines are considerably

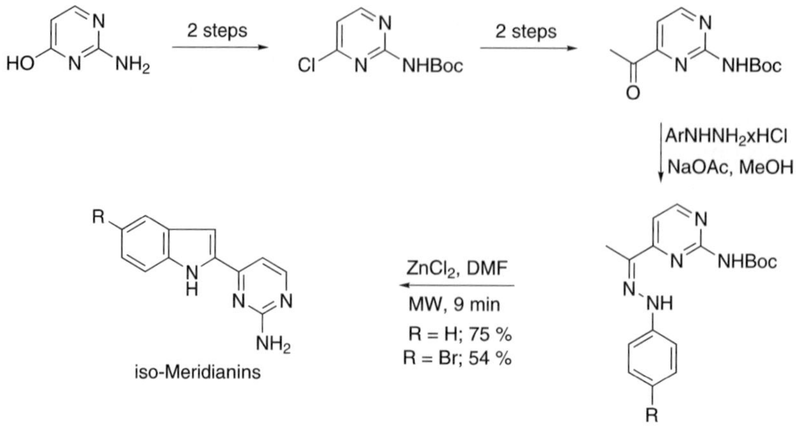

Scheme 12 Microwave-assisted synthesis of iso-Meridianins

cheaper and more easily available than the corresponding functionalized indoles, the authors reasoned that a diversity oriented and flexible synthesis could be achieved (Scheme 12). They started the synthesis from the commercially available isocytosine, which was converted to the corresponding N-Boc-4-chloro analogue by known transformations. The methyl ketone functionality was introduced at the C-4 position via Pd-catalyzed cross-coupling of the chloro derivative with tri-n-butyl(1-ethoxyvinyl)tin followed by an acidic hydrolysis of the intermediate. The phenylhydrazones of the compounds were then generated following standard protocols.

However, the Fischer indole synthesis met with failure under classical heating conditions, using a variety of solvents and catalysts. The authors then explored $ZnCl_2$-mediated microwave-assisted cyclization in a domestic microwave oven using the full irradiation power of c.a. 1500 W. The cyclization was found to proceed smoothly in DMF in a mere 9 min and the iso-Meridianins were isolated in good yields.

Another interesting example of the influence of microwave irradiation is illustrated in the total synthesis of the indoloquinoline alkaloids Cryptotackieine and Cryptosanguinolentine, isolated from *Cryptolepis sanguinolenta*, a shrub indigenous to West Africa. These alkaloids exhibit various interesting biological activities such as antimuscarinic, antibacterial, antiviral, antimycotic and antihyperglycemic activities, considerable *in vitro* antitumor activity, and strong antiplasmodial activity of Cryptotackieine against chloroquine-resistant strains of *P. falciparum* [54–57]. P.M. Fresneda et al. reported an elegant microwave-assisted synthesis of Cryptotackieine and Cryptosanguinolentine [58], based on a divergent approach for the generation of the key 1-methyl-(*o*-azidophenyl)-quinoline-2-one followed by its selective indolization (Scheme 13). 2-Azidobenzaldehyde was converted into the key isocyanate intermediate following known synthetic manipulations and the

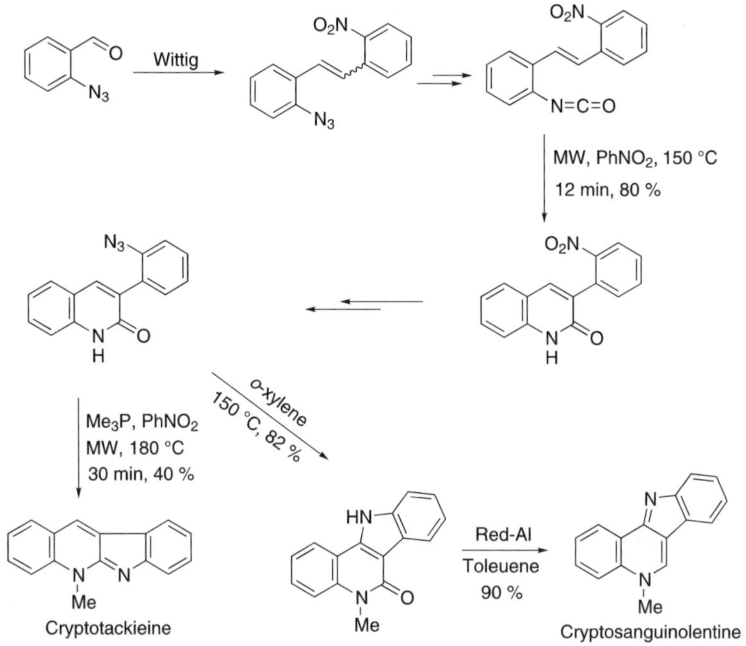

Scheme 13 Microwave-assisted synthesis of Cryptotackieine and Cryptosanguinolentine

quinolin-2-one skeleton was generated via the microwave-assisted cyclization of the corresponding isocyanate intermediate at 150 °C, using a dedicated microwave reactor (Scheme 13).

The quinoline-2-one was then converted into the required 1-methyl-(*o*-azidophenyl)-quinoline-2-one via known synthetic manipulations. Cyclization of this intermediate to the target Cryptotackieine was performed under microwave irradiation at 180 °C for 30 min, in the presence of trimethyl phosphine. It is noteworthy that the same aza-Wittig reaction, when carried out under classical heating for 24 h, furnished inferior yields. The quinolin-2-one intermediate, when heated at 150 °C in *o*-xylene under conventional heating conditions, underwent a nitrene insertion at the C-4 position of the pyridinone ring. Subsequent Red-Al reduction of the carbonyl group furnished the Cryptosanguinolentine.

S. Hibino et al. have recently reported an interesting total synthesis of a new furo[3,2-*h*]isoquinoline alkaloid TMC 120B [59], starting from 2,4-bismethoxymethyloxybenzaldehyde (Scheme 14). TMC 120A, B and C were isolated in 1999 by J. Kohno et al. from a fermentation broth of *Aspergillus ustus* TC 1118. TMC 120B shows moderate inhibitory activity against the interleukin-5 mediated prolongation of eosinophil survival with an IC_{50} value of 2.0 μM [60–62]. The key 3,7,8-trisubstituted isoquinoline intermediate was obtained starting from the *o*-alkenylbenzaldoxime methyl ester,

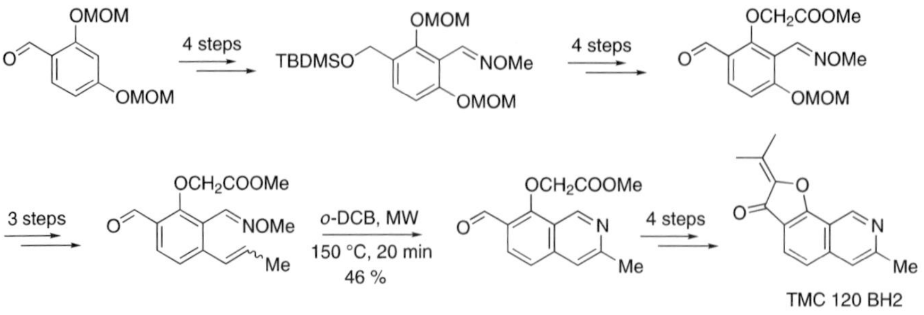

Scheme 14 Microwave-assisted synthesis of alkaloid TMC 120B

which was synthesized from 2,4-bismethoxymethyloxybenzaldehyde in 11 steps via known synthetic manipulations.

The electrocyclic cyclization was investigated by heating the sample in *o*-dichlorobenzene (*o*-DCB), both under conventional heating conditions and under microwave irradiation in a dedicated monomode apparatus. It was found that the microwave-assisted cyclization provided slightly elevated yields at the lower reaction temperature of 150 °C, in comparison with the conventional heating experiment conducted at 180 °C (Scheme 14). The intermediate isoquinoline was then converted to the final alkaloid using known synthetic manipulations to complete the synthesis.

Y. Langlois et al. have recently reported the total synthesis of the marine alkaloid Bengacarboline [63], using a microwave-assisted imination and subsequent Pictet–Spengler cyclization [64]. Bengacarboline, extracted from the Fijian ascidian *Didemnum* sp. exhibits interesting inhibitory activity of topoisomerase II and significant in vitro activity against tumor cell lines [65].

The authors started their explorations from indole-3-carboxylic acid, which was easily converted into the corresponding amide using known synthetic manipulations. The di-indolyl ketone intermediate was then developed in two steps via a Bischler–Napieralski reaction with $POCl_3$ followed by nucleophilic attack of a hydroxyl anion and spontaneous ring opening (Scheme 15). The imination was then explored under a variety of conventional heating conditions, albeit without favorable results. Microwave-assisted conditions in the presence of $ZnCl_2$ in *p*-xylene at 130 °C for 1 h solved this problem. The imine intermediate was then converted into the target Bengacarboline structure in two steps.

The pyrroloquinazolinoquinoline alkaloids Luotonin A and B were isolated recently from the aerial parts of *Peganum nigellastrum*, a plant that has been used against rheumatism, abscesses and inflammation in Chinese traditional medicine. The alkaloid Luotonin A was found to possess a pentacyclic skeleton closely resembling Camptothecin, derivatives of which are clinically useful anticancer agents [66, 67]. The compound was proven to be cytotoxic against the P-388 cell line of murine leukemia. J.S. Yadav et al. elab-

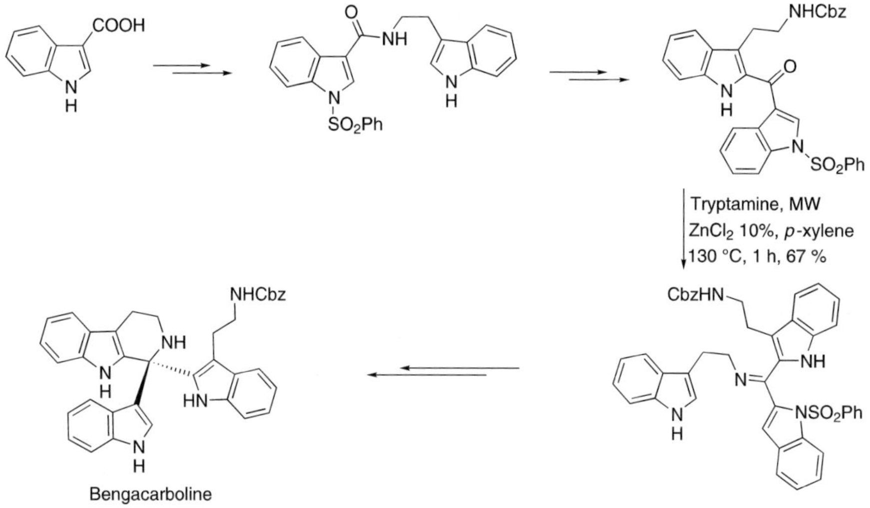

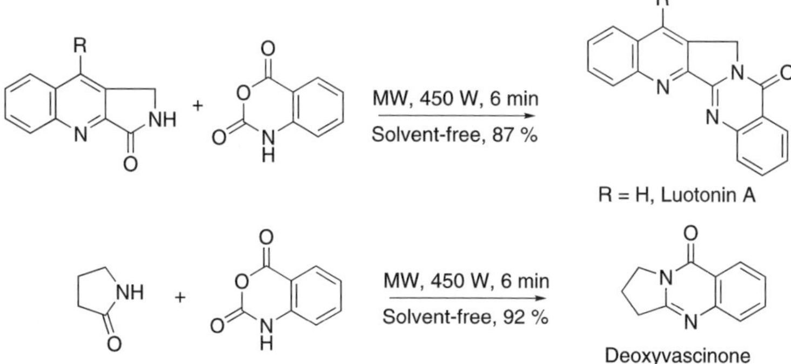

Scheme 15 Microwave-assisted total synthesis of Bengacarboline

orated a rapid microwave-assisted synthesis of these cytotoxic alkaloids [68], albeit using a domestic microwave oven (Scheme 16). The authors devised a two-component condensation reaction for the generation of the pentacyclic system of the target molecule. Thus, 3-oxo-1H-pyrrolo[3,4-b]quinoline, easily generated in three steps following known procedures, was condensed with isatoic anhydride under solvent-free microwave-assisted conditions. Condensation proceeded smoothly in 6 min at an irradiation power of 450 W, and Luotonin A was isolated in the high yield of 87% (Scheme 16). Furthermore, the antitumor active alkaloid deoxyvascinone, also isolated from *Peganum nigellastrum*, was obtained in a similar fashion in an excellent yield of 92%. It is noteworthy that this condensation furnished a much lower yield of 60%

Scheme 16 Microwave-assisted synthesis of Luotonin A and Deoxyvascinone

in a considerably longer reaction time of 5–8 h when carried out under conventional heating conditions at a maximum temperature of 120 °C.

T. Lipinska et al. have recently reported an ingenious microwave-assisted synthesis of E-ring modified Sempervirine analogues [69] (Scheme 17). Since the first isolation from *Gelsenium sempervirens* in 1916, Sempervine and its structural analogues have gained considerable synthetic interest due to their impressive range of biological potential [70, 71]. Some of the known members of this family such as Flavopereirine, Serpentine and Alstonine exhibit a variety of biological activities, for example anti-HIV, antipsychotic, sedative and immunostimulant activities together with notable cytostatic effects [72]. Even though the synthesis of these molecules has been targeted in the past, a major issue to be addressed was to find synthetic routes towards Sempervirine analogues with a modified core structure. An Inverse Electron-Demand Diels–Alder reaction (IEDDA) [73] between a suitably functionalized 1,2,4-triazene and a 1-pyrrolidino-1-cycloalkene was initially explored to generate the cyclic intermediate, which upon loss of nitrogen furnished the required bicyclic intermediate for the synthesis of the target molecules (Scheme 17). After converting the intermediate to the corresponding phenylhydrazone quantitatively in refluxing ethanol with traces of acetic acid, the authors carried out a microwave-assisted, solvent-free Fischer indole synthesis with morillonite K10 modified with $ZnCl_2$ at the very high temperatures of 165–190 °C for 1.5–2.5 min (Scheme 17). The thus formed 2-substituted indole-analogues were then converted into the final targets by a three-step protocol based on known synthetic manipulations, furnishing the Sempervirine analogues in high yields.

T. Lipinska further used this methodology to generate a series of 9-methoxy analogues of Sempervirines [74, 75], by simply choosing 4-methoxyphenylhydrazine as the starting material. This case study is particularly interesting due to the fact that a variety of methoxy analogues of indole alkaloids are known to be more effective on the nervous system than the parent compounds. The Fischer indolizations were performed under con-

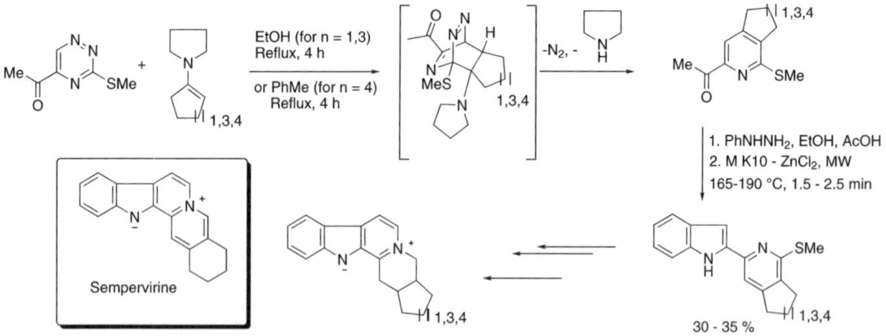

Scheme 17 Microwave-assisted synthesis of Sempervirine analogues

trolled microwave-assisted conditions in an open vial, applying the optimized MK10/ZnCl$_2$ system at 150 °C for 2 min.

S.E. Wolkenberg et al. have recently investigated the microwave-assisted synthesis of richly decorated imidazoles from 1,2-diketones [76] in the presence of ammonium acetate as the nitrogen source (Scheme 18). The reactions were carried out at 180 °C for 5 min in a dedicated microwave apparatus using acetic acid as the solvent, and the corresponding imidazoles were isolated in very high yields and purities. Furthermore, the authors extended the strategy for the synthesis of Lepidiline B [77], an alkaloid isolated from the root extracts of *Lepidium meyenii* of Peruvian origin, which exhibits potent micromolar cytotoxicity against several human cancer cell lines.

An interesting investigation regarding the microwave-assisted synthesis of biologically potent alkaloids has recently been reported by J.-F. Liu et al. [78]. The authors initially explored a microwave-assisted one-pot synthesis of the 2,3-disubstituted 3*H*-quinazolin-4-one skeleton, a privileged structure present in many biologically interesting alkaloids such as the sedative-hypnotic Methaqualone, antitussive Chloroqualone and anticonvulsant Piriqualone [79–81]. The quinazolinone core is found in a variety of natural products possessing interesting bioactivity, such as anti-inflammatory, cardiovascular, CNS, antimalarial and antiviral effects [82]. The authors developed a highly efficient, microwave-assisted three-component procedure in view of circumventing the known drawback that most of the conventional methodologies were only applicable to the *N*-aryl analogues (Scheme 19). This strategy involved the one-pot reaction between a suitably substituted anthranilic acid, a functionalized acid chloride, carboxylic acid or aldehyde and a suitable aryl or alkyl amine in the presence of P(OPh)$_3$ as the coupling agent. The synthesis was performed using a monomode microwave station integrated in the solution-phase high-throughput automated platform, with the aim of combining the rapid microwave-assisted synthesis and the novel high-throughput methods for library design. It is noteworthy that the same

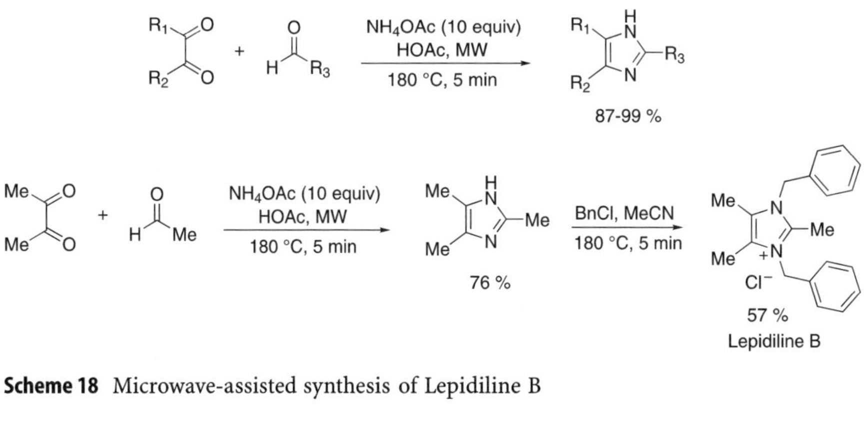

Scheme 18 Microwave-assisted synthesis of Lepidiline B

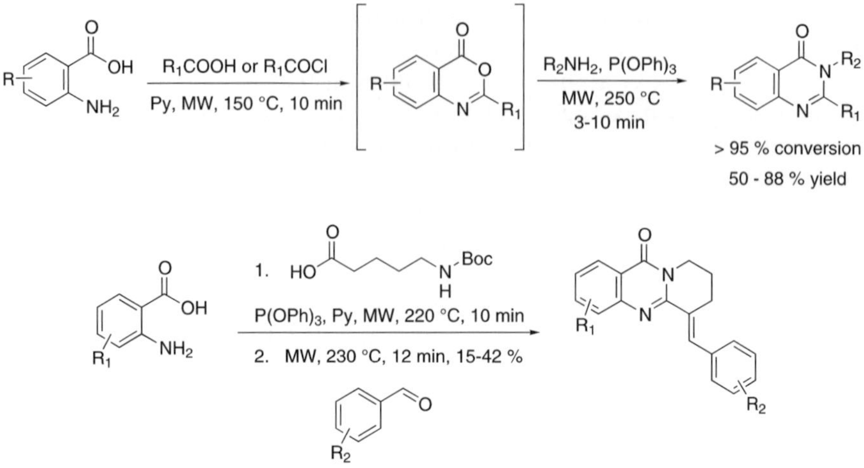

Scheme 19 Microwave-assisted one-pot synthesis of 2,3-disubstituted 3*H*-quinazolin-4-ones

protocol, under conventional heating conditions, provided inferior conversions of $\leq 50\%$ (Scheme 19).

J.-F. Liu et al. further explored this novel three-component one-pot strategy to generate a variety of alkaloids bearing a pyrazino[2,1-*b*]quinazoline-3,6-dione skeleton. The synthesis and post-synthetic manipulations of these natural products had been previously explored owing to the large variety of biological and pharmacological activities they exhibit. Some notable members of these families are Fumiquinazolines F and G, and Glyantrypine [83–85]. J.-F. Liu et al. applied their previously developed microwave-assisted one-pot protocol for the synthesis of Glyantrypine, Fumiquinazoline F and Fiscaline B (Scheme 20) [86]. The cleavage of the Boc protecting group was found to be strongly temperature dependent. When the temperature was raised from 180 °C to 220 °C, the deprotection of the *N*-Boc to NH was found

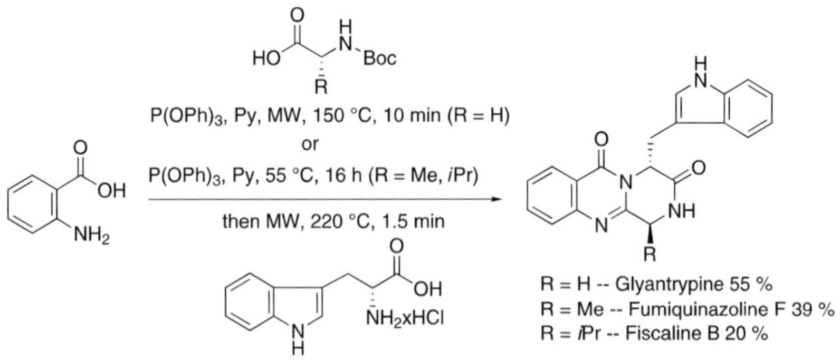

Scheme 20 Synthesis of Glyantrypine, Fumiquinazolines F and Fiscaline B

to increase from 0 to 100%. The strategy was successfully applied for the synthesis of these alkaloids.

J.-F. Liu et al. have explored a highly efficient, microwave-assisted, domino reaction protocol [87, 88] for the synthesis of a variety of 2,3-disubstituted-quinazolin-4-ones starting from readily available compounds (Scheme 21) [89]. A one-pot procedure was applied using P(OPh)$_3$ as the coupling reagent in pyridine at 200 °C for 10 min, using a monomode microwave apparatus. The target alkaloids Deoxyvasicinone, Mackinazolinone and 8-Hydroxydeoxyvasicinone were isolated in high yields of 72–89% in a mere 10 min irradiation time. The authors used the thus optimized procedure to generate a small library of structurally similar alkaloid analogues with high yields and purity.

This optimized, microwave-assisted, domino protocol was used for the synthesis of a variety of quinazolinobenzadiazepine alkaloid natural products [90]. The members of this broad family of naturally occurring alkaloids feature a broad spectrum of very intriguing biological potential: Sclerotigenin exhibits high anti-insectant activity, Aspercilin C and E show high Cholecystokinin (CCK) antagonist activity and Benzomalvin A features inhibitory activity against substance P at the neurokinin NK1 receptor [91, 92]. The microwave-assisted domino reaction between anthranilic acid and a suitable N-Boc amino acid in the presence of P(OPh)$_3$ was explored, as in the previous cases, to generate the target natural products (Scheme 22).

However, the reaction was found to proceed slowly, taking 20 min under microwave irradiation, resulting in comparatively lower yields, probably owing to the seven-membered ring in the target compounds. In a similar fashion, the authors generated the chemotaxonomic marker analogues for the quinazolinobenzadiazepine alkaloids (Circumdatin E derivatives) by employing N-Boc-proline as the amino acid coupling partner.

Another interesting investigation of the use of microwave irradiation for the synthesis of alkaloids was recently reported by M. Álvarez et al., where the authors investigated a modular total synthesis of Lamellarin D [93]. After 20 years of the first isolation of this compound from the marine prosobranch mollusk *Lamellaria* sp, this family currently comprises more than 30 members, isolated from various natural sources. With its recent identification as

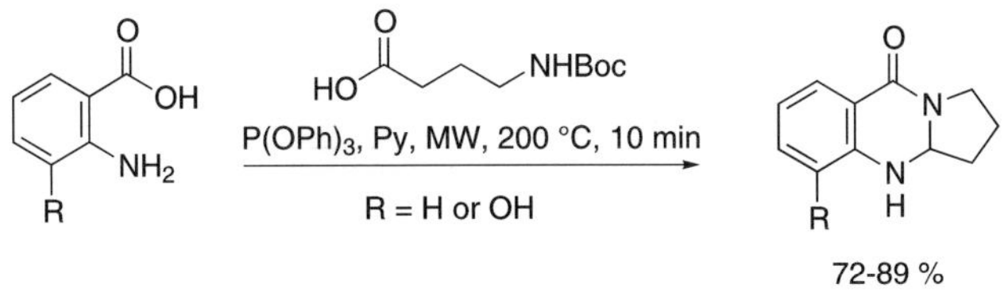

Scheme 21 Microwave-assisted domino reaction protocol for the synthesis of 2,3-disubstituted-quinazolin-4-ones

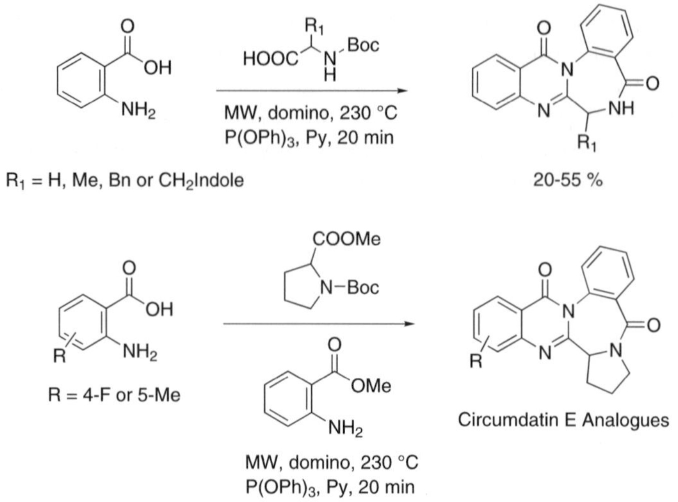

Scheme 22 chemical reaction:

R₁ = H, Me, Bn or CH₂Indole

MW, domino, 230 °C
P(OPh)₃, Py, 20 min

20-55 %

R = 4-F or 5-Me

MW, domino, 230 °C
P(OPh)₃, Py, 20 min

Circumdatin E Analogues

Scheme 22 Synthesis of the quinazolinobenzadiazepine alkaloids

a potent topoisomerase I inhibitor as well as its candidature for in vivo pre-clinical development as an antitumor agent, there has been a lot of current interest in the synthesis and structural manipulation of Lamellarin D [94, 95]. The authors started the synthesis with the *N*-alkylation of methyl pyrrole-2-carboxylate applying a suitable tosylate under conventional heating conditions, followed by an intramolecular Heck-type cyclization to access the basic pyrroloisoquinoline skeleton (Scheme 23). Then a regioselective bromination and subsequent Suzuki–Miyaura reaction was performed to generate the intermediate 3-aryl pyrrole and the sequence was repeated to furnish the 3,4-diarylpyrrole intermediate.

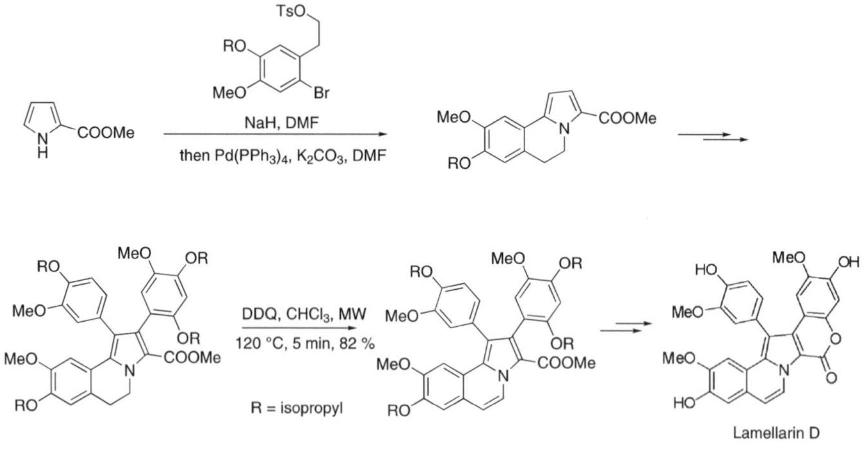

Scheme 23 reactions:

NaH, DMF
then Pd(PPh₃)₄, K₂CO₃, DMF

DDQ, CHCl₃, MW
120 °C, 5 min, 82 %

R = isopropyl

Lamellarin D

Scheme 23 Total synthesis of Lamellarin D

The aromatization of the dihydroisoquinoline was achieved under microwave-assisted conditions, irradiating the intermediate in $CHCl_3$ at 120 °C for 5 min with DDQ as the oxidizing agent (Scheme 23). It is noteworthy that the use of DDQ, MnO_2 or Pd – C in various solvents under conventional heating conditions was not successful. The authors completed the total synthesis by an $AlCl_3$-mediated lactonization to furnish lamellarin D in a total of eight steps with 18% overall yield.

4
Modifications of Natural Products

Boosted by the demand for novel molecules for the purpose of biological screening, recent research in both organic synthesis and medicinal chemistry have resulted in a plethora of recent literature related to the modification of natural products. Altering the structures of existing natural products for the purpose of increasing their biological potential has therefore received widespread current interest. Consequently, microwave irradiation has been investigated for the fast and easy modifications of biologically interesting molecules.

4.1
Modifications of Steroids

An efficient preparation of 1,5-diketones as precursors to D-ring annulated heterosteroids was elaborated by R.C. Boruah et al. [96] (Scheme 24). Readily available 16-dehydropregnenolone acetate (16-DPA) was used in a Michael reaction with enamines.

Thus, freshly prepared 1-morpholino-1-cyclohexene was mixed thoroughly with 16-DPA and basic alumina and this was irradiated in a dedicated

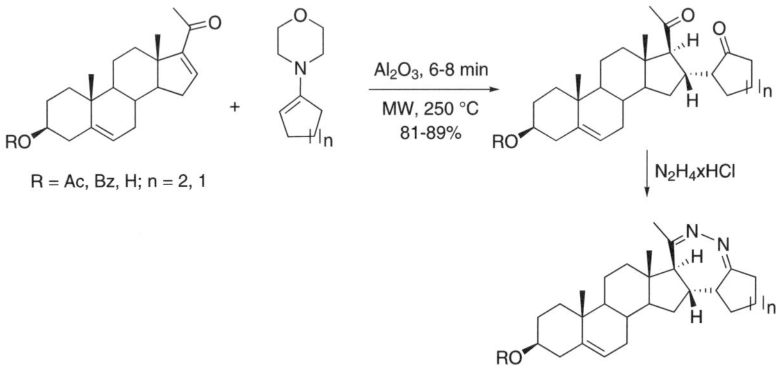

Scheme 24 Synthesis of a novel class of 1′,2′-diazepino(17,16-d') steroids

microwave instrument for 6 min. The 3β-acetoxy-16-(2'-cyclohexanoyl)preg-nenolone was isolated in 79% yield. In a similar manner, related Michael adducts were obtained in high yields. It was observed that the reactions were sluggish in the absence of basic alumina, which indicated the catalytic role of alumina in this solid-phase reaction. Interestingly, the reactions did not work under conventional heating conditions, even with prolonged heating. The obtained compounds were then cyclized upon reaction with hydrazine hydrochloride, affording the 1',2'-diazepino(17,16-d') steroids.

R.C. Boruah et al. described the microwave-mediated [2 + 2] cycloaddition of steroidal formyl enamides and alkynes under solvent-free conditions [97] (Scheme 25). Although much attention has been paid to the development of synthetic methodologies for oxetanes, the corresponding unsaturated ana-logue oxetene has received only limited interest. In the course of a program directed towards the synthesis of D-ring annulated heterosteroids, the authors investigated the reaction of a mixture of 3β-acetoxy-16-formyl-17-acetamido-androst-5,16-diene, triphenylphosphine and various acetylenes, which were thoroughly mixed with basic alumina. This mixture was irradiated using a microwave reactor for 5 min. Steroidal oxetane derivatives were obtained in good yields (72–80%). Interestingly, in the absence of alumina, the reaction was found to be sluggish. Also, the conventional thermal heating of formyl enamides with alkynes in the presence of PPh$_3$ in a protic solvent under pro-longed heating (72 h) led to very poor yields of oxetenes (15–23%).

R. Skoda-Földes et al. synthesized steroidal dienes and investigated them as starting materials for further functionalization of the steroidal skeleton via cycloaddition reactions [98] (Scheme 26). In a previous work the authors de-scribed the synthesis of these dienes via Stille coupling of steroidal alkenyl halides with vinyltributyltin [99, 100]. However, under conventional heat-ing conditions, these reactions were rather slow, requiring a reaction time of many hours. Rapid reactions could here be achieved under microwave irradiation using a domestic oven. 17-Iodo-androst-16-ene, 17-iodo-4-aza-4-methyl-androst-16-en-3-one and 17-iodo-4-aza-androst-16-en-3-one were reacted with vinyltributyltin in the presence of Pd(PPh$_3$)$_4$. As toluene does

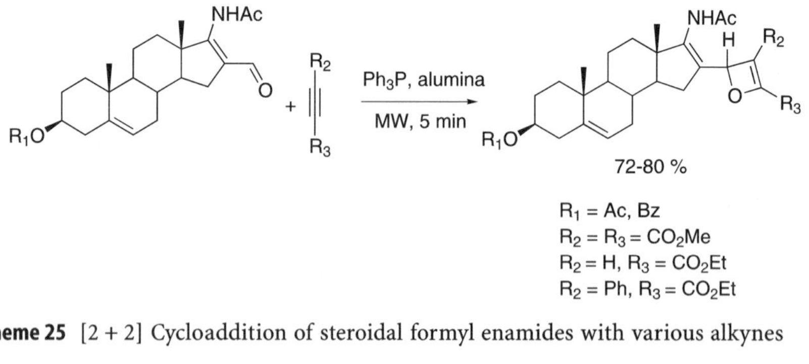

Scheme 25 [2 + 2] Cycloaddition of steroidal formyl enamides with various alkynes

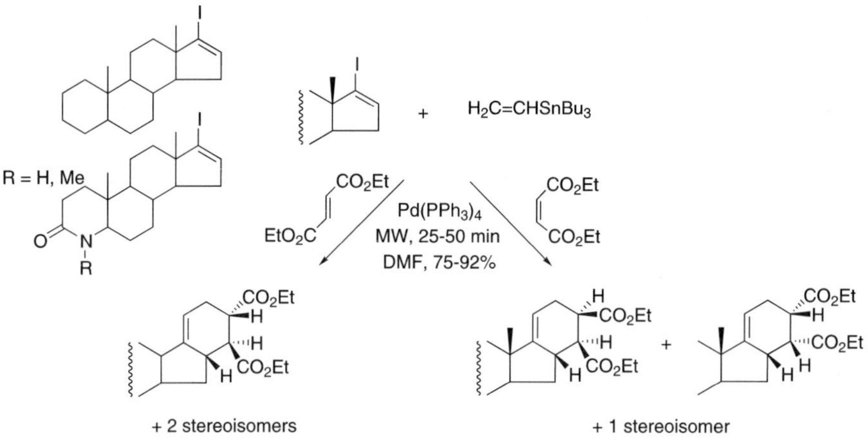

Scheme 26 One-pot Stille coupling/Diels–Alder reaction of steroidal alkenyl halides

not couple efficiently with microwaves, DMF was chosen as the solvent. Reactions were complete in minutes rather than in hours. As it was possible to increase the substrate/catalyst ratio, the products precipitated on cooling and could be isolated by simple filtration. Moreover, the authors found that the subsequent Diels–Alder reaction with diethyl fumarate or diethyl maleate could be run as a one-pot reaction under microwave irradiation. In some cases the stereoselectivity of this reaction could be greatly improved.

4.2
Modifications of Non-Steroidal Compounds

B. Das et al. have demonstrated a microwave-assisted modification of Parthenin [101]. This compound is a major sesquiterpinoid content of the weed *Parthenium hysterophorus L* (compositae) showing a variety of potent biological effects such as anticancer, antibacterial, antiamoebic and anti-malarial activities [102–106]. However, Parthenin is also known to be highly toxic and to cause allergy, which prevents its usage for any medical purposes. In view of increasing its biological potential and decreasing its toxicity, B. Das et al. carried out a microwave-assisted reduction protocol, applying Zn – AcOH as the reducing media (Scheme 27).

The reduction was carried out applying a household microwave oven. In contrast to the Zn – AcOH reduction of Parthenin, which was performed under conventional heating conditions, the microwave-assisted protocol was found to be useful in preventing the deoxygenation of the compound and the reduction of the exocyclic methylene group.

B. Das et al. have also elaborated the microwave-assisted synthesis of Mappicine ketone, a potent antiviral lead compound derived from the natural product Camptothecin. The naturally occurring pyrrolo[3,4-*b*]quinoline

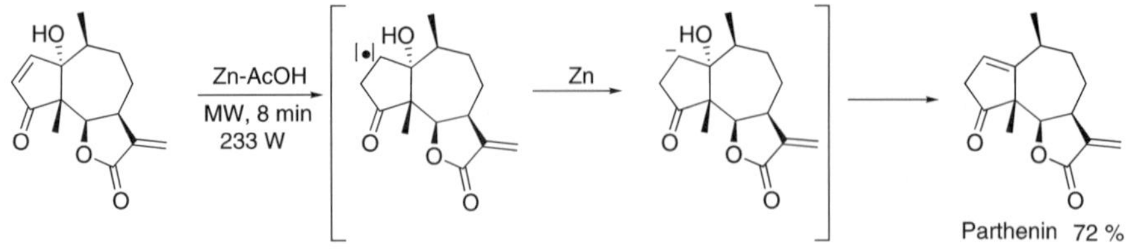

Scheme 27 Microwave-assisted modification of Parthenin

alkaloid Camptothecin, isolated from *Camptotheca acuminate* Decne (Nyssaceae), shows promising antineoplastic activity against animal tumor models [107–110]. The decarboxylated E-ring analogue of Camptothecin, known as Mappicine ketone, has been shown to exhibit potent activity against herpes viruses (HSV1 and HSV2) and human cytomegalovirus (HCMV) [107–110]. B. Das et al. have elaborated a successful conversion of Camptothecin into Mappicine ketone [111] using a commercial household microwave oven (Scheme 28).

Camptothecin was irradiated under solvent-free conditions for 7 min at the full power of the microwave oven (Scheme 28). The product, Mappicine ketone, was isolated in 96% yield without a trace of undesired side products, which clearly exhibits the potential of microwave-assisted chemistry. In comparison, when the reaction was run at rt in THF and in the presence of $BF_3 \times Et_2O$, Mappicine ketone was isolated in a mere 65% yield.

B. Das et al. have further demonstrated the use of microwave irradiation in the modification of natural products, by converting Berberine into Berberrubine (Scheme 29) [112]. Berberine, the quarternary alkaloid from *Tinospora*

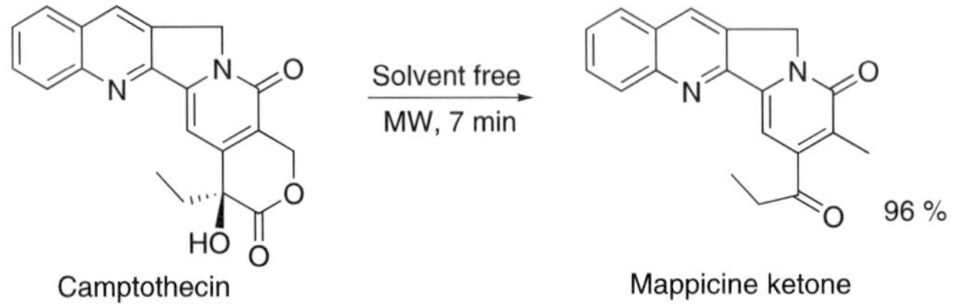

Scheme 28 Microwave-assisted conversion of Camptothecin into Mappicine ketone

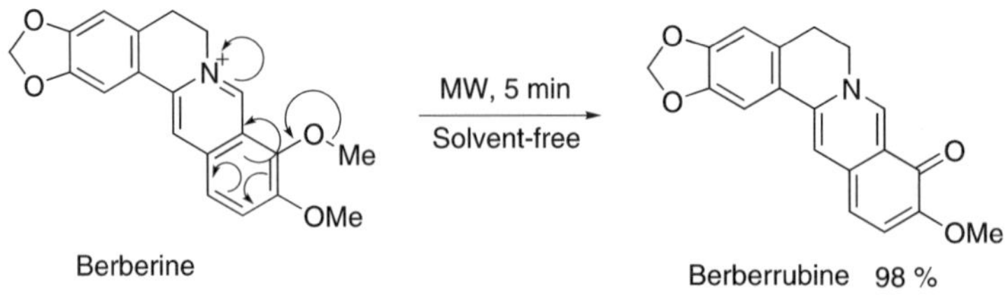

Scheme 29 Microwave-assisted conversion of Berberine to Berberrubine

cardifolia, exhibits potent antianaemic and antineoplastic activities, besides featuring interesting pharmacological effects including respiratory stimulation, transient hypotension and convulsion [113, 114]. The authors converted the naturally occurring alkaloid Berberine into Berberrubine by irradiating the former under solvent-free conditions for 5 min, applying a domestic microwave oven.

B. Das et al. have also investigated the microwave-assisted generation of 7-cyano camptothecins and 7-cyanomappicine ketones [115], using a direct conversion of a formyl group into a nitrile functionality (Scheme 30). The 7-Cyanocamptothecin derivatives were irradiated with hydroxylamine hydrochloride using $NaHSO_4 \times SiO_2$ as a catalyst to perform the desired conversions. Microwave-irradiation experiments were carried out under solvent-free conditions for 2 min using a domestic microwave oven and the 7-cyano derivatives of Camptothecin and Mappicine ketone were isolated in good yields.

An interesting exploration related to the microwave-assisted modification of a naturally occurring bioactive molecule was recently demonstrated by P.S. Baran et al., where the authors converted the natural product Sceptrin into Ageliferin under microwave-assisted conditions [116]. While attempting to rationalize the presence of Sceptrin as the major product during the biosynthesis of Ageliferin, the authors managed an ingenious interconversion of Sceptrin to Ageliferin simply by shifting the thermodynamic equilibrium of the reaction (Scheme 31).

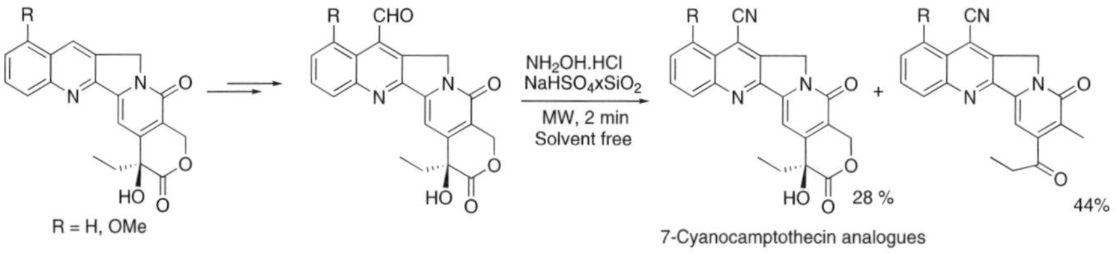

Scheme 30 Microwave-assisted synthesis of 7-Cyanocamptothecin analogues

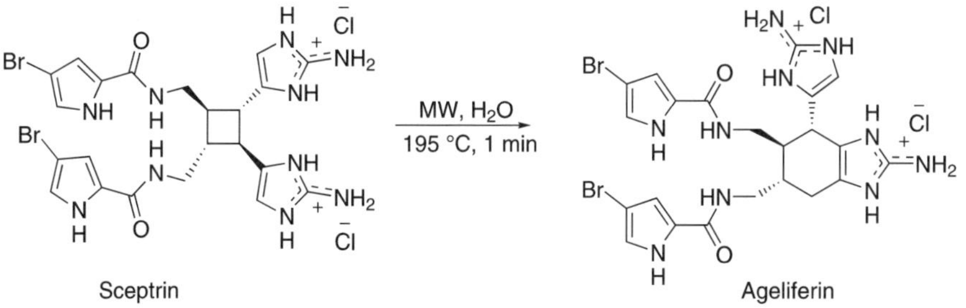

Scheme 31 Microwave-assisted formation of Ageliferin from Sceptrin

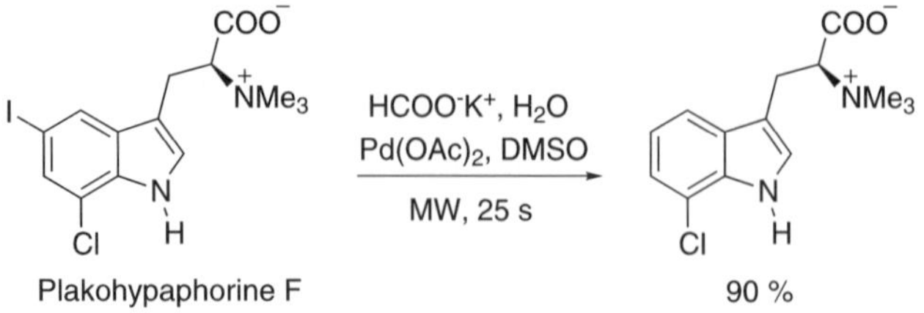

Scheme 32 Microwave-assisted selective dehalogenation of Plakohypaphorine F

The authors irradiated a sample of Sceptrin in water at 195 °C for 1 min, using a dedicated monomode microwave apparatus. They investigated the process in detail and found that after a series of tandem tautomerizations and [1,2] shifts, Scerptin did indeed convert into the structural analogue Ageliferin. They further confirmed the reaction and mechanistic details by deuterium-labeling.

Another interesting application of microwave irradiation is found in the selective dehalogenation of the iodinated indole alkaloid Plakohypaphorine F, described by E. Fattorusso et al. [117, 118] (Scheme 32). The bis-halogenated compound was treated with potassium formate and palladium acetate under controlled microwave irradiation, resulting in selective deiodination. The choice of the solvent, in this case DMSO, was found to be crucial.

5
Synthesis of Unnatural Analogues of Natural Products

In parallel with the developments in microwave-assisted synthesis and modification of known natural products, there has been some progress in the use of microwave irradiation in the synthesis of unnatural analogues of known natural product skeletons. A typical example of such an attempt is the synthesis of unnatural β-carboline alkaloids by C.W. Lindsey et al. [119]. The authors devised a microwave-assisted one-pot procedure to easily access the basic Canthine skeleton [120]. Canthines represent a tetracyclic subclass of β-carboline alkaloids, bearing an additional D-ring. Almost forty members have been reported in the literature up to the present day from the first isolation in 1952 of the parent compound Canthine from *Pentaceras australis*. Several members of this family have been known to possess very intriguing anti-fungal, antiviral and antitumor activities [121–127]. The authors applied the ingenious Inverse Electron-Demand Diels–Alder (IEDDA) strategy developed by S. Snyder et al. [128] (Scheme 33), based on the formation of an indole derived triazine followed by intramolecular IEDDA at high temperature, while the indole was acting as the dienophile.

While trying to improve the yields and diversity of the reaction, the authors studied the reaction under microwave-assisted conditions. They ini-

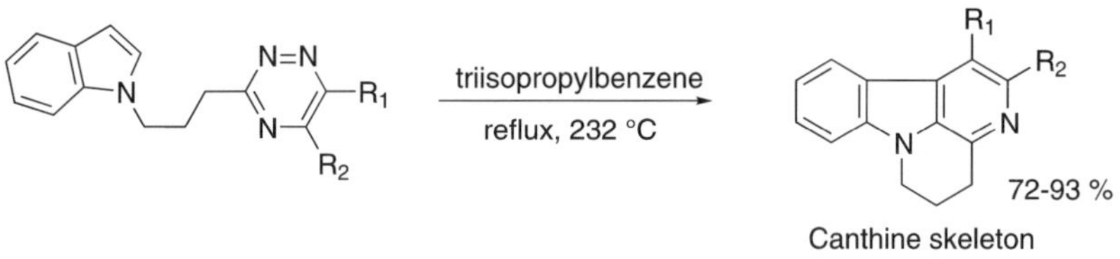

Scheme 33 Intramolecular IEDDA reaction in the synthesis of the Canthine core

tially explored the synthesis of the hydrazide intermediate needed for the formation of the triazine derivative, applying a dedicated microwave apparatus. The indole-derived ester (Scheme 34) was thus irradiated together with hydrazine in THF at 150 °C and the corresponding hydrazides were generated in high yields in a mere 2 min irradiation time. The authors further explored a microwave-assisted one-pot strategy for the formation of the triazine derivatives, followed by the IEDDA reaction to generate the corresponding Canthine analogues in high yields and purity. The condensation of the acyl hydrazide with the appropriate diketone was found to be temperature sensitive, and resulted in high yields at 220 °C in 40 min irradiation time (Scheme 34).

The authors further developed this strategy for the synthesis of unsymmmetrically substituted canthine analogues, starting from the synthesis of an unsymmetrical benzil, followed by the synthesis of the corresponding acylhydrazide and subsequently the Canthine analogues [129]. This rapid microwave-assisted one-pot procedure was then utilized in the synthesis of more complex Canthine analogues (Fig. 3) featuring potent Akt (Protein Kinase B, a serine/threonine kinase) kinase inhibitory activity, equipotent against both Akt1- and Akt2-type kinases.

Another interesting exploration was recently published by J.P. Bazureau et al., where the authors investigated a microwave-assisted, solventless procedure for the synthesis of the 2-aminoimidazolone core [130] (Fig. 4). This cyclic guanidine skeleton represents an interesting pharmacophore that displays a wide range of pharmacological activities such as hypoglycemic and hypotensive activities, protein kinase C activity and inhibitory activity for the Nuclear Factor Kappa B (NF-κB) activation (Fig. 4). Among the interesting marine sponge-derived alkaloids comprising the guanidine moiety can Dispacamide, exhibiting potent antihistamine activity and Leucettamine B,

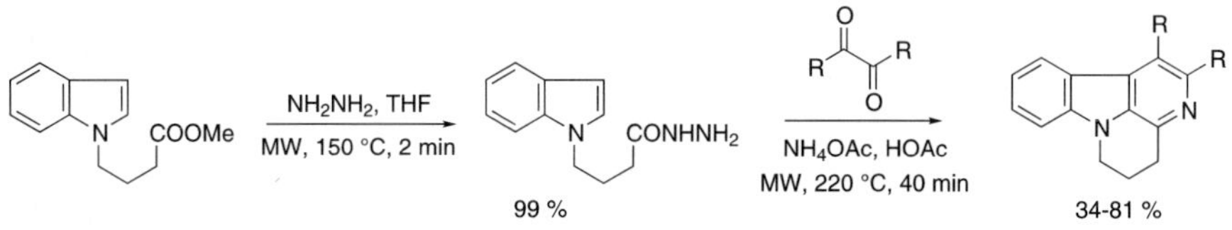

Scheme 34 Microwave-assisted synthesis of Canthine analogues

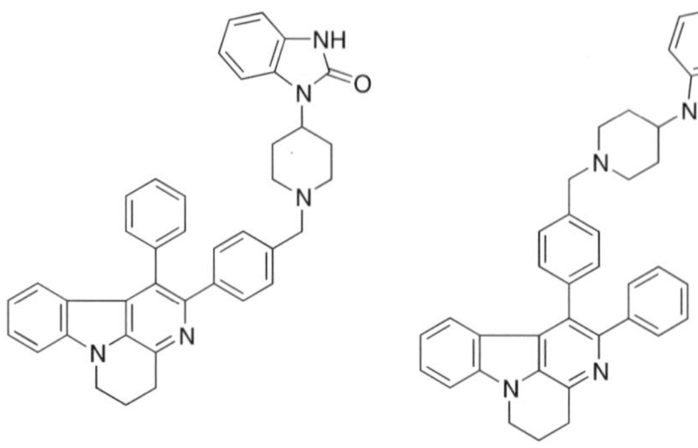

Fig. 3 Two Akt(PKB) kinase inhibitors

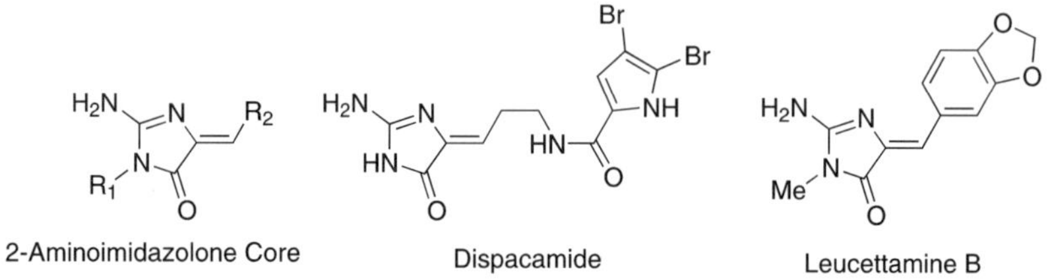

2-Aminoimidazolone Core　　　Dispacamide　　　Leucettamine B

Fig. 4 The 2-aminoimidazolone core

which acts as a mediator of inflammation, be mentioned [131, 132]. Even though there are a number of known methods for the synthesis of the basic skeleton, there are almost no literature precedents dealing with the construction of the 2-alkylamino derivatives of the 2-aminoimidazolones.

The authors explored two high yielding convergent approaches for the synthesis of the 2-alkylamino derivatives of Leucettamine B, applying microwave-assisted conditions using a dedicated reactor (Scheme 35). The 2-thiohydantoine intermediates were easily generated by the base-mediated condensation of the corresponding isothiocyanates with the required amino esters. These compounds were then converted to the corresponding 5-benzo-[1,3]-dioxo-5-ylmethylene-2-alkylsulfanyl-3,5-dihydroimidazol-4-ones under microwave-assisted conditions, as can be viewed in Scheme 35. It is noteworthy that the microwave irradiation provided the required intermediates in high yields and purity, while the reactions were performed under solvent-free conditions at 70–80 °C for 1 h. The compounds were then converted into the corresponding Leucettamine B analogues by reacting them with the corresponding alkylamines at 50 °C.

Y.-H. Chu et al. described the application of a microwave-accelerated Pictet–Spengler reaction for the preparation of 1,1-disubstituted indole alkaloids [133] (Scheme 36). Most of the synthetic pathways concerning the Pictet–Spengler cyclization were performed with aldehydes or activated ke-

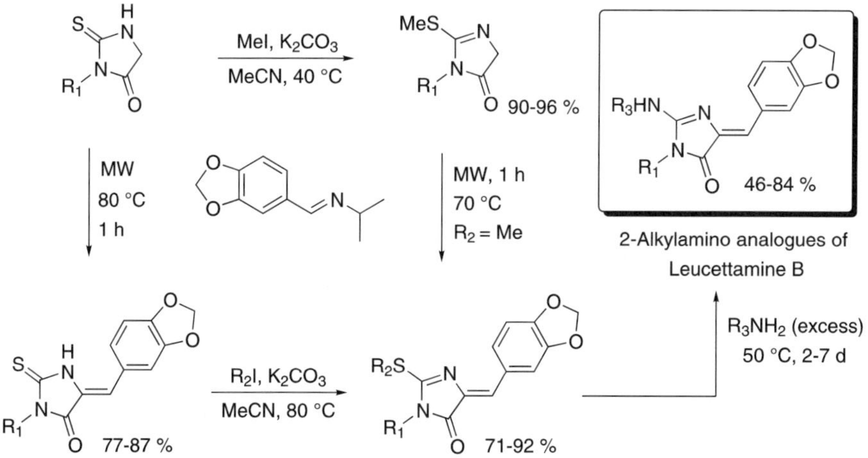

Scheme 35 Microwave-assisted synthesis of Leucettamine B analogues

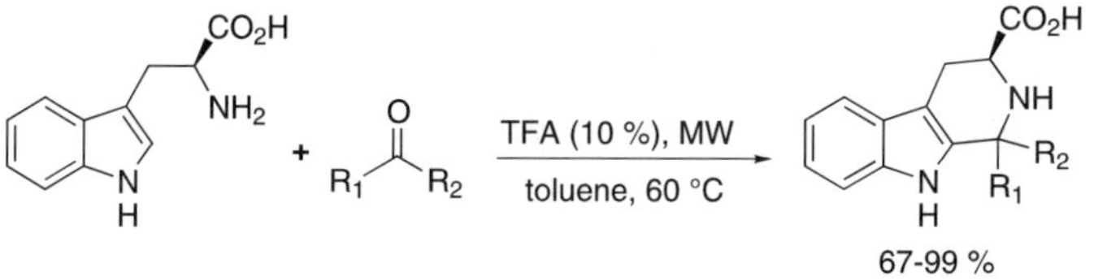

Scheme 36 Pictet–Spengler reaction of tryptophan with various ketones

tones such as 1,2-dicarbonyl compounds. Reactions with ketones at room temperature mostly require days and are sluggish even under reflux conditions. Many arylketones are known to be unreactive and consequently these ketone reactions have never been addressed [134–138]. The authors found that at ambient temperature, Pictet–Spengler reactions of tryptophan with ketones could be sped up if a large excess (12 equiv) of ketone was used, although the average reaction time was in the order of days. However, upon the application of controlled microwave irradiation, the reactions proceeded remarkably faster and cleaner with both aliphatic and aromatic ketones, though the latter were much less reactive [139].

To demonstrate the usefulness of this microwave methodology the authors carried out a multistep synthesis of a new, ketone-derived Demethoxyfumitremorgin C analogue and tetrahydro-β-carbolinehydantoins [140] (Scheme 37). The former was synthesized starting from the L-tryptophan methyl ester, in an overall yield of 11%, via a three-step synthesis comprising a microwave assisted Pictet–Spengler reaction, a Schotten–Baumann acylation, deprotection and an intramolecular cyclization. In addition, good overall yields (5–70%) were achieved for the latter compound applying microwave ir-

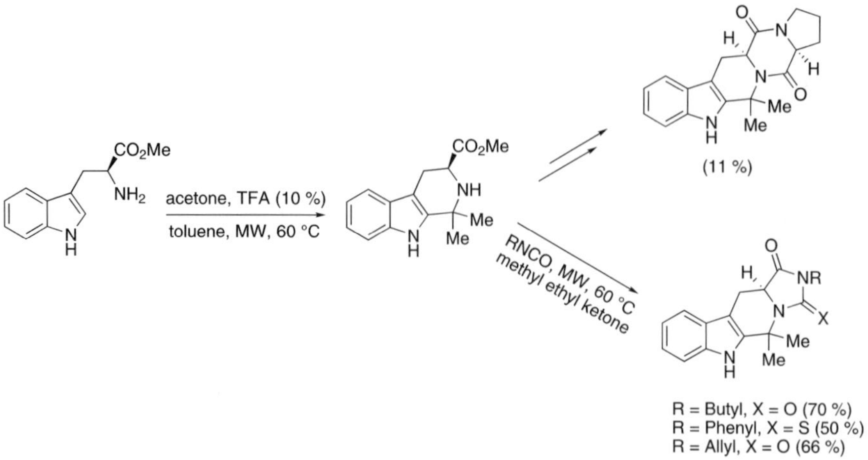

Scheme 37 Microwave-assisted synthesis of a new, ketone-derived Demethoxyfumitremorgin C analogue and Tetrahydro-β-carbolinehydantoins

radiation on both the Pictet–Spengler cyclization and the following hydantoin formation.

In a route towards new estrogens which bind to the β-unit of the K^+-channel located on the surface of the endothelium, L.F. Tietze et al. described the synthesis of a novel enantiopure B-Nor-steroid, applying multiple Pd-catalyzed transformations [141] (Scheme 38). A combination of a Suzuki–Miyaura and a Heck reaction using a 2-bromobenzylchloride derivative and a boronic ester, derived from the enantiopure Hajos–Wiechert ketone [142–

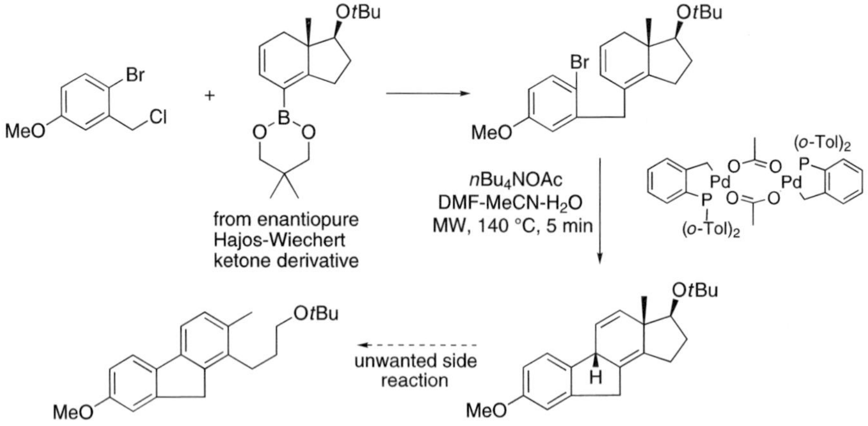

Scheme 38 Synthesis of the B-Nor-estradiol analogue

144] was used. The Suzuki–Miyaura reaction was performed in refluxing THF using Pd(PPh$_3$)$_4$ as the catalyst and sodium hydroxide as the base yielding the *seco*-B-Nor-steroid in 72% with high regioselectivity. Ring B could be formed at 140 °C in an intramolecular Heck reaction using a suitable palladacycle as the catalyst, yielding the B-Nor-estradiol in 70% yield. As expected, reaction took place from below, *anti* to the angular methyl group, to form a single diastereomer. Under high-density microwave irradiation the reaction time could be shortened to 5 min and the aromatization of ring C with opening of ring D, which is an unwanted side reaction under standard conditions, could be suppressed.

E. Van der Eycken et al. have recently explored an interesting microwave-assisted synthesis of the 1,2,3-triazole-derived aza-analogues of the naturally occurring lignan lactones (–)-Steganacin and (–)-Steganone (Fig. 5). They were isolated from *Steganotaenia Araliacea* by the late S.M. Kupchan and have been demonstrated to possess significant *in vivo* activity against P-388 leukemia in mice and *in vitro* activity against cells derived from human carcinoma of the nasopharynx [145–148]. In order to increase the biological potential and to solve the problems associated with stereoselection, K. Koga et al. proposed the synthesis of unnatural (–)-Steganacin 7-aza-analogues [149, 150]. Some of these compounds have been shown to exhibit anti-tumor activity even higher than the corresponding natural lignan lactones.

The authors explored a microwave-assisted nine-step protocol for the synthesis of the unnatural Steganacin and Steganone analogues [151], where the fused oxazolidinone ring was replaced by a 1,2,3-triazole ring system (Scheme 39). They used a microwave-assisted Suzuki–Miyaura cross-coupling reaction [152–157] to generate the key biaryl intermediates in high yields and purity. It is noteworthy that these cross-couplings, owing to the difficult oxidative addition of the transition-metal catalyst to the highly electron-rich aryl bromide and consequently the increase of competitive proto-deboronation of the boronic acid, often furnish inferior yields under conventional heating conditions. In this case, the authors observed a drop

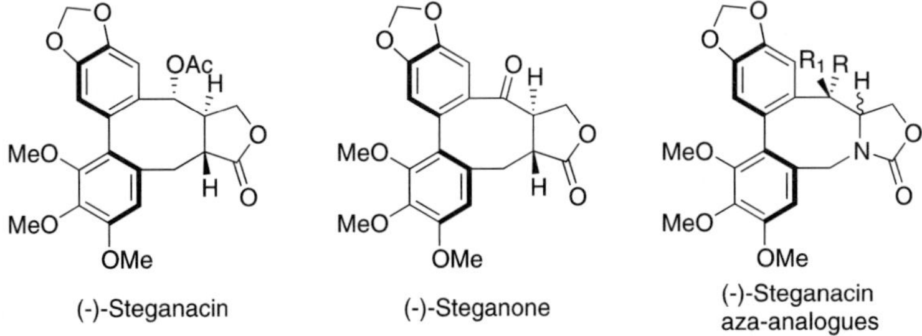

Fig. 5 Naturally occurring lignan lactones and their 7-aza-analogues

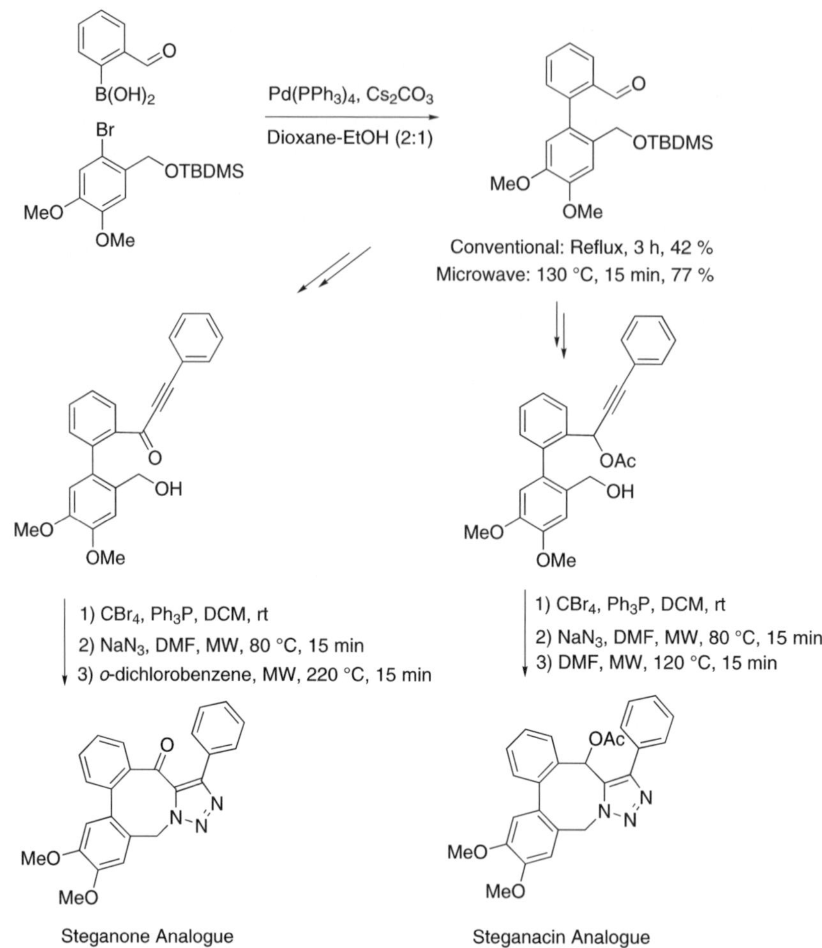

Scheme 39 Microwave-assisted synthesis of triazole-derived analogues

in yield from 77% when the microwave irradiation was used to a mere 42% under conventional heating conditions.

The required intermediate alcohols were generated by known synthetic manipulations. The intramolecular cyclization into the target molecules was then carried out following a three-step one-pot procedure, applying a microwave-assisted intramolecular Huisgen 1,3-dipolar cycloaddition as the key step. As expected, the ketone requires a much higher temperature for cyclization compared to the acetate precursor. It is also noteworthy that the intramolecular cycloaddition reaction, when carried out under conventional heating conditions, completely failed to promote the reaction.

E. Van der Eycken et al. explored a microwave-assisted Suzuki–Miyaura reaction to generate structural analogues of the apogalanthamine family of *Amaryllidaceae alkaloids* [158]. The Apogalanthamine analogues (Fig. 6) represent an intriguing class of natural products, featuring a rare 5,6,7,8-tetrahydrobenzo[*c,e*]azocine skeleton. Buflavine, isolated from *Boophane flava*, an endemic *Amaryllidaceae* alkaloid species from South Africa, is a typical member of the family featuring potent biological activities such as alpha-adrenolytic and serotonin antagonistic activities [159–164].

In search of a suitable protocol for the selective generation of the biaryl axis of these compounds, the authors explored the Suzuki–Miyaura reaction of highly electron-rich phenethylamines (PEA) with a variety of boronic acids, including hindered boronic acids and acids bearing electron-withdrawing substituents, promoting proto-deboronations during cross-coupling. Despite the non-aqueous conditions advocated in the literature for conventional heating protocols, the authors were successful in developing a small library of 2-(hetero)aryl PEAs, performing the cross-coupling in a 1 : 1 mixture of DMF and water with $NaHCO_3$ or Cs_2CO_3 as the base, under focussed microwave irradiation at 150 °C for 15 min (Scheme 40). Furthermore, a microwave-assisted pseudo one-pot three-step procedure was developed for the construction of the medium-sized ring to generate a Buflavine analogue in high yield and purity.

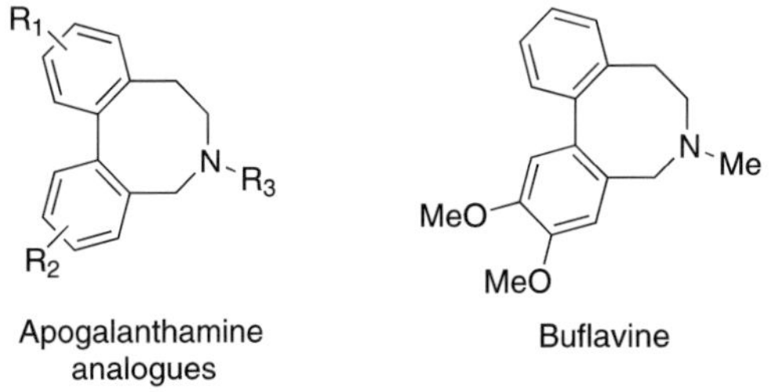

Fig. 6 Apogalanthamine analogues and Buflavine

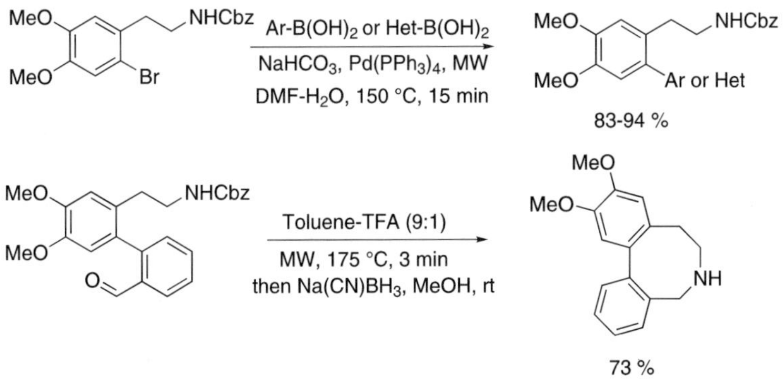

Scheme 40 Microwave-assisted biaryl coupling and synthesis of a Buflavine analogue

It is noteworthy that the microwave-assisted Suzuki–Miyaura reactions were found to be far superior to the reactions under conventional heating, as can be demonstrated by yield improvements from 22% to 84% in the case of cross-coupling of PEA derivatives. Microwave-irradiation also provided smooth conversions in water as the sole solvent.

E. Van der Eycken et al. further explored the microwave-assisted transition-metal catalyzed reactions [165–170] for the synthesis of structural analogues of Buflavine, using a combination of microwave-assisted Suzuki–Miyaura cross-coupling and Ring-Closing Metathesis (RCM). The authors developed a novel nine-step protocol for the synthesis of *N*-shifted Buflavine analogues [171], making use of their previously optimized cross-coupling conditions. Thus, highly electron-rich aryl bromides were cross-coupled with the hindered 2-pivaloylamino phenylboronic acid applying a monomode reactor to generate the required biaryl intermediates (Scheme 41).

The reactions were performed in a 1 : 1 mixture of DMF and water for 15 min at a temperature of 150 °C to furnish the biaryl intermediates in excellent yields. After installing the allyl-"handle" under standard conditions, the RCM reaction was investigated under microwave irradiation. The application of Grubbs' 2nd-generation catalyst (G-2 cat) was found to be highly beneficial, irradiating the samples for 5 min in toluene at a temperature of 150 °C (Scheme 41). It is worth mentioning that microwave-irradiation was found to be highly beneficial in generating the highly strained ring system of the target compounds, as the yields under conventional heating conditions were lower.

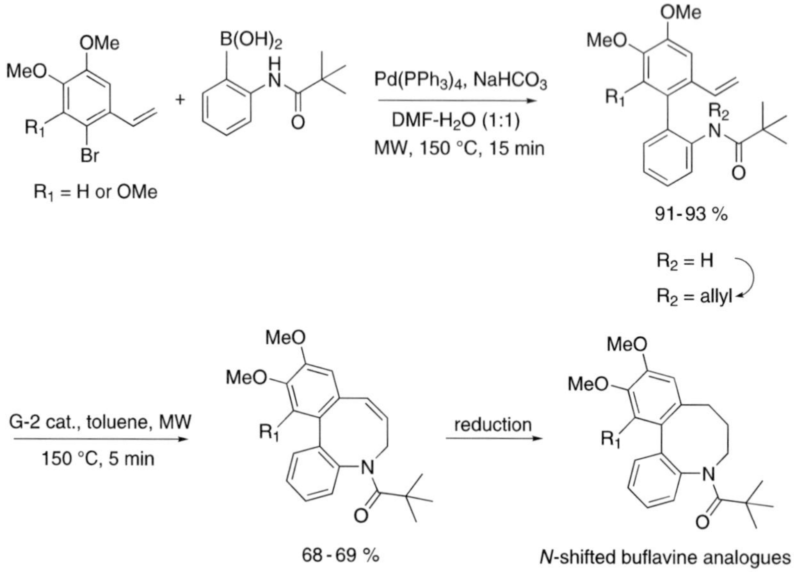

Scheme 41 Microwave-assisted synthesis of *N*-shifted Buflavine analogues

The newly generated double bond was then reduced under Pd-mediated hydrogenation conditions to successfully complete the synthesis.

6
Synthesis of Interesting Building Blocks for Natural Product Synthesis

Apart from valuable applications in the synthesis of natural products and analogues, microwave irradiation has been applied for the generation of generally applicable building blocks for natural product synthesis. A typical example was recently reported by J.C. Menéndez et al. [172]. The authors described the synthesis of a small library of molecules possessing the pyrazino[2,1-*b*]quinazoline-3,6-dione skeleton, which is the structural core of a number of biologically interesting natural products such as Glyantrypine, the Fiscalins and Fumiquinazolines (Fig. 7). More complex natural products bearing this skeleton have been demonstrated to possess important biological properties. As an example, Ardeemin is a very potent inhibitor of multi-drug resistance (MDR) to antitumor drugs, which is an important obstacle to antitumor chemotherapy, and it has been shown that the pharmacophore unit of *N*-acetylardeemin is the pyrazino[2,1-*b*]quinazoline-3,6-dione moiety. Additionally, Spiroquinazoline inhibits the binding of substance P to the human NK-1 receptor [173, 174].

The authors explored an ingenious synthesis involving the direct cyclocondensation of the lactim ethers of the corresponding diketopiperazine

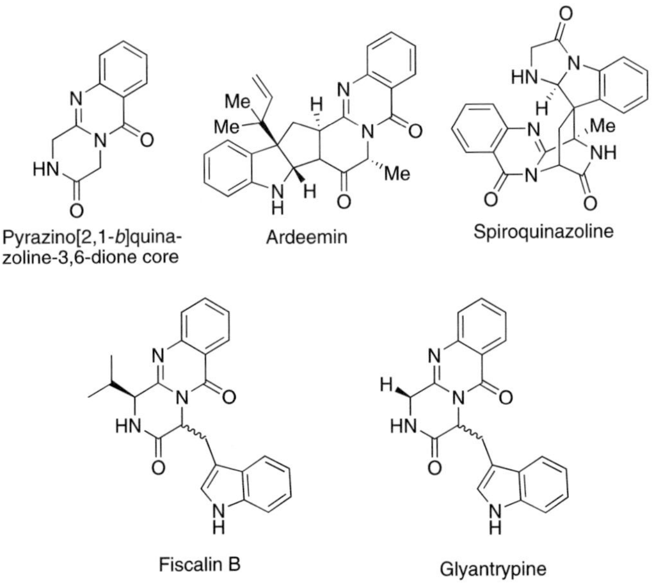

Fig. 7 Natural products with a pyrazino[2,1-*b*]quinazoline-3,6-dione skeleton

(DKP) building blocks with a suitable anthranilic acid derivative (Scheme 42). Even though this method has been explored previously under conventional heating conditions, it has been rather neglected owing to the long and harsh reaction conditions as well as inferior product yields. An additional problem observed during conventional heating was the epimerization of the stereocenters adjacent to the carbonyl groups. The authors carried out the reaction in a domestic microwave oven at 600 W maximum irradiation power with three 1 min heating-2 min cooling cycles with the reactants absorbed on alumina (Scheme 42). Reactions under microwave irradiation proved to be beneficial both in terms of rates and yields in comparison with the conventional heating conditions.

Another interesting example of the use of microwave irradiation in generating interesting building blocks for natural product synthesis, was reported by A.G. Falliset al. [175] regarding the synthesis of the functionalized tricycle[9,3,1,0]pentadecene system of Taxanes, the core skeleton that can be found in Taxol (Fig. 8). Taxol has elicited considerable interest [176–178] due

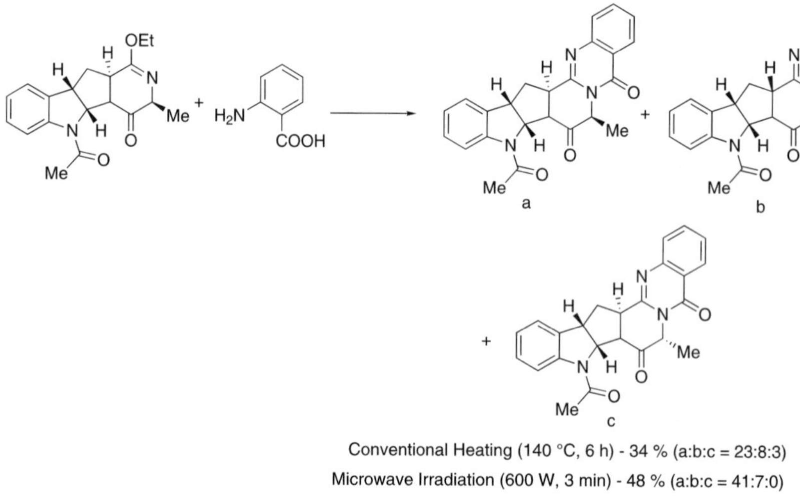

Conventional Heating (140 °C, 6 h) - 34 % (a:b:c = 23:8:3)

Microwave Irradiation (600 W, 3 min) - 48 % (a:b:c = 41:7:0)

Scheme 42 Microwave-assisted synthesis of de-Prenylardeemins

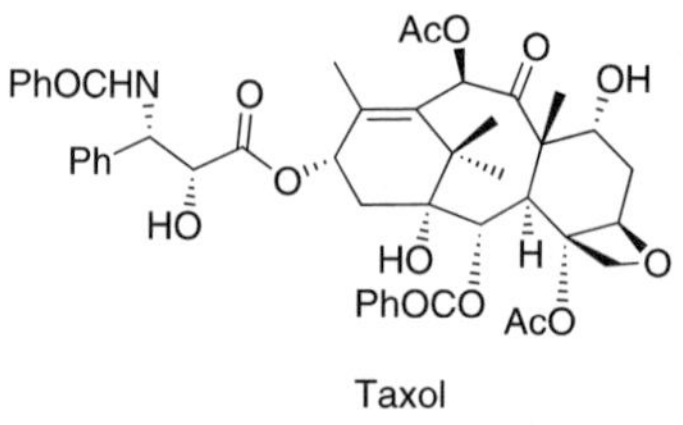

Fig. 8 Taxol skeleton

to its high biological potential as an antitumor agent and the tremendous synthetic challenges in generating this complex ring system (Fig. 8). Diversely functionalized synthetic Taxane cores could serve as building blocks for the synthesis of Taxol analogues for the purpose of drug discovery. As a consequence, a number of synthetic studies have been reported on the synthesis and manipulation of Taxol analogues under conventional heating conditions.

The authors elaborated an innovative synthesis of the Taxane core based on an intramolecular Diels–Alder reaction as the key step. The TBDMS-protected cylohexenone-derived alcohol was converted into the corresponding nitrile intermediate in five steps by known synthetic manipulations (Scheme 43), mainly based on transmetallation protocols. The diene handle for the Diels–Alder reaction was then introduced following a simple but highly efficient four-step procedure. The dienophile for the cycloaddition, the terminal acetylene moiety, was incorporated via lithiation chemistry to furnish the substrate for the intramolecular Diels–Alder cyclization (Scheme 43).

The reaction was carried out in a sealed glass tube applying a "modified" microwave oven. A 0.05 M solution of the substrate together with 1 mol equiv of hydroquinone in toluene was irradiated for 10 h to furnish the target Taxane core in 40% yield. When conventional heating conditions were used, the authors found that the double bond of the cyclohexene ring shifted into conjugation with the acetylenic ketone, preventing the cyclization from taking effect. This example clearly illustrates the advantage of microwave irradiation over conventional heating.

Another interesting investigation was reported by J.A. Prieto et al. describing the microwave-assisted VO(acac)$_2$-catalyzed epoxidation [179] of hindered homoallylic alcohols. From a synthetic point of view this is of particular interest as the literature abounds with examples of syntheses of 3,4-epoxy alcohols, owing to their potential in the generation of valuable intermedi-

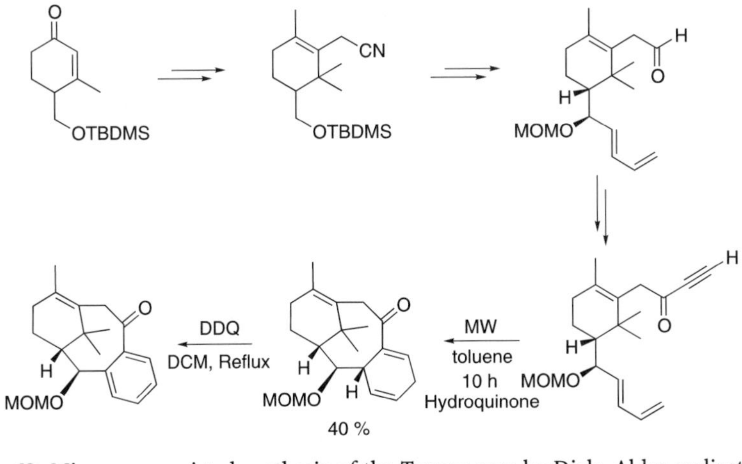

Scheme 43 Microwave-assisted synthesis of the Taxane core by Diels–Alder cyclization

ates such as 1,3-diols [180–186]. The 3,4-epoxy alcohols are usually generated via epoxidation of the corresponding homoallylic alcohols, albeit with minor control of the stereochemistry of the resultant epoxides. The most common method for this purpose is iodocyclization after carbonylation. However, this procedure has been shown to be rather slow and the application of very strong bases, such as *n*-BuLi, is required. Complex mixtures are mostly obtained when *trans* homoallylic alcohols are employed.

The reagent VO(acac)$_2$/*t*-butyl hydroperoxide has been extensively used for the epoxidation of homo-allylic alcohols. While the *cis*-homoallylic alcohols generally give high yields and stereoselectivity, the *trans*-analogues are known to furnish poor selectivities and sluggish reactions, particularly as the steric demand of the alkenol system increases. Therefore, the authors sought to increase the reaction rates by performing the epoxidation under microwave irradiation applying a multimode apparatus. Dramatic accelerations and yield enhancements were observed under these conditions. The authors demonstrated that the reaction times were shortened from 6–10 days, to only 45 min–3 h under microwave-assisted conditions (Scheme 44). In addition, the yields were improved considerably. Further attempts to improve the synthesis, such as the use of solvent-free conditions or doping with an ionic liquid failed to improve the outcome even under microwave irradiation. The diastereoselectivities remained unchanged in comparison with the reactions under conventional heating.

Furthermore, the authors explored the epoxidation of more complex homoallylic alcohols, in view of their application as precursors for the synthesis of the polypropionate chains of Streptovaricins D and U (Scheme 45) [187–189]. These complex alcohols were found to undergo the VO(acac)$_2$-catalyzed epoxidations smoothly under microwave-assisted conditions and the products were isolated in very high yields in only 10 min, after being protected as the corresponding acetonides (Scheme 45). Moreover, in this specific case, the authors were successful in obtaining exclusive diastereoselectivity, as only the *syn*-isomers were isolated as the products of epoxidation.

Another interesting investigation was reported by A.K. Bose et al. [190], where the authors successfully carried out Catalytic Transfer Hydrogena-

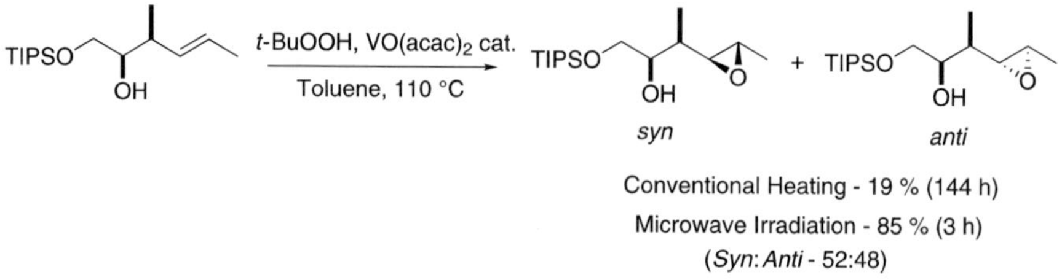

Scheme 44 VO(acac)$_2$-catalyzed epoxidation; conventional heating versus microwave irradiation

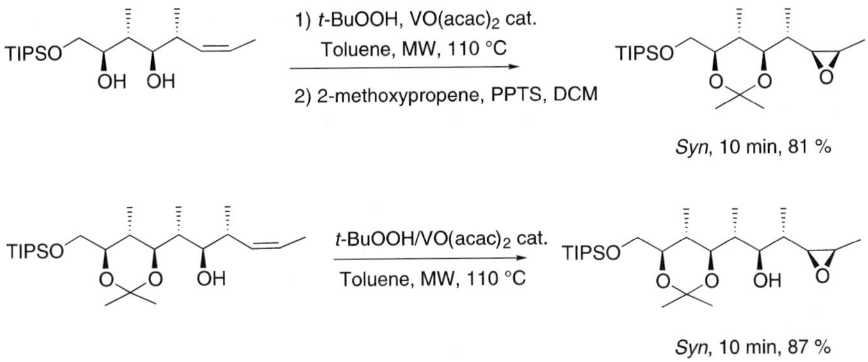

Scheme 45 Microwave-assisted VO(acac)$_2$-catalyzed diastereoselective epoxidations

tion (CTH) under microwave-irradiation [191–194] (Scheme 46). They investigated the reduction via CTH of ring substituents of diverse β-lactams, containing alkene or alkylidene groups or conjugated esters, applying an unmodified domestic microwave oven. A beaker or an Erlenmeyer flask was used as the reaction vessel, in the presence of a beaker containing water as the heat sink. The authors used this simple yet effective set-up for the CTH reactions of differently substituted β-lactams in ethylene glycol using an irradiation power of 600–1000 W. Ammonium formate or hydrazine hydrate were used as the hydrogen transfer sources, and the reactions were catalyzed either by Pd – C or Raney nickel. The reactions were found to proceed with the very high yields of 80–90% in only 2–3 min of irradiation times.

Curiously, the *N*-benzyl moiety on a variety of β-lactams was untouched by the microwave-assisted CTH reactions. The authors further elaborated this interesting strategy for the reduction of the double bond in an unsaturated sugar moiety, without inducing the ring fission of the attached β-lactam side chains (Scheme 46).

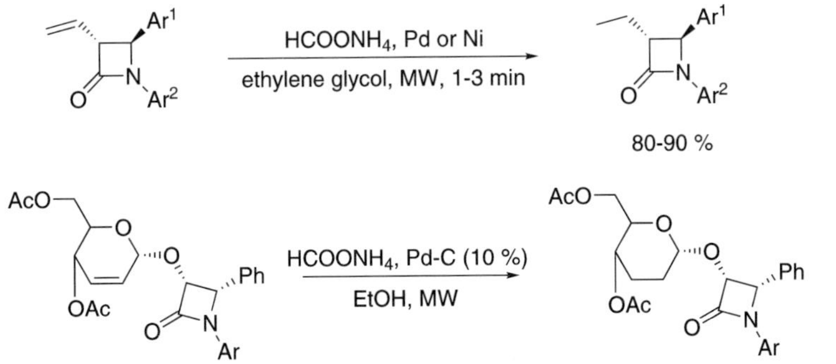

Scheme 46 Microwave-assisted CTH reactions of β-lactams

7
Summary and Overview

Natural product synthesis is undoubtedly one of the most challenging topics for synthetic organic chemists and has received much attention during the past decades. In contrast, the application of microwave irradiation in natural product synthesis has only been fragmentarily explored and has so far not been systematically applied for the majority of possible conversions in synthetic sequences. Nevertheless, its usefulness has already been demonstrated in the synthesis of complex systems such as the taxol skeleton. Among the advantages of microwave irradiation can the dramatic shortening of reaction time and in many cases also higher yields be mentioned. Sometimes a different outcome of the reaction can be observed, for example an induction of a more favorable reaction ratio. It is clear that shorter heating periods, caused by the application of microwave irradiation, might be advantageous for these complex and sometimes fragile and labile systems.

Concomitant with the tremendously increased application of microwave irradiation in synthesis during the last 5 years, we might expect a rise in interest in its application for natural product chemistry, as more and more natural product chemists are discovering the benefits of this powerful technique.

References

1. Ridley DD, Ritchie E, Taylor WC (1970) Aust J Chem 23:147
2. Lytollis W, Scannell RT, An H, Murty VS, Reddy KS, Barr JR, Hecht SM (1995) J Am Chem Soc 117:12683
3. Singh US, Scannell RT, An H, Carter BJ, Hecht SM (1995) J Am Chem Soc 117:12691
4. Scannell RT, Barr JR, Murty VS, Reddy KS, Hecht SM (1988) J Am Chem Soc 110:3650
5. Fürstner A (2002) Chem Eur J 8:1856
6. Agrofoglio LA, Nolan SP (2005) Topics Curr Med Chem 5:1541
7. Nicolaou KC, Bulger PG, Sarlah D (2005) Angew Chem Int Ed 44:4490
8. Grubbs RH (2004) Tetrahedron 60:7117
9. Schmidt B, Hermanns J (2004) Topics Organometallic Chem 13:223
10. Deiters A, Martin SF (2004) Chem Rev 104:2199 and references cited therein
11. Mulzer J, Oehler E (2004) Topics Organometallic Chem 13:269
12. Diver ST, Giessert AJ (2004) Chem Rev 104:1317
13. Furstner A, Davies PW (2005) Chem Commun, p 2307
14. Thiel OR (2004) In: Transition Metals for Organic Synthesis. 2nd Ed. Wiley-VCH, Weinheim 1:321
15. Michell SS, Rhodes D, Bushman FD, Faulkner D (2000) Org Lett 2:1605
16. Kiyota H, Dixon DJ, Luscombe CK, Hettstedt S, Ley SV (2002) Org Lett 4:3223
17. Dixon DJ, Ley SV, Reynolds DJ (2002) Chem Eur J 8:1621
18. Gonzalez N, Rodriguez J, Jiminez C (1999) J Org Chem 64:5705
19. Kashiwada Y, Yamazaki K, Ikeshiro Y, Yamasishi T, Fujioka T, Mihashi K, Mizuki K, Cosentino LM, Fowke K, Morris-Natschke SL, Lee K-H (2001) Tetrahedron 57:1559

20. Mamoru E, Hiroshi E (1982) Jpn Kokai Tokko Koho JP 82–28:080
21. Kang Y, Mei Y, Du Y, Jin Z (2003) Org Lett 5:4481
22. Saimoto H, Yoshida K, Murakami T, Morimoto M, Sashiwa H, Shigemasa Y (1996) J Org Chem 61:6768
23. Narkhede DD, Iyer PR, Iyer CSR (1989) J Nat Prod 52:502
24. Brink AJ, Rall GJH, Breytenbach JC (1977) Phytochemistry 16:273
25. Subburaj K, Katoch R, Murugesh MG, Trivedi GK (1997) Tetrahedron 53:12621
26. Volkmann RA, Andrews GC, Johnson WS (1975) J Am Chem Soc 97:4777
27. Corey EJ, Ohno M, Mitra RB, Vatakencherry PA (1964) J Am Chem Soc 86:478
28. Lei B, Fallis AG (1993) J Org Chem 58:2186
29. Abegaz BM (2002) Phytochem Rev 1:299
30. Aguilar R, Benavides A, Tamariz J (2004) Synthetic Commun 34:2719
31. Kad GL, Singh V, Khurana A, Singh J (1998) J Nat Prod 61:297
32. Grainger RS, Patel A (2003) Chem Commun, p 1072
33. Lewis FD, Reddy GD, Bassani DM (1993) J Am Chem Soc 115:6468
34. Lewis FD, Reddy GD, Bassani DM, Schneider S, Gahr M (1994) J Am Chem Soc 116:597
35. Subbaraju GV, Manhas MS, Bose AK (1991) Tetrahedron Lett 32:4871
36. Ezaki M, Iwami M, Yamashita M, Hashimoto S, Komori T, Umehara K, Mine Y, Kohsaka M, Aoiki H, Imanaka H (1985) J Antibiot 38:1453
37. Uchida I, Shigematsu N, Ezaki M, Hashimoto M, Hatsuo A, Hiroshi I (1985) J Antibiot 38:1462
38. Uchida I, Ezaki M, Shigematsu N, Hashimoto M (1985) J Org Chem 50:1341
39. Chang CC, Morton GO, James JC, Siegel MM, Kuck NA, Testa RT, Borders DB (1991) J Antibiot 44:674
40. Lépine R, Zhu J (2005) Org Lett 7:2981
41. Kappe CO (2004) Angew Chem Int Ed 43:6250
42. Kappe CO (2002) Curr Opin Chem Biol 6:314
43. Kappe CO, Dallinger D (2006) Drug Discov 5:51
44. Ersmark K, Larhed M, Wannberg J (2004) Curr Opin Drug Discov Dev 7:417
45. Mavandadi F, Lidstroem P (2004) Curr Top Med Chem 4:773
46. Alexandre F-R, Domon L, Frere S, Testard A, Thiery V, Besson T (2003) Mol Diver 7:273
47. Somei M, Yamada F (2005) Nat Prod Rep 22:73
48. Somei M, Yamada F (2004) Nat Prod Rep 21:278
49. Toyota M, Ihara M (1998) Nat Prod Rep 15:327, and references cited therein
50. Franco LH, Palermo JA (2003) Chem Pharm Bull 51:975
51. Hernández FL, Bal de Kier JE, Puricelli L, Tatian M, Seldes AM, Palermo JA (1998) J Nat Prod 61:1130
52. Butler MS, Capon RJ, Lu CC (1992) Aust J Chem 45:1871
53. Perry NB, Ettouati L, Litaudon M, Blunt JW, Munro MHG, Parkin S, Hope H (1994) Tetrahedron 50:3987
54. Rauwald HW, Kobar M, Mutschler E, Lambrecht G (1992) Planta Med 58:486
55. Sharaf MH M, Schiff PL Jr, Tackie AN, Phoebe CH Jr, Martin GE (1996) J Heterocycl Chem 33:239
56. Cimanga K, De Bruyne T, Pieters L, Claeys M, Vlietinck A (1996) Tetrahedron Lett 37:1703
57. Cimanga K, De Bruyne T, Pieters L, Claeys M, Vlietinck A (1997) J Nat Prod 60:688
58. Fresneda PM, Molina P, Delgado S (2001) Tetrahedron 51:6197
59. Kumemura T, Choshi T, Hirata A, Sera M, Takahashi Y, Nobuhiro J, Hibino S (2005) Chem Pharm Bull 53:393

60. Kohno J, Hiramatsu H, Nishio M, Sakurai M, Okuda T, Komatsubara S (1999) Tetrahedron 55:11247
61. Kohno J, Sakurai M, Kameda N, Nishio M, Kawano K, Kishi N, Okuda T, Komatsubara S (1999) J Antibiot 52:913
62. Kumemura T, Choshi T, Hirata A, Sera M, Takahashi Y, Nobuhiro J, Hibino S (2003) Heterocycles 61:13
63. Pouìlhes A, Langlois Y, Chiaroni A (2003) Synlett, p 1488
64. Cox ED, Cook JM (1995) Chem Rev 95:1797
65. Foderato TA, Barrows LR, Lassota P, Ireland CM (1997) J Org Chem 62:6064
66. Ma ZZ, Hano Y, Nomura T, Chen JJ (1997) Heterocycles 46:541
67. Slichenmyer WJ, Rowinsky EK, Donehower RC, Kaufmann SH (1993) J Nat Cancer Inst 85:271
68. Yadav JS, Reddy BVS (2002) Tetrahedron Lett 43:1905
69. Lipińska T (2002) Tetrahedron Lett 43:9565
70. Costa-Campos L, Lara DR, Nunes DS, Elisabetsky E (1998) Pharmacol Biochem Behav 60:133
71. Lipińska T, Guibe-Jampel E, Petit A, Loupy A (1999) Synth Commun 29:1349
72. Beljanski M, Crochet S (1996) Int J Oncol 8:1143 and references cited therein
73. Rykowski A, Lipińska T (1997) Pol J Chem 71:83
74. Lipińska T (2004) Tetrahedron Lett 45:8831
75. Lipińska T (2005) Tetrahedron 61:8148
76. Wolkenberg SE, Wisnoski DD, Leister WH, Wang Y, Zhao ZJ, Lindsley CW (2004) Org Lett 6:1453
77. Cui B, Zheng BL, He K, Zheng QY J (2003) Nat Prod 66:1101
78. Liu J-F, Lee J, Dalton AM, Bi G, Yu L, Baldino CM, McElory E, Brown M (2005) Tetrahedron Lett 46:1241
79. Wolfe JF, Rathman TL, Sleevi MC, Campbell JA, Greenwood TD (1990) J Med Chem 33:161
80. Buzas A, Hoffmann C (1959) Bull Soc Chim Fr, p 1889
81. Welch WM, Ewing FE, Huang J, Menniti FS, Pagnozzi MJ, Kelly K, Seymour PA, Guanowsky V, Guhan S, Guinn MR, Critchett D, Lazzaro J, Ganong AH, DeVries KM, Staigers TL, Chenard BL (2001) Bioorg Med Chem Lett, p 177
82. Liu J-F, Kaselj M, Isome Y, Ye P, Sargent K, Sprague K, Cherrak D, Wilson CJ, Si Y, Yohannes D, Ng S-C (2006) J Comb Chem 8:7
83. Penn J, Mantle PG, Bilton JN, Sheppard RN (1992) J Chem Soc, Perkin Trans 1, p 1495
84. Numata A, Takahashi C, Matsushita T, Miyamoto T, Kawai K, Usami Y, Matsumura E, Inoue M, Ohishi H, Shingu T (1992) Tetrahedron Lett 33:1621
85. Takahashi C, Matsushita T, Doi M, Minoura K, Shingu T, Kumeda Y, Numata A (1995) J Chem Soc, Perkin Trans 1, p 2345
86. Liu J-F, Ye P, Zhang B, Bi G, Sargent K, Yu L, Yohannes D, Baldino CM (2005) J Org Chem 70:6339
87. Tietze LF (1996) Chem Rev 96:115
88. Tietze LF, Haunert F (2000) In: Shibasaki M, Stoddart JF, Vögtle F (eds) Stimulating Concepts in Chemistry. Wiley, Weinheim, p 39
89. Liu J-F, Ye P, Sprague K, Sargent K, Yohannes D, Baldino CM, Wilson CJ, Ng S-C (2005) Org Lett 7:3363
90. Liu J-F, Kaselj M, Isome Y, Chapnick J, Zhang B, Bi G, Yohannes D, Yu L, Baldino CM (2005) J Org Chem 70:10488
91. Goetz MA, Monaghan RL, Chang RSL, Ondeyka J, Chen TB, Lotti VJ (1988) J Antibiot 41:875

92. Sun HH, Barrow CJ, Sedlock DM, Gillum AM, Cooper R (1994) J Antibiot 47:515
93. Pla D, Marchal A, Olsen CA, Albericio F, Alvarez M (2005) J Org Chem 70:8231
94. Cironi P, Albericio F, Alvarez M (2004) In: Gribble GW, Joule JA (eds) Progress in Heterocyclic Chemistry. Pergamon, Oxford, UK, vol 16, p 1
95. Facompré M, Tardy C, Bal-Mahieu C, Colson P, Pérez C, Manzanares I, Cuevas C, Bailly C (2003) Cancer Res 63:7392
96. Sharma U, Bora U, Boruah RC, Sandhu JS (2002) Tetrahedron Lett 43:143
97. Longchar M, Bora U, Boruah RC, Sandhu JS (2002) Synthetic Commun 32:3611
98. Skoda-Földes R, Pfeiffer P, Horváth J, Tuba Z, Kollár L (2002) Steroids 67:709
99. Skoda-Földes R, Kollár L, Heil B, Gálik G, Tuba Z, Arcadi A (1991) Tetrahedron Asymm 2:633
100. Skoda-Földes R, Csákai Z, Kollár L, Horváth J, Tuba Z (1995) Steroids 60:812
101. Das B, Venkataiah B, Kashinatham A (1999) Tetrahedron 55:6585
102. Kupchan SM, Eakin MA, Thomas AM (1971) J Med Chem 14:1147
103. Mew D, Baiza F, Towers GHN, Levy JG (1982) Planta Med 45:23
104. Picman AK, Towers GHN (1983) Biochem Syst Ecol 11:321
105. Sharma GL, Bhutani KK (1988) Planta Med 54:120
106. Hopper M, Kirby GC, Kulkarni MM, Kulkarni SN, Nagasampagi BA, O'Neill MJ, Philipson JD, Rojatkar SR, Warhurs DC (1990) Eur J Med Chem 25:717
107. Gallo ILC, Whang-peng J, Adamson ILK (1971) J Nat Cancer Inst 46:789
108. Wall ME, Wani MC (1977) Ann Rev Pharmacol Toxicol 17:117
109. Pendrak I, Wittrock R, Kingsbury WD (1995) J Org Chem 60:2912
110. Pendrark I, Berney S, Wittrock R, Lambert DM, Kingsbury WD (1994) J Org Chem 59:2623
111. Das B, Madhusudhan P, Kashinatham A (1998) Tetrahedron Lett 39:431
112. Das B, Srinivas KVNS (2002) Synth Commun 32:3027
113. Buckingham J (ed) (1994) Dictionary of Natural Products. 1st Ed. Chapman and Hall, London, vol 1, p 640
114. Gharbi SA, Beal JL, Doskotch RW, Mitscher LA (1973) Lloydia 36:349
115. Das B, Srinivas KVNS (2004) Synth Commun 34:199
116. Baran PS, O'Malley DP, Zografos AL (2004) Angew Chem Int Ed 43:2674
117. Borrelli F, Campagnuolo C, Capasso R, Fattorusso E, Taglialatela-Scafati O (2004) Eur J Org Chem 15:3227
118. Campagnuolo C, Fattorusso E, Taglialatela-Scafati O (2003) Eur J Org Chem 2:284
119. Lindsley CW, Wisnoski DD, Wang Y, Leister WH, Zhao Z (2003) Tetrahedron Lett 44:4495
120. Ohmoto T, Koike K (1989) In: Brossi A (ed) The Alkaloids. Academic Press, New York, vol 36, p 135
121. Haynes HF, Nelson ER, Price JR (1952) Aust J Sci Res Ser A 5:387
122. Nelson ER, Price JR (1952) Aust J Sci Res Ser A 5:563
123. Nelson ER, Price JR (1952) Aust J Sci Res Ser A 5:68
124. Anderson LA, Harris A, Phillipson JD (1983) J Nat Prod 46:374
125. Arisawa M, Kinghorn AD, Cordell GA, Farnsworth NR (1983) J Nat Prod 46:222
126. Ohmoto T, Nikaido T, Koide K, Kohda K, Sankawa U (1988) Chem Pharm Bull 36:4588
127. Ouyang Y, Koide K, Ohmoto T (1994) Phytochemistry 36:1543
128. Benson SC, Li J-H, Snyder JK (1992) J Org Chem 57:5285
129. Lindsley CW, Bogusky MJ, Leister WH, McClain RT, Robinson RG, Barnett SF, Defeo-Jones D, Ross CW, Hartman GD (2005) Tetrahedron Lett 46:2779
130. Chérouvrier J-R, Carreaux F, Bazureau JP (2002) Tetrahedron Lett 43:3581

131. Cafieri F, Fattorusso E, Mangani A, Tagliakatela-Scafati O (1996) Tetrahedron Lett 37:3587
132. Boehm JC, Gleason JG, Pendrak I, Sarau HM, Schmidt B, Foley JJ, Kingsbury WD (1993) J Med Chem 36:3333
133. Kuo F-M, Tseng M-C, Yen Y-H, Chu Y-H (2004) Tetrahedron 60:12075
134. Horiguchi Y, Nakamura M, Saitoh T, Sano T (2003) Chem Pharm Bull 51:1368
135. Horiguchi Y, Nakamura M, Kida A, Kodama H, Saitoh T, Sano T (2003) Heterocycles 59:691
136. Cheng CC, Chu Y-H (1999) J Comb Chem 1:461
137. Papadopoulou D, Spyros IP, Varvoglis A (1998) Tetrahedron Lett 39:2865
138. Mayer JP, Bankaitis-Davis D, Zhang J, Beaton G, Bjergarde K, Anderson CM, Goodman BA, Herrera CJ (1996) Tetrahedron Lett 37:5633
139. Horiguchi Y, Kodama H, Nakamura M, Yoshimura T, Hanezi K, Hamada H, Saitoh T, Sano T (2002) Chem Pharm Bull 50:253
140. Tseng M-C, Liang Y-M, Chu Y-H (2005) Tetrahedron Lett 46:6131
141. Tietze LF, Wiegand JM, Vock CA (2004) Eur J Org Chem, p 4107
142. Micheli RA, Hajos ZG, Cohen N, Parrish DR, Portland LA, Sciamanna W, Scott MA, Wehrli PA (1975) J Org Chem 40:675
143. Eder U, Sauer G, Wiechert R (1971) Angew Chem 83:492
144. Eder U, Sauer G, Wiechert R (1971) Angew Chem Int Ed Engl 10:496
145. Kupchan SM, Britton RW, Ziegler MF, Gilmore CJ, Restivo RJ, Bryan RF (1973) J Am Chem Soc 95:1335
146. Wang RW, Rebhum LI, Kupchan SM (1977) Cancer Res 37:3071
147. Sackett DL (1993) Pharmacol Ther 59:163
148. Zavala F, Guenard D, Robin J-P, Brown E (1980) J Med Chem 23:546
149. Tomioka K, Kubota Y, Kawasaki H, Koga K (1989) Tetrahedron Lett 30:2949
150. Kubota Y, Kawasaki H, Tomioka K, Koga K (1993) Tetrahedron 49:3081
151. Beryozkina T, Appukkuttan P, Mont N, Van der Eycken E (2006) Org Lett 8:487
152. Miyaura N, Suzuki A (1995) Chem Rev 95:2457
153. Diederich F, Stang PJ (eds) (1998) Metal-Catalyzed Cross-coupling Reactions. Wiley, Weinheim, p 517
154. Suzuki A (1999) J Organometallic Chem 576:147
155. Kotha S, Lahiri K, Kashinath D (2002) Tetrahedron 58:9633
156. Leadbeater NE (2005) Chem Commun 23:2881
157. Apukkuttan P, Van der Eycken E, Dehaen W (2005) Synlett, p 127
158. Appukkuttan P, Orts AB, Chandran RP, Goeman JL, Van der Eycken J, Dehaen W, Van der Eycken E (2004) Eur J Org Chem, p 3277
159. Ishida Y, Sasaki Y, Kimura Y, Watanabe K (1985) J Pharmacobiodyn 8:917
160. Ishida Y, Sadamune K, Kobayashi S, Kihara M (1983) J Pharmacobiodyn 6:391
161. Viladomat F, Bastida J, Codina C, Campbell WE, Mathee S (1995) Phytochemistry 40:307
162. Hoarau C, Couture A, Deniau E, Grandclaudon P (2002) J Org Chem 67:5846
163. Patil PA, Snieckus V (1998) Tetrahedron Lett 39:1325
164. Sahakitpichan P, Ruchiravat S (2003) Tetrahedron Lett 44:5239
165. Lidström P, Tierney J, Wathey B, Westman J (2001) Tetrahedron 57:9225
166. Lidström P, Westman J, Lewis A (2002) Comb Chem High Throughput Screening 6:441
167. Larhed M, Moberg C, Hallberg A (2002) Acc Chem Res 35:717
168. Loupy A (2002) Microwaves in Organic Synthesis. Wiley, Weinheim, Germany, and references therein

169. Kappe CO (2004) Angew Chem Int Ed 43:6250
170. Hayes BL (2004) Aldrichim Acta 37:66
171. Appukkuttan P, Dehaen W, Van der Eycken E (2005) Org Lett 7:2723
172. Cledera P, Sánchez JD, Caballero E, Avendaño C, Ramos MT, Menéndez JC (2004) Synlett, p 803
173. Hochlowski JE, Mullally MM, Spanton SG, Whittern DN, Hill P, McAlpine JB (1993) J Antibiotics 46:380
174. Barrow CJ, Sun HH (1994) J Nat Prod 57:471
175. Lu Y-F, Fallis AG (1993) Tetrahedron Lett 34:3367
176. Wani MC, Taylor HL, Wall ME, Coggan P, McPhail AT (1971) J Am Chem Soc 93:2325
177. Pamess J, Horwitz SB (1981) J Cell Biol 91:479
178. Chapin SJ, Bulinski JC, Gundersen GG (1991) Nature 349:24
179. Torres G, Torres W, Prieto JA (2004) Tetrahedron 60:10245
180. Corey EJ, Hase T (1979) Tetrahedron Lett 20:335
181. Lipshutz BH, Kozlowski J (1984) J Org Chem 49:1147
182. Lipshutz BH, Barton JC (1988) J Org Chem 53:4495
183. Mohr P (1992) Tetrahedron Lett 33:2455
184. Burova SA, McDonald FE (2002) J Am Chem Soc 124:8188
185. Martin HJ, Drescher M, Mulzer J (2000) Angew Chem Int Ed 39:581
186. Wang Z, Schreiber SL (1990) Tetrahedron Lett 31:312
187. Rinehart K, Maheshwari ML, Antosz FJ, Mathur HH, Sasaki K, Schacht RJ (1971) J Am Chem Soc 93:6273
188. Rinehart KL, Antosz FJ, Sasaki K, Martin PK, Maheshwari ML, Reussre F, Li HL, Moran D, Wiley PF (1974) Biochemistry 13:861
189. Miyashita M, Shiratani T, Kawamine K, Hatakeyama S, Miyazawa M, Irie H (1996) Chem Commun, p 1027
190. Banik BK, Barakat KJ, Wagle DR, Manhas MS, Bose AK (1999) J Org Chem 64:5746
191. Sharma A, Kumar V, Sinha AK (2006) Adv Synth Catal 348:354
192. Arefalk A, Larhed M, Hallberg A (2005) J Org Chem 70:938
193. Berthold H, Schotten T, Honig H (2002) Synthesis 11:1607
194. Daga MC, Taddei M, Varchi G (2001) Tetrahedron Lett 42:5191

Top Curr Chem (2006) 266: 49–101
DOI 10.1007/128_058
© Springer-Verlag Berlin Heidelberg 2006
Published online: 11 August 2006

Microwave-assisted Heterocyclic Chemistry

Edgars Suna (✉) · Ilga Mutule

Latvian Institute of Organic Synthesis, Aizkraukles 21, 1006 Riga, Latvia
edgars@osi.lv

Abstract Recent developments in the microwave-assisted synthesis of heterocycles are surveyed with the focus on diversity-oriented multi-component and multi-step one-pot procedures. Both solution- and solid-phase as well as polymer-supported methodologies for the preparation of libraries of heterocycles are reviewed. Advantages of microwave dielectric heating are highlighted by comparison with conventional thermal conditions.

Keywords Heterocycles · High-throughput synthesis · Microwaves ·
Multi-component reactions

Abbreviations

BEMP	2-tert-butylimino-2-dimethylamino-1,3-dimethylperhydro-1,3,2-diazaphosphorine
DBU	1,8-diazabicyclo[5.4.0]undec-7-ene

DHP   dihydropyridine
DIEA   diisopropylethylamine
DMAP   4-(dimethylamino)pyridine
DMF N,N-dimethylformamide
DMF–DMA N,N-dimethylformamide dimethyl acetal
HBTU O-benzotriazol-1-yl-N,N,N',N'-tetramethyluronium hexafluorophosphate
HMDS   hexamethyldisilazane
IL   ionic liquids
MCR   multi-component reaction
MW   microwaves
NBS N-bromosuccinimide
NMP N-methyl-2-pyrrolidinone
PEG   poly(ethylene glycol)
PPTS   pyridinium p-toluenesulfonate
PS   polymer supported
RCM   ring-closing metathesis
SPE   solid-phase extraction
SPOS   solid-phase organic synthesis
TFA   trifluoroacetic acid
THF   tetrahydrofuran
TMS   trimethylsilyl
Ts   tosyl (p-toluenesulfonyl)

1
Introduction

Heterocycles are among the most frequently encountered scaffolds in drugs
and pharmaceutically relevant substances. Because of the drug-like character and considerable range of structural diversity, large collections or libraries of diverse heterocycles are routinely employed in high-throughput
screening at early stages of drug discovery programs. Furthermore, a heterocyclic core is propitious for variations of substitution patterns during
structure-activity relationship (SAR) studies. Consequently, relatively small
(< 300 membered) focused libraries of heterocycles are frequently generated
for SAR studies during the development and optimization of lead structures [1]. Among the different synthetic methodologies available for the production of compound libraries, a multi-component synthesis of heterocycles
is especially suitable because high structural diversity in the desired scaffolds can be introduced in a single synthetic step simply by proper variation
of precursors. One-pot multi-step synthesis constitutes a convenient alternative as it can avoid workup and purification of the intermediates between
different reaction steps. Similarly, domino reactions where "two or more
bond-forming transformations take place under the same reaction conditions
without adding additional reagents and catalysts, and in which the subsequent reactions result as a consequence of the functionality formed in the

previous step" [2] are also perfectly suited for the generation of libraries of heterocyclic compounds.

The competition in the field of drug discovery has helped to identify speed of synthesis as a top priority in drug development. Consequently, technologies that could accelerate and facilitate both synthesis and screening of substances have become highly desirable. The advent of microwave technology enabled organic and medicinal chemists to reduce the time of synthesis of heterocycles from days and hours to minutes and even seconds. In addition, suppressed formation of side-products and improved yields have frequently been observed under microwave heating conditions. Moreover, the extremely fast and efficient dielectric heating mode has sometimes identified unusual reactivities that could not be achieved by conventional heating. The frequently observed acceleration of the reaction speed by microwave heating could be rationalized by considering the Arrhenius law [$k = A \exp(- E_a/RT)$]. Thus, Baghurst and Mingos calculated $t_{9/10}$ lifetimes of a typical first-order reaction for a given activation energy E_a and pre-exponential factor A [3]. Accordingly, 90% completion of the reaction would require 13.4 h in a solvent refluxing at 77 °C, and just 1.61 s (!) if performed in a closed vessel where the temperature is maintained at 227 °C. Thus, a reduction of the reaction time by five orders of magnitude could be achieved by performing reactions 150 °C above the boiling point of the solvent. In the meantime, Strauss and co-workers demonstrated that for the thermally homogeneous reactions at known temperatures the kinetics of microwave-heated and conventionally heated reactions do not differ significantly [4]. Consequently, the authors concluded that "if the microwave conditions can be adequately mimicked, conventional heating will produce a comparable outcome" [5]. On the other hand, reproduction of the dielectric heating conditions is usually a challenging task given the extremely rapid dielectric heating mode originating from the direct coupling of microwave energy with molecules (solvents, reagents, catalysts) that are present in the reaction mixture. Moreover, in-core volumetric heating can suppress the formation of side-products on the hot surface of a reaction vessel, resulting in cleaner reactions. Eventually, mimicking the dielectric heating conditions is virtually impossible in the case of thermally heterogeneous conditions, which operate when inorganic supports such as relatively microwave transparent silica, alumina and clay or strongly absorbing graphite are employed. In this case, selective heating of strongly microwave absorbing reagents or heterogeneous catalysts in a less polar reaction medium (specific microwave effects) presumably accounts for the acceleration of the reaction speed.

Although being essentially reproducible in an oil-bath, thermally homogeneous microwave dielectric heating conditions nevertheless are the preferred methodology for the synthesis of heterocycles on a laboratory scale, especially when high temperatures and long reaction times are necessary for the reaction to occur. Not surprisingly, microwave-assisted synthesis of heterocycles has been a subject of numerous reviews during the last 5 years [6–10]. The operational simplicity of the microwave technology has led to the design of purpose-built microwave equipment possessing high levels of automation for use in drug development and, particularly, for synthesis of heterocycle libraries. Therefore, the goal of the present review is to demonstrate applications of microwave dielectric heating to facilitate rapid parallel or sequential synthesis of solution- and solid-phase libraries of heterocycles. Consequently, the review focuses preferentially on diversity-oriented (multi-component, multi-step one-pot as well as domino) microwave-assisted synthesis of heterocycles [11]. Generally, procedures are surveyed only where microwave technology is employed in a ring-forming step and examples dealing with modification of pre-existing heterocyclic scaffolds are not discussed.

Last but not least, workup of the reaction mixture and product isolation have frequently been "rate limiting steps" in rapid and automated production of heterocycle libraries. Clearly, acceleration of the reaction speed by microwaves is of little advantage given the subsequent laborious and time-consuming isolation and purification of multi-membered libraries. Therefore, rapid synthesis is to be accompanied by simple and efficient workup-purification sequences, preferentially amenable to automation. In the simplest cases, proper choice of solvents allowed the precipitation of the desired products in acceptable purity directly from the reaction mixture. Routine workup and purification approaches are based on standard phase separation techniques, such as attachment of reagents or catalysts to a solid (or soluble) polymer matrix or the introduction of fluorous tags followed by fluorous-phase extraction. Notably, under microwave flash heating conditions slow heterogeneous solid-phase reactions could be efficiently accelerated without any degradation of the polymer backbone. Consequently, the workup and purification issues are also considered in the review.

Rather than being comprehensive, this review is aimed at demonstrating tendencies and approaches of diversity oriented synthesis of heterocycles, that appeared in the literature from 2002 through January 2006. Synthesis of heterocycles in dedicated microwave systems is discussed preferentially as this type of equipment renders higher reproducibility and is more amenable to automation than custom-designed or household ovens. The survey of heterocycle synthesis is arranged according to ring-size and number of heteroatoms in the ring.

2
Five-membered Heterocycles

2.1
Pyrrolidines, Pyrrolines and Pyrroles

A series of pyrrolidines was conveniently prepared in a microwave-assisted double alkylation of aniline derivatives with alkyl dihalides in water in the presence of K_2CO_3 as a base (Scheme 1) [12, 13]. Although the reaction mixture could be regarded as a multi-phase system, as neither reactant was soluble in the mildly basic aqueous medium, the microwave-assisted reaction proceeded readily without the use of phase-transfer reagents. The amount of side-reactions such as hydrolysis of bromides to alcohols in an alkaline reaction medium was substantially suppressed compared to the conventional thermal conditions. The reaction conditions were sufficiently mild to tolerate a variety of functional groups in anilines such as hydroxyls, ketones and esters. Alkyl bromides and tosylates were equally efficient as alkylating agents. Notably, isolation and purification comprised simply of phase separations (filtration or decantation) of the desired product from the aqueous media.

The use of microwave dielectric heating led to the reduction of reaction time from hours to minutes (Scheme 1, pyrrolidine **1**). As the magni-

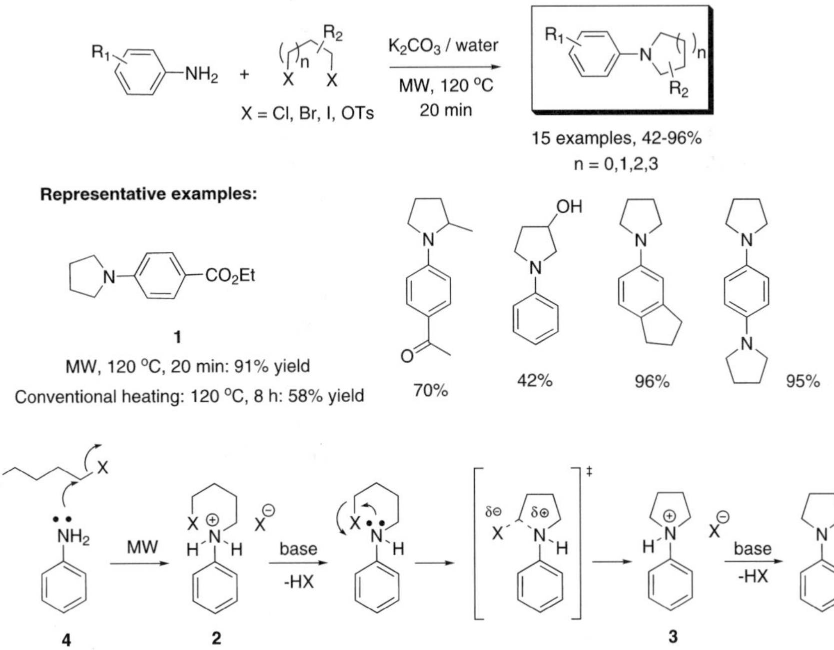

Scheme 1 Microwave-promoted synthesis of *N*-aryl pyrrolidines in neat water

tude of heating by microwaves depends on the dielectric properties of the molecules, the greater the polarity of the reacting species, the more efficient is the absorbance of the microwave energy. Consequently, acceleration of the reaction by microwaves has been attributed to a greater stabilization of the charged intermediates **2** and **3** by the dipole–dipole interaction with the microwave electric field when compared to the less polar ground state **4** (Scheme 1).

One of the most efficient approaches towards pyrrolines is ring-closing metathesis (RCM) of suitably substituted dienes. The ring closure of *N,N*-diallyl *p*-toluensulfonamide, yielding *N*-tosyl-2,5-dihydropyrrole has been often used as a model reaction in the evaluation of novel metathesis catalysts. While in the case of simple dienes the RCM reaction works sufficiently well even at room temperature, tri- and tetrasubstituted olefins carrying electron-withdrawing groups require elevated temperatures (40 °C; refluxing in CH_2Cl_2) to enter the RCM reaction. The advent of the air- and thermally stable Grubbs "2nd generation" ruthenium metathesis catalyst **5** (Scheme 2) enabled the use of microwave heating as a tool to enhance the effectiveness of the RCM method. Subsequently, several groups have reported successful applications of dielectric heating to accelerate the synthesis of different heterocycles, including pyrrolines [14–17]. Starting dienes for the RCM reaction were conveniently synthesized in a three component aza-Baylis–Hillman reaction, followed by *N*-allylation (Scheme 2) [16, 17]. A cyclization reaction in the presence of Grubbs catalyst **5** was completed in 1–2 minutes under dielectric heating conditions at 100 °C [16]. The 2,5-dihydropyrrole products were isolated in high yields after simple filtration of the reaction

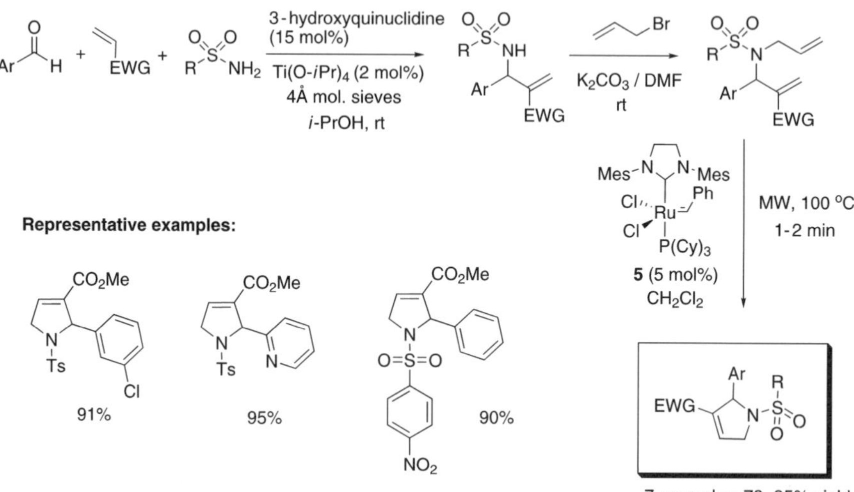

Scheme 2 Rapid preparation of functionalized dihydropyrroles using RCM reaction

mixture to remove the catalyst, followed by evaporation of the solvent. Overall, the three-step reaction sequence could be conveniently employed for the preparation of a library of trisubstituted pyrroles, with diversity being introduced in the first step (aza-Baylis–Hillman reaction) (Scheme 2). The use of the N-2-trimethylsilylethylsulfonyl protecting group instead of the N-tosyl moiety (not shown) furnished a straightforward conversion of pyrrolines to pyrroles via a base-promoted dehydrodesulfinylation–aromatization reaction [17].

Selective heating-activation of the polar reagents (catalyst and/or olefin) in a non-polar and microwave transparent reaction medium [18–20] was claimed to be responsible for the observed acceleration of the reaction rate (specific microwave effects) [14, 15]. Accordingly, the heating energy is delivered directly to the reacting species and the bulk reaction medium acts as a thermostat. However, careful comparison studies between microwave and thermal (oil bath) experiments indicated that the reaction proceeded equally well regardless of the type of heating employed. Consequently, the observed rate enhancements can be rationalized solely by thermal effects (the Arrhenius equation) rather than by specific microwave effects [21].

One of the most common approaches to pyrrole synthesis is the Paal–Knorr reaction. The microwave-assisted Paal–Knorr cyclization was successfully carried out on various 1,4-diketoesters in the presence of amines, affording the desired pyrroles in good yields [22]. For example, pyrrole **6** was obtained in 70% yield after dielectric heating at 150 °C for 5 min (Scheme 3). When the reaction was conducted under traditional heating, less than 15% of the pyrrole **6** was formed after 12 h of heating at 110 °C (external oil bath 150 °C). A number of polysubstituted pyrrole derivatives were prepared in a one-pot reaction from 1,4-diaryl but-2-ene-1,4-diones and but-2-yne-1,4-diones via hydrogenation of the carbon–carbon double bond (or triple bond) followed by amination–cyclization. Ammonium formate or aryl/alkyl ammonium formates were employed both as reducing agents in the palladium-catalyzed transfer hydrogenation and also as sources of ammonia. Poly(ethylene glycol)-200 (PEG-200) was identified as the most convenient solvent for the microwave-assisted reaction owing to its high dielectric constant ($\varepsilon = 20$), high boiling point (> 250 °C) and excellent water miscibility (Scheme 4) [23].

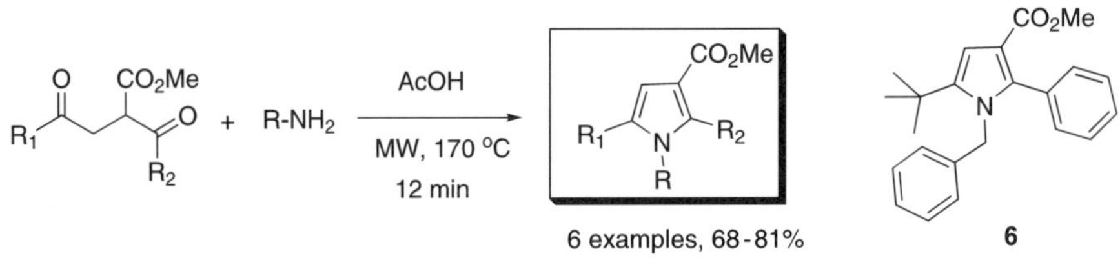

Scheme 3 Paal–Knorr cyclization under microwave irradiation

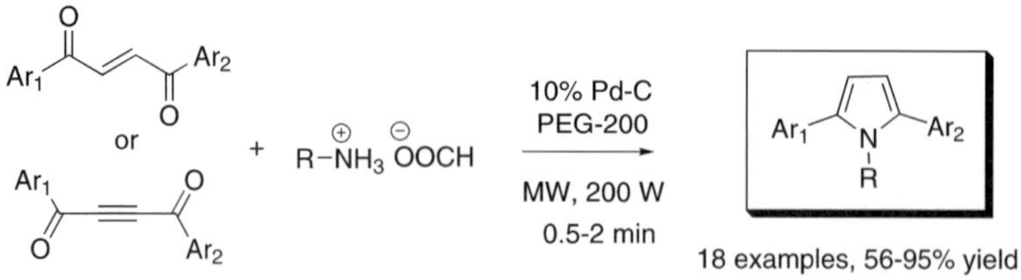

Scheme 4 Preparation of pyrroles in a one-pot hydrogenation–cyclization sequence

Tetrasubstituted pyrroles could be obtained by skeletal rearrangement of 1,3-oxazolidines, a reaction that is substantially accelerated by microwave irradiation. Dielectric heating of a 1,3-oxazolidine **7**, absorbed on silica gel (1 g silica gel/mmol) for 5 min in a household MW oven (900 W power) cleanly afforded the 1,2,3,4-tetrasubstituted pyrrole **8** in 78% yield, thus reducing the reaction time from hours to minutes (Scheme 5) [24]. 1,3-Oxazolidines are accessible in one-pot, two-step, solvent-free domino processes (see also Sect. 2.6). The first domino process, a multi-component reaction (MCR) between 2 equivalents of alkyl propiolate and 1 equivalent of aldehyde furnished enol ethers **9** (Scheme 5). Subsequent microwave-accelerated solvent-free reactions of enol ethers **9** with primary amines on silica support afforded intermediate 1,3-oxazolidines, which in situ rearranged to the tetrasubstituted pyrroles (2nd domino process). Performed in a one-pot format, these

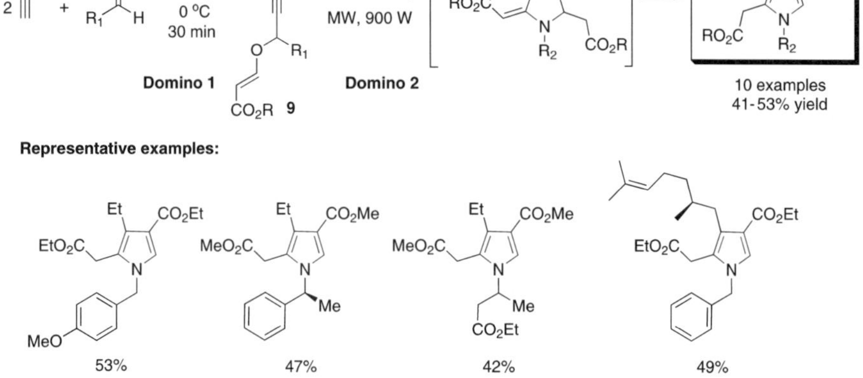

Scheme 5 Sequential domino synthesis of pyrroles

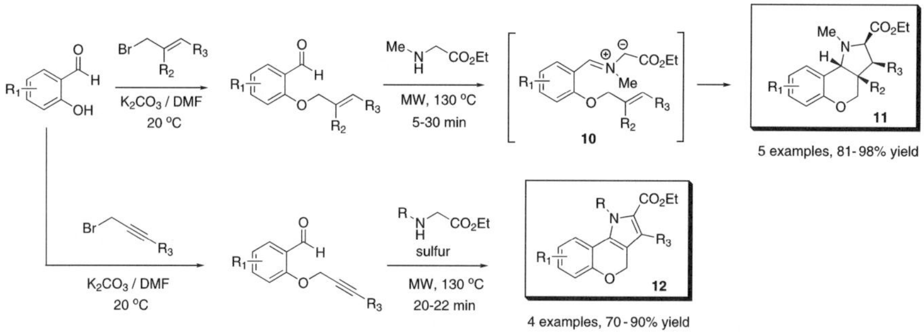

Scheme 6 Microwave accelerated intramolecular [3 + 2] cycloaddition

two sequential domino reactions resulted in an efficient, diversity-oriented synthesis of 1,2,3,4-tetrasubstituted pyrroles from simple and commercially available components [24].

The superiority of microwave dielectric heating was demonstrated in an intramolecular [3 + 2] cycloaddition reaction leading to fused pyrrolidines **11** and complex tricyclic benzopyrano-pyrroles **12** [25]. The reaction evidently proceeded via formation of charged intermediates—for example, the 1,3-dipolar azomethine ylides **10**. As the dielectric heating was proposed to be especially advantageous in the case of polar reaction mechanisms (when the polarity is increased during the reaction from the ground state towards the transition state), acceleration of the reaction speed could be anticipated [26]. Indeed, microwave dielectric heating at 130 °C for 15 minutes afforded the fused pyrrolidine **11** in 93% yield. In contrast, it took 1.5 hours to achieve comparable yields under conventional thermal conditions (130 °C, pre-heated oil bath). Access to fused pyrroles **12** required the [3 + 2] cycloaddition with the *O*-propargyl moiety and subsequent in situ oxidation of the cycloaddition product with sulfur. Both reactions were performed as one-pot procedures, thus establishing a novel one-pot MCR towards complex heterocycles (Scheme 6).

2.2
Indoles, Carbazoles and Phthalimides

The widely employed Leimgruber–Batcho protocol for indole synthesis comprises two consecutive steps—the formation of enamines followed by a reductive cyclization. The formation of enamines (such as **14**, Scheme 7) presumably required an initial deprotonation of the methyl *ortho* to the aromatic nitro-group by methoxide generated from DMF–DMA under elevated temperatures (overnight heating in DMF) [27]. The use of microwave irradiation at 180 °C allowed the reduction of the time of formation of enamines such as **14** from 22 h (at 110 °C) to 4.5 h (and even to 40 min in several

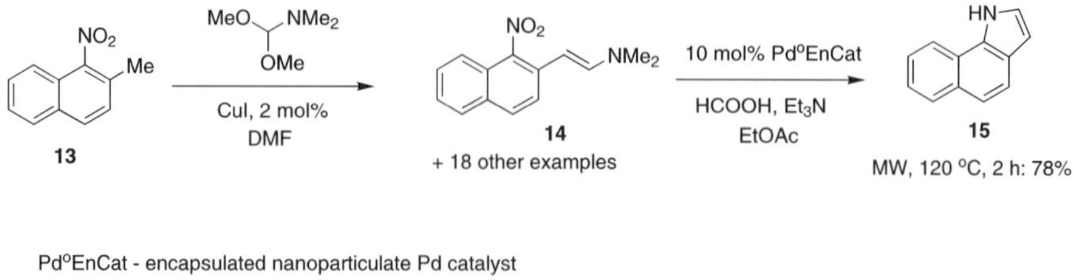

Pd^{o}EnCat - encapsulated nanoparticulate Pd catalyst

Scheme 7 Leimgruber–Batcho synthesis of indoles

cases) [28]. For example, the enamine **14** was prepared in a microwave-accelerated condensation of 2-methyl-1-nitronaphthalene **13** with DMF–DMA in the presence of catalytic amounts of CuI. A subsequent reductive cyclization of enamine **14** leading to 1*H*-benz[*g*]indole **15** was achieved by transfer hydrogenation in the presence of an encapsulated nanoparticulate Pd catalyst and HCOOH/Et$_3$N as the reducing agents (Scheme 7). Thus, the combination of microwave-accelerated enamine formation with the use of a recyclable catalyst for reductive cyclization under dielectric heating conditions diminished the reaction times from days to hours.

The Fisher indole synthesis (the Lewis or protic acid catalyzed rearrangement of arylhydrazones into indoles) is among the most widely used approaches towards the indole heterocycle. The construction of 2-(2-pyridyl)indoles **17** from 2-acetylpyridines **16** required forced conditions and was consequently regarded as a difficult example of the Fisher indolization and microwave heating was anticipated to be of value for this synthesis. Indeed, several microwave-assisted methods towards indoles **17** have been reported. An initially developed solvent-free protocol employed montmorillonite K10 clay modified with ZnCl$_2$ [29]. Higher yields, however, were obtained by using catalytic amounts (10 mol %) of ZnCl$_2$ in triethylene glycol as a high-boiling, polar solvent [30]. Thus, dielectric heating of a mixture of 2-acetylpyridine **16** ($n = 1$) and phenylhydrazine (R = H) at 180 °C afforded indole **17** in 52% yield after 7 min. Notably, the reaction furnished indole **17** in only 12% yield after 3 hours under otherwise identical conditions in an oil-bath at 180 °C, which clearly demonstrates the advantages of microwave flash heating (Scheme 8).

A two-step, one-pot process has been developed for the synthesis of aza-indoles under dielectric heating conditions [31]. In the first step, aminopyridines and ketones were condensed either at room temperature (in the case of aza-indoles **19–20**) or under dielectric heating at 160–220 °C to yield intermediate enamines **18**. Subsequent microwave-assisted intramolecular Heck reactions furnished the corresponding 4-, 5-, 6- or 7-azaindoles in moderate to good yields (Scheme 9).

Microwave flash heating was successfully employed in another palladium catalyzed two-step process leading to *N*-substituted oxindoles [32]. The

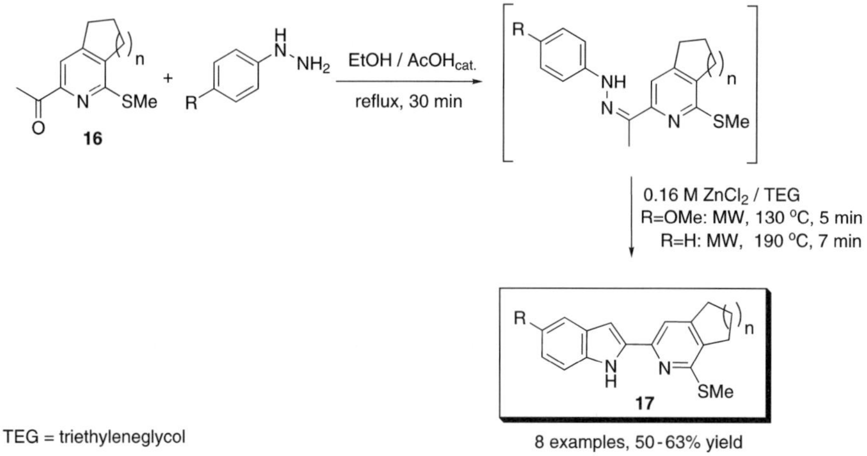

Scheme 8 Microwave-assisted Fisher synthesis

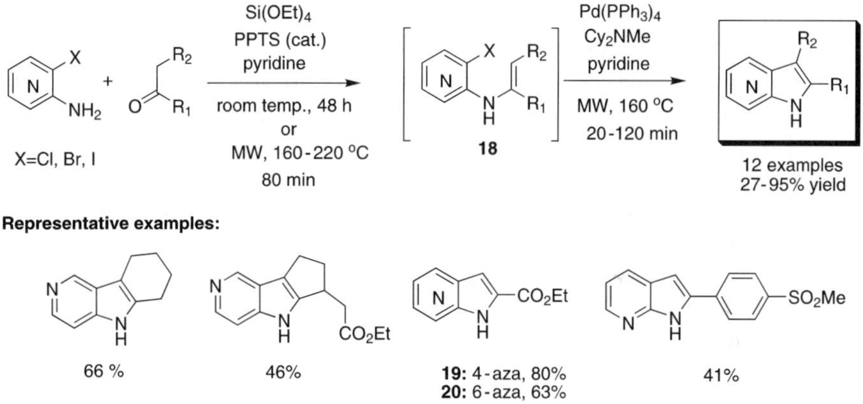

TEG = triethyleneglycol

8 examples, 50-63% yield

Representative examples:

66 % 46% **19:** 4-aza, 80% 41%
20: 6-aza, 63%

Scheme 9 A two-step, one-pot synthesis of aza-indoles

method involves initial microwave-assisted amide bond formation between 2-halo-arylacetic acids and various alkylamines and anilines. Subsequent palladium-catalyzed intramolecular amidations afforded oxindoles. In the case of alkylamines, the procedure can be carried out under aqueous conditions as a one-pot process without isolation of the intermediate amide **21** (Scheme 10). A number of *N*-substituted oxindoles were also synthesized by microwave-assisted radical cyclization on solid support in DMF as the solvent [33].

A novel microwave-accelerated three-component coupling of α-acyl bromides, pyridine and internal alkynes under solvent-free conditions afforded a collection of indolizines [34]. It was proposed that basic alumina catalyzed the in situ formation of a 1,3-dipole from the *N*-acyl pyridinium salt. Subse-

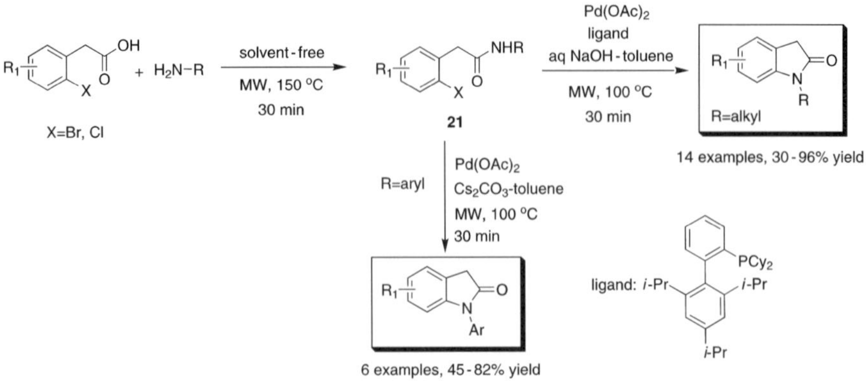

Scheme 10 Preparation of *N*-substituted oxindoles via a condensation–arylation sequence

quent [3 + 2] cycloaddition reaction of the 1,3-dipole with alkyne afforded the indolizine core. The reaction products were easily separated from alumina by extraction with dichloromethane or ethyl acetate (Scheme 11).

A rapid parallel solvent-free synthesis of a representative 28-membered library of phthalimides was achieved utilizing a household microwave oven under highly optimized conditions [35]. Thus, the highest irradiation area inside the household microwave oven was determined to ensure better reproducibility of results (Scheme 12).

Sequential Suzuki–Miyaura cross-couplings and Cadogan cyclizations were developed under microwave dielectric heating conditions to access a variety of 2-substituted carbazoles and other fused heterocyclic systems (Scheme 13). The use of microwave irradiation not only minimized the proto-

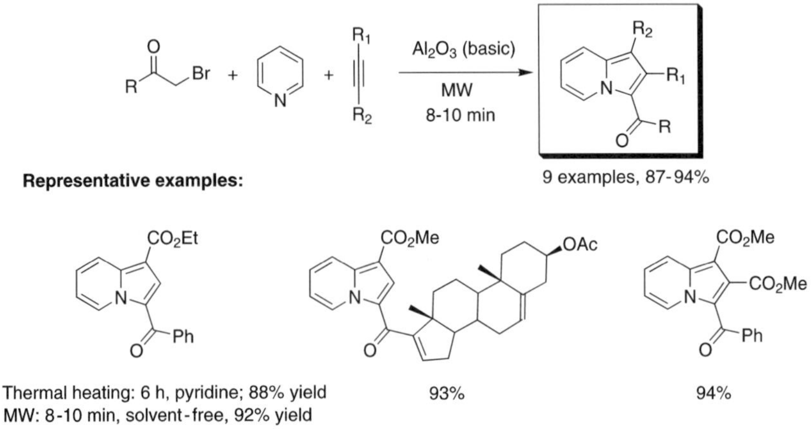

Scheme 11 Microwave-mediated synthesis of indolizines via a three-component cyclocondensation

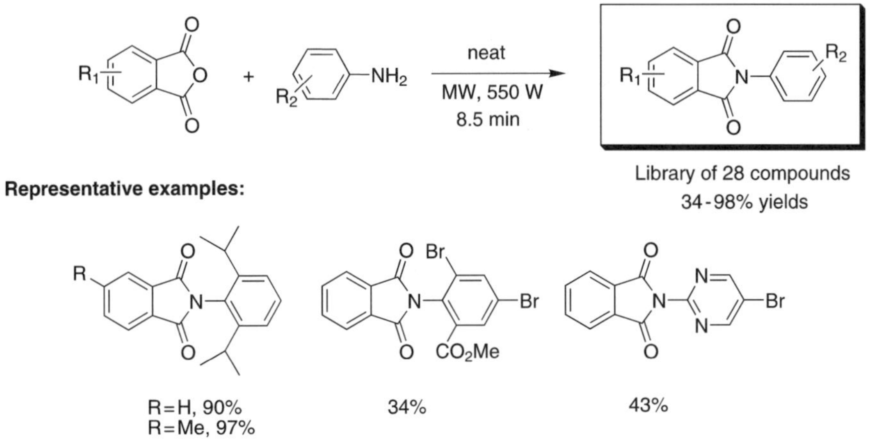

Scheme 12 A rapid parallel solvent-free synthesis of a library of phthalimides

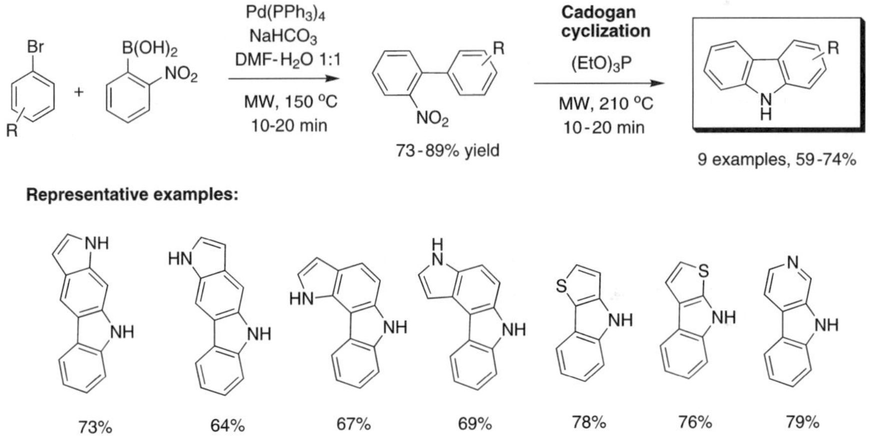

Scheme 13 Preparation of carbazoles via microwave-enhanced Cadogan cyclization

deboronation side reaction in the Suzuki–Miyaura cross-coupling step, but also accelerated the Cadogan cyclization reaction [36]. Thus, carbazoles were formed from 2-nitro-biaryls within 10–20 min of microwave heating at 210 °C. The Cadogan cyclization under conventional thermal heating required up to 24 hours to go to completion [37].

2.3
Thiophenes

Multi-component condensation of ketones (or aldehydes), α-active methylene nitriles and elementary sulfur (the Gewald reaction) is an efficient methodol-

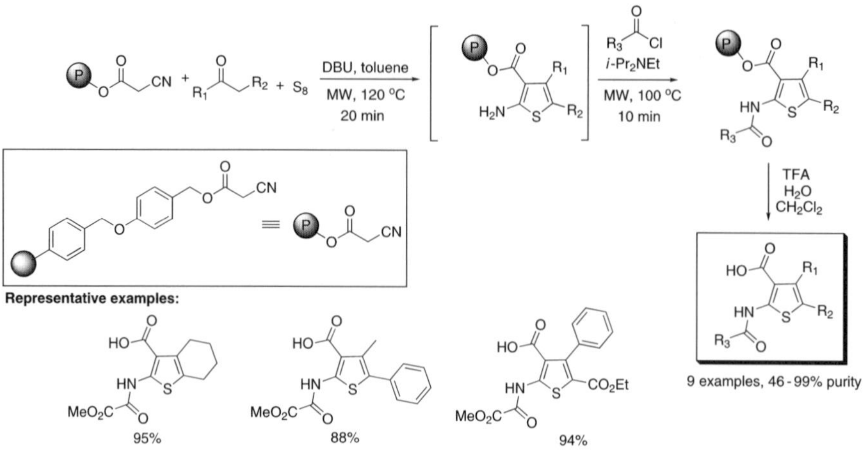

Scheme 14 Microwave-assisted Gewald reaction on solid support

ogy to access diverse 2-aminothiophenes. The Gewald reaction, however, suffers from long reaction times (8–48 hours) and laborious purification of the desired products. To address these disadvantages, the reaction was performed on solid support under microwave dielectric heating conditions, furnishing 2-aminothiophenes within 20 min [38]. Moreover, additional diversity was introduced by a one-pot N-acylation of the initially formed 2-aminothiophene within 10 min. The use of a solid support facilitated the workup substantially and the desired heterocycles were obtained after cleavage from the polymer support with 46–99% HPLC purity (Scheme 14).

2.4
Imidazoles, Benzimidazoles

Condensation of 1,2-diketones with aldehydes in the presence of NH_4OH constitutes an efficient approach towards trisubstituted imidazoles. Impressively, under dielectric heating conditions (180 °C), the cyclocondensation reaction required merely 5 min to go to completion [39]. Moreover, analytically pure products were easily isolated from the reaction mixture by a neutralization–filtration sequence. The high speed of the reaction combined with the ease of product isolation rendered the cyclocondensation especially suitable for the generation of imidazole libraries (Scheme 15).

Benzoin **23** could be used instead of benzil **22**, provided that the microwave-assisted cyclocondensation is performed on inorganic support (silica gel or alumina) under solvent-free conditions [40]. A related diversity-oriented approach towards imidazoles utilized the cyclocondensation of unsymmetrical keto-oximes with aldehydes in the presence of NH_4OAc [41] (Scheme 16). Hydroxyimidazoles **24** were formed in the cyclization step and

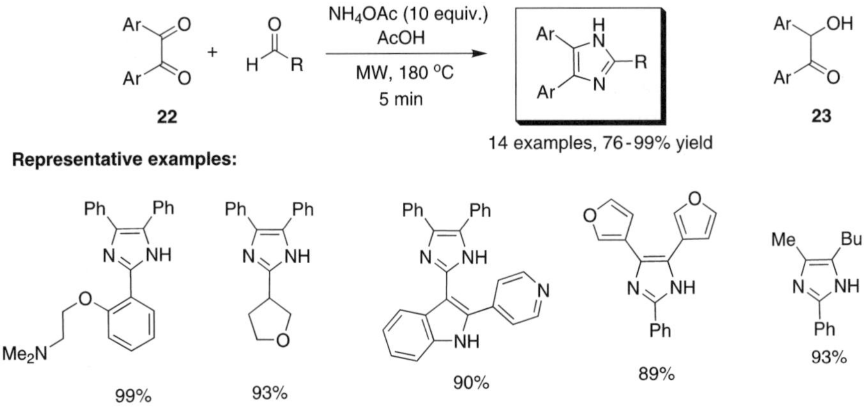

Representative examples:

99% 93% 90% 89% 93%

Scheme 15 Fast three-component synthesis of imidazoles

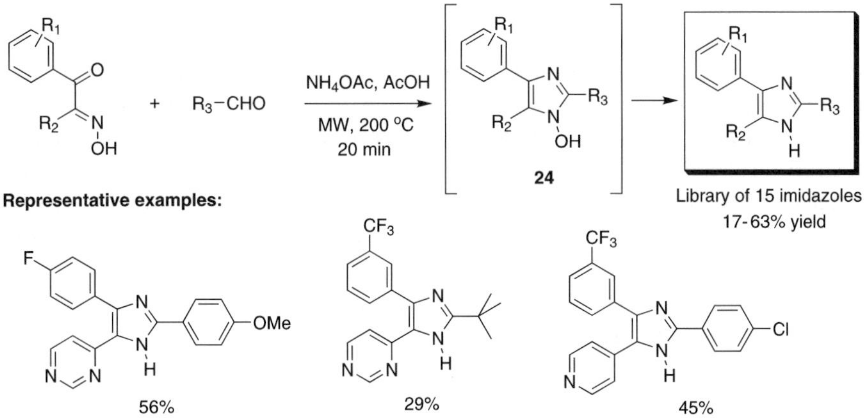

Representative examples:

56% 29% 45%

Scheme 16 Preparation of a library of imidazoles under microwave irradiation

the use of microwave dielectric heating at 160 °C diminished the reaction time from 3 hours (reflux in AcOH) to 20 min. Subsequent reduction of the *N*-hydroxyimidazoles with TiCl$_3$ was also accelerated by microwave irradiation (5 min at 120 °C vs. 16 h at room temperature). It was noteworthy that the use of higher temperatures (200–210 °C) in the cyclocondensation step unexpectedly brought about the concomitant cleavage of the N – O bond, thus leading directly to the desired imidazoles in a one-pot two-step process. The crude products were filtered and purified by injection directly onto a preparative LCMS, thus setting up a high throughput methodology for generation of an imidazole library (Scheme 16). A related one-pot synthesis of tetrasubstituted imidazoles from 1,2-diketones, substituted benzonitriles and primary amines has been achieved in a household microwave oven in the presence of silica gel [42].

A diverse 24-membered library of sulfanyl-imidazoles was prepared in a four-component coupling of 21 different aldehydes, two 2-oxothioacetamides, 12 alkyl bromides and NH_4OAc [43]. Upon completion of the reaction, the products were precipitated from the reaction mixture in an overall average purity of 76% and average yields of 68%. The same representative library was generated under conventional thermal conditions. While purity and yields of the products obtained by the two different methods were comparable, the application of the parallel microwave processing technique reduced the time needed for the library generation dramatically from 12 hours to 16 minutes. Notably, repetitive pulses of microwave heating (2 min, four times) were employed instead of a single prolonged heating profile (8 min; Scheme 17) (the superiority of multiple irradiation cycles compared to prolonged heating for the same time period was also demonstrated by [44]).

To address the purification issue, which frequently is a bottle-neck in the fast microwave chemistry, a solid-phase "catch and release" methodology was utilized in a two-component, two-step synthesis of 1-alkyl-4-imidazolecarboxylates [45]. In the first step, a collection of isonitriles 25 was immobilized onto a solid support by the reaction with the commercially available N-methyl aminomethylated polystyrene 26. Subsequent treatment with various amines brought about simultaneous derivatization and release of the desired imidazoles 27 back into solution. Significantly, only derivatized material was released from the resin, thus ensuring high purity of the desired product. Both steps of the reaction were substantially accelerated by microwave dielectric heating, resulting in the overall reaction time reduction from 60 hours to 70 minutes (Scheme 18).

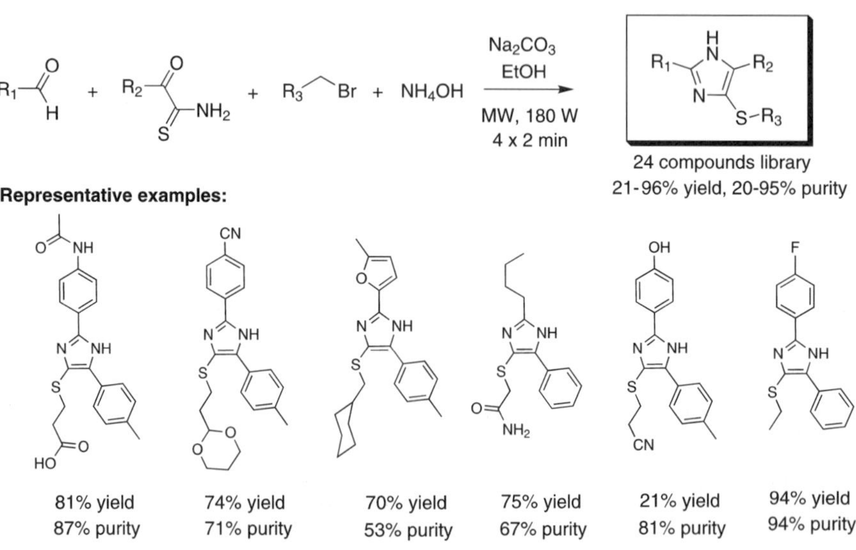

Representative examples:

81% yield	74% yield	70% yield	75% yield	21% yield	94% yield
87% purity	71% purity	53% purity	67% purity	81% purity	94% purity

Scheme 17 Microwave-accelerated parallel synthesis of 4(5)-sulfanyl-1H-imidazole library

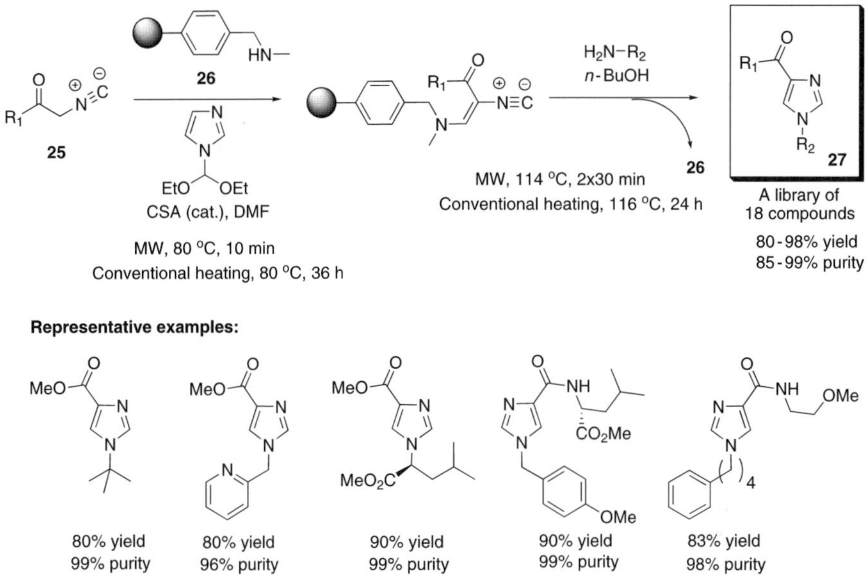

26
25
EtO OEt
CSA (cat.), DMF
MW, 80 °C, 10 min
Conventional heating, 80 °C, 36 h
H2N–R2
n-BuOH
MW, 114 °C, 2x30 min
Conventional heating, 116 °C, 24 h
27
A library of
18 compounds
80-98% yield
85-99% purity
Representative examples:
80% yield
99% purity
80% yield
96% purity
90% yield
99% purity
90% yield
99% purity
83% yield
98% purity

Scheme 18 Solid-phase synthesis of 1-alkyl-4-imidazolecarboxylates using catch and release methodology

A related synthesis of 1-substituted 4-imidazolecarboxylates under microwave dielectric heating conditions employed Wang resin-bound 3-*N*,*N*-(dimethylamino)-isocyanoacrylate [46]. Imidazolidine-2-ones (cyclic ureas) were synthesized by a microwave-accelerated coupling of urea (a cheap and convenient carbonyl source) with aliphatic and aromatic diamines in the presence of ZnO as a catalyst [47]. The use of microwave irradiation in the synthesis of benzimidazoles from *ortho*-phenylenediamine and glycolic acid accelerated the reaction 80 times (from 120 h to 90 min) compared to the standard thermal heating [48]. Other microwave-assisted approaches towards benzimidazoles employed the in situ reduction of *ortho*-nitro anilines to diamines [49] or reaction of dianilines with aldehydes [50]. Purification using a strongly acidic resin was utilized to avoid chromatography in the preparation of *N*-functionalized benzimidazoles via Pd-catalyzed intramolecular arylations of amidines. Notably, the intramolecular aryl-amination was performed under aqueous conditions and the reaction was completed after 20 min of microwave heating at 200 °C [51].

Rapid access to an array of fused 3-aminoimidazoles was conveniently achieved by a $Sc(OTf)_3$ catalyzed three-component cyclocondensation of heterocyclic amidines (such as 2-aminopyridine) and aldehydes with isocyanides (Ugi MCR) [52] or, alternatively, with trimethylsilylcyanide (TMSCN) [53]. *N*-Unsubstituted 3-aminoimidazo[1,2-*a*]pyridines **28** were formed in the latter case (Scheme 19). Microwave dielectric heating of the methanolic

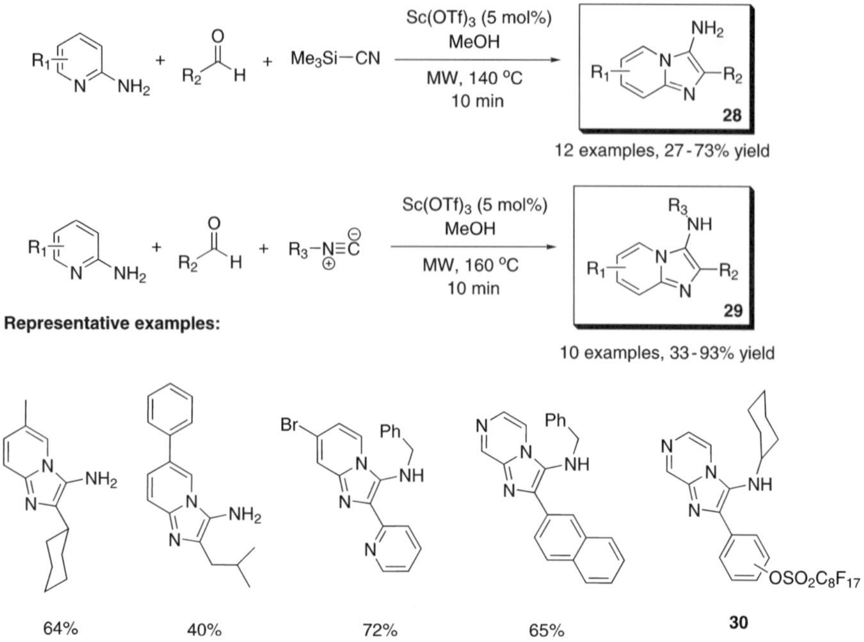

Scheme 19 Sc(OTf)$_3$-catalyzed preparation of fused 3-aminoimidazoles

reaction mixture at 140–160 °C in sealed reaction tubes (75–95 °C above the boiling point) rendered substantial acceleration of the reaction speed. Thus, the desired heterocycles **28** and **29** were formed within merely 10 min. Purification of the target heterocycles was achieved by flash chromatography or supercritical fluid chromatography (Scheme 19).

The time consuming chromatographical purification of heterocycles **28** and **29** slowed down the rate of library production. A phase separation using fluorous chemistry was employed by Zhang and Lu to address the workup and purification of fused 3-aminoimidazo[1,2-*a*]pyridines (such as **30**) [54]. Thus, attachment of a perfluorooctanesulfonyl tag to aldehydes and subsequent Ugi three-component microwave-assisted condensations with 2-aminopyridines and isocyanides furnished the desired heterocycles **30**, which were conveniently isolated by fluorous solid-phase extraction. The fluorous tag could be subsequently used as an activating group in the post-condensation modifications, such as Suzuki–Miyaura cross-coupling reactions.

A library of 2-(arylamino)benzimidazoles was prepared in a microwave-accelerated cyclocondensation of PEG-supported *ortho*-phenylenediamines **31** with isothiocyanates, followed by the product cleavage from polymer support [55, 56]. The quantitative cyclization required either microwave heating (in an open vessel system) for 10 min or reflux in MeOH for 4 hours (Scheme 20). The use of a soluble PEG matrix substantially simplified the

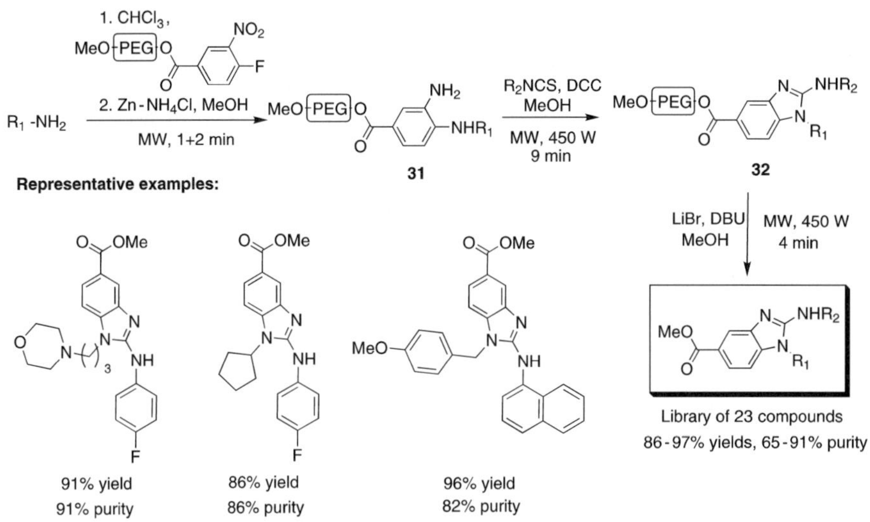

Scheme 20 Synthesis of benzimidazoles using soluble PEG matrix

isolation of products as PEG-bound products were precipitated selectively from a suitable combination of solvents. Similarly, after microwave-assisted cleavage of the substituted benzimidazoles from the polymer support, MeO – PEG – OH was removed from the homogeneous solution by precipitation and filtration. Notably, all polymer-supported intermediates and the polymer support itself remained stable under microwave exposure. Unlike the solid-phase synthesis, soluble polymer-supported reactions could be easily monitored by conventional analytical methods. In later works, Sun and co-workers demonstrated that the microwave-assisted conversion of diamine **31** to benzimidazoles **32** (via the intermediate N,N'-disubstituted thiourea) could be facilitated by $HgCl_2$ [57] and $BiCl_3$ [58] catalysis. A soluble PEG-matrix was also employed in a microwave-assisted combinatorial synthesis of libraries of hydantoins [59, 60] and thiohydantoins [61]. A parallel synthesis of 1,5-disubstituted hydantoins and thiohydantoins was reported [62] under solvent free conditions in the presence of polyphosphoric ester. A collection of 3,5,5-trisubstituted hydantoins was also prepared in a microwave-assisted condensation of substituted benzils with ureas, followed by N^3-alkylation [63].

2.5
Pyrazoles, Isoxazoles, Indazoles

A small library of pyrazoles was synthesized in a MCR of cyclic 1,3-diketones, DMF–DMA and hydrazines [64]. Although the reaction presumably proceeds via cyclodehydration of the intermediate hydrazonoketone **34**, the use of water as a solvent nevertheless was found to be beneficial. Microwave heating of the reaction at 200 °C for 2 min in water was equal in terms of yields to the reflux reaction in an oil-bath for 4 hours. Moreover, the desired pyrazoles were easily isolated by simple filtration from the aqueous reaction mixture, thus simplifying the workup and rendering the procedure especially useful for the production of compound libraries (Scheme 21). The use of hydroxylamine instead of hydrazine brought about the formation of isoxazole **35**. The cyclocondensation of monosubstituted hydrazines with *N,N*-dialkyl enaminones (such as **33**) or *N,N*-dialkyl enamino-*β*-keto esters (derived from ketones or *β*-keto-esters, respectively, and DMF–DMA) has been widely employed both in the solution phase [65, 66] and in the polymer-supported synthesis [67] of substituted pyrazoles. Thus, the attachment of enamino-*β*-keto esters to cellulose beads [68] was accomplished through the enamine moiety, which is a leaving group in the subsequent reaction with phenylhydrazine or hydroxylamine. As a result, concomitant release of the desired pyrazoles or imidazoles from the polymer-supported intermediate **36** into solution occurred upon the cyclization reaction. The catch and release approach delivered a library of trisubstituted pyrazoles with minimal purification and without the need for cleavage of the final product from the resin. The use of microwave dielectric heating both in the cellulose-bound enaminone formation step and

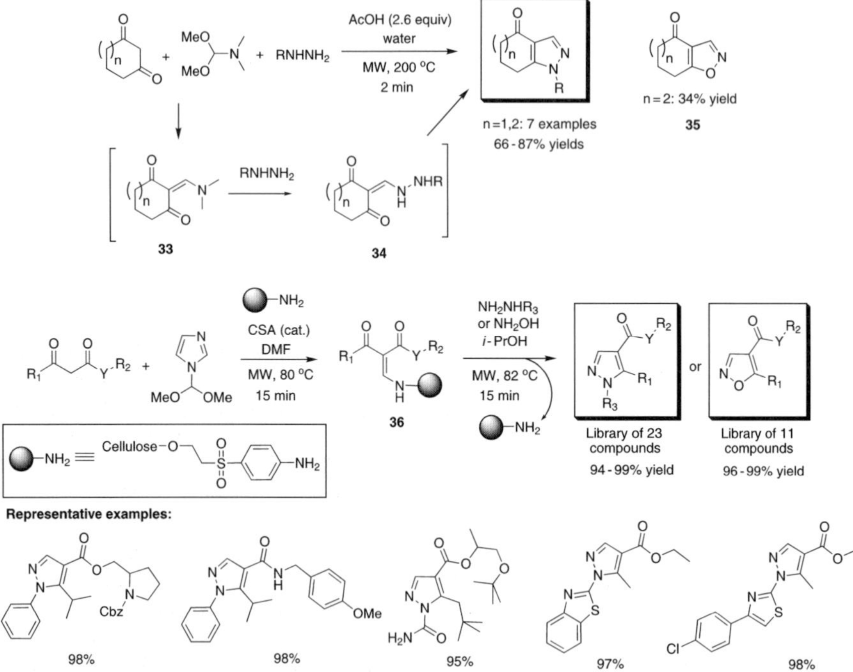

Scheme 21 Microwave-promoted multicomponent synthesis of pyrazoles and isoxazoles

in the subsequent cyclization step dramatically reduced the overall reaction time from 49–53 hours to 30 minutes. The cellulose-supported aniline could be recycled up to 10 times without any reduction in yields and purity. Similarly, the use of hydroxylamine instead of hydrazine delivered a 11-membered library of isoxazoles (Scheme 21).

1,3-Dipolar cycloaddition of nitrile oxides to olefins and acetylenes is among the most widely exploited synthetic routes towards isoxazoles and isoxazolines. It is well-known that microwave irradiation in cycloaddition reactions considerably reduces reaction times. Indeed, the use of dielectric heating (microwave-heated reactions were performed in a flask with a reflux condenser mounted outside the apparatus) allowed for a remarkable reduction of the cycloaddition reaction time from 6–12 hours to merely 3 minutes [69]. Simple aqueous workup provided the target isoxazoles and isoxazolines. The requisite nitrile oxides for the cycloaddition reaction were generated in situ from the corresponding nitroalkanes, 4-(4,6-dimethoxy [1, 3, 5]triazin-2-yl)-4-methylmorpholinium chloride (DMTMM) and 4-dimethylaminopyridine (DMAP) (Scheme 22).

A remarkable improvement of the cycloaddition process was observed also in the synthesis of isoxazolidine **37** under microwave irradiation conditions [70] (Scheme 23). Thus, microwave dielectric heating of nitrone and allyl alcohol for merely 1 hour was equivalent in terms of yields to the oil-bath heating for as long as 15 days. The addition of Zn(OTf)$_2$ not only improved the diastereoselectivity of the process, but also resulted in a substantial reduction of the reaction time. Still, microwave irradiation at 120 °C delivered the diastereomerically pure isoxazolidine **37-cis** in just 15 minutes,

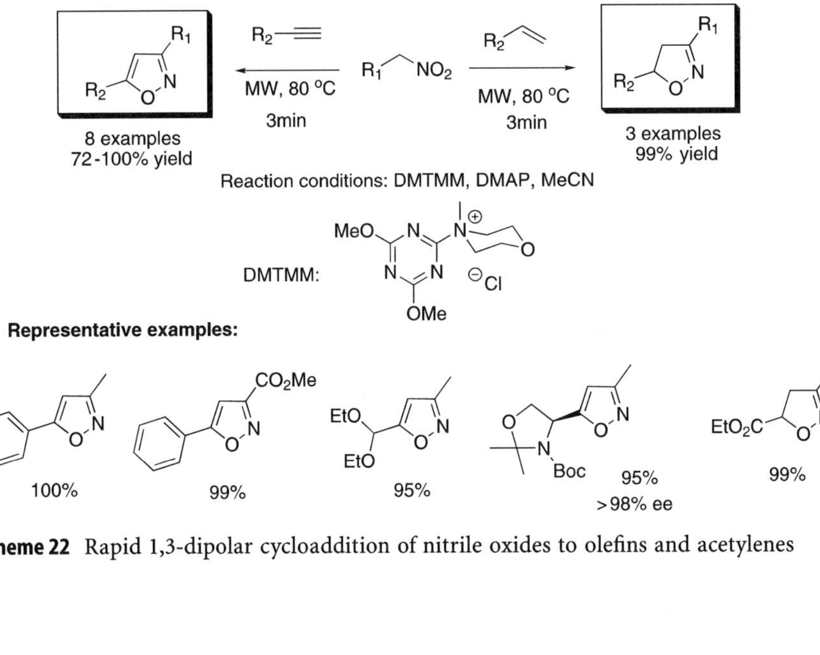

Scheme 22 Rapid 1,3-dipolar cycloaddition of nitrile oxides to olefins and acetylenes

while the corresponding oil-bath reaction at 80 °C was remarkably slower (4 days) (Scheme 23).

A 63-membered library of pyrazoloquinazolinones was prepared from nine hydrazinobenzoic acids and seven cyanoketones [71] in a microwave-assisted tandem α-aminopyrazole formation—an amide bond-forming ring closure reaction. In many cases, the products precipitated out of the reaction mixture and simple washing with diethyl ether furnished the products with purities greater than 95%. Alternatively, heterocycles were isolated in a high-throughput fashion via preparative HPLC (Scheme 24).

A microwave-assisted one-pot two-step approach towards 1-arylindazoles relied on a copper-catalyzed intramolecular *N*-arylation–cyclization of in situ formed arylhydrazones [72]. Traditionally, the CuI-diamine-complex catalyzed *N*-arylation reactions occurred at elevated temperatures (up to 110 °C) and required prolonged time (up to 24 h) to go to completion [73]. The com-

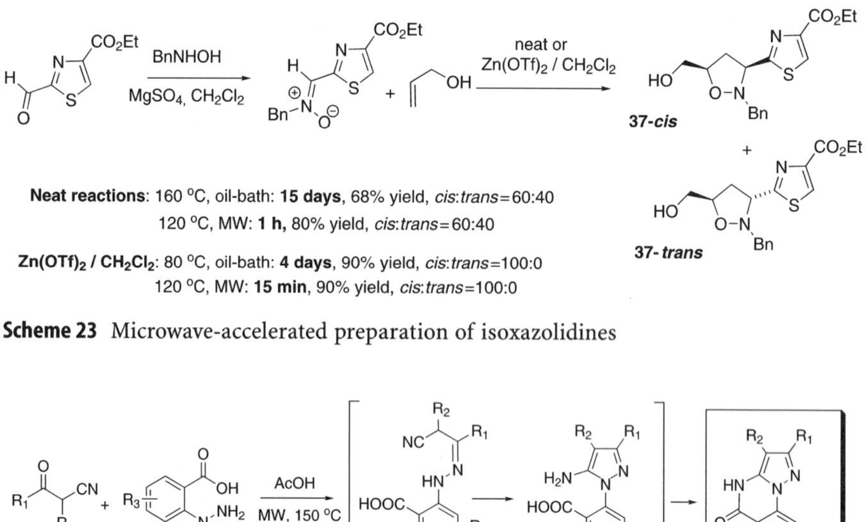
BnNHOH
MgSO4, CH2Cl2
neat or
Zn(OTf)2 / CH2Cl2
37-cis
37-trans
Neat reactions: 160 °C, oil-bath: 15 days, 68% yield, cis:trans=60:40
120 °C, MW: 1 h, 80% yield, cis:trans=60:40
Zn(OTf)2 / CH2Cl2: 80 °C, oil-bath: 4 days, 90% yield, cis:trans=100:0
120 °C, MW: 15 min, 90% yield, cis:trans=100:0

Scheme 23 Microwave-accelerated preparation of isoxazolidines

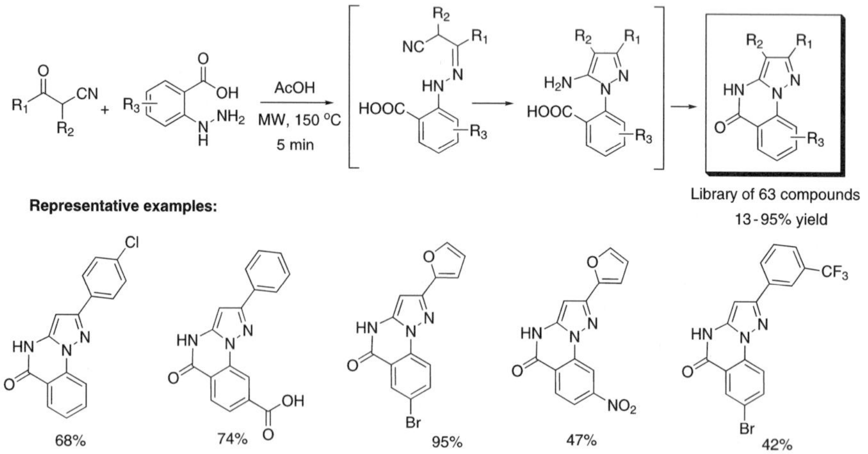
AcOH
MW, 150 °C
5 min
Representative examples:
Library of 63 compounds
13-95% yield
68%
74%
95%
47%
42%

Scheme 24 One-pot synthesis of a library of pyrazoloquinazolinones under microwave irradiation

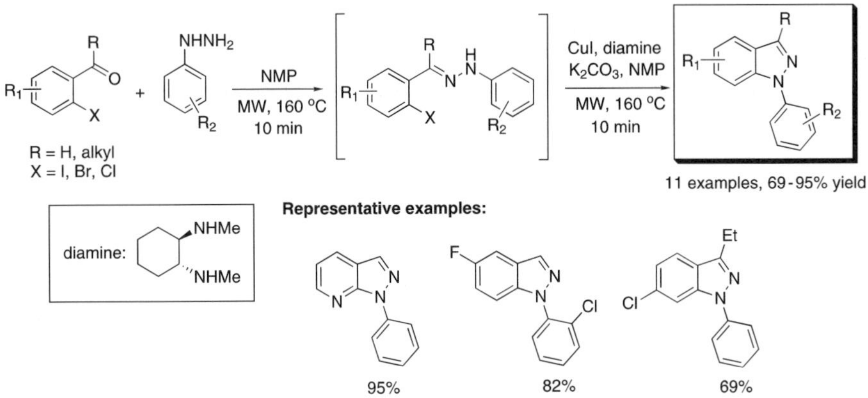

Scheme 25 Cu-catalyzed cyclization of transient arylhydrazones

bination of microwave dielectric heating at 160 °C with the use of a polar high boiling solvent—NMP ($bp = 202$ °C) reduced the *N*-arylation reaction time substantially to merely 10 minutes. It is noteworthy that not only aryl iodides and bromides, but also aryl chlorides were reactive in the intramolecular *N*-arylation reaction (Scheme 25).

2.6
Oxazoles, Benzoxazoles

Polysubstituted 1,3-oxazolidines were prepared in a one-pot diversity oriented four-component reaction (4-MCR), comprising two linked domino processes. Thus, domino synthesis of enol ethers **9** was followed by a sequential amine addition–cyclization sequence [74]. While strong microwave irradiation (900 W) of silica-gel absorbed conjugated alkynoates **9** and amines afforded tetrasubstituted pyrroles (via the skeletal rearrangement of 1,3-oxazolidines, see Sect. 2.1 and Scheme 5) [24], the use of milder microwave conditions (160 W power, 90 min) furnished 1,3-oxazolidines. Under these mild conditions the 1,3-oxazolidines did not rearrange to pyrroles and with respect to diastereoselectivity, the 1,3-oxazolidines were obtained as mixtures of syn/anti isomers. Overall, the formation of one C – C bond, one C – O bond, two C – N bonds and a ring in this MCR required less than 3 hours and utilized simple and commercially available reagents (Scheme 26).

A microwave-assisted one-pot approach towards 2,4,5-trisubstituted oxazoles employed a hypervalent iodine (III) catalyst to bring about the reaction of ketones, 1,3-diketones and β-keto-carboxylic acid derivatives with amides [75]. Microwave dielectric heating was also successfully utilized in a solid-supported, solvent-free synthesis of 2-phenyl-oxazol-5-ones (azlactones) [76] as well as in a solution phase synthesis of isomeric 2-phenyl-oxazol-4-ones (oxalactims) [77].

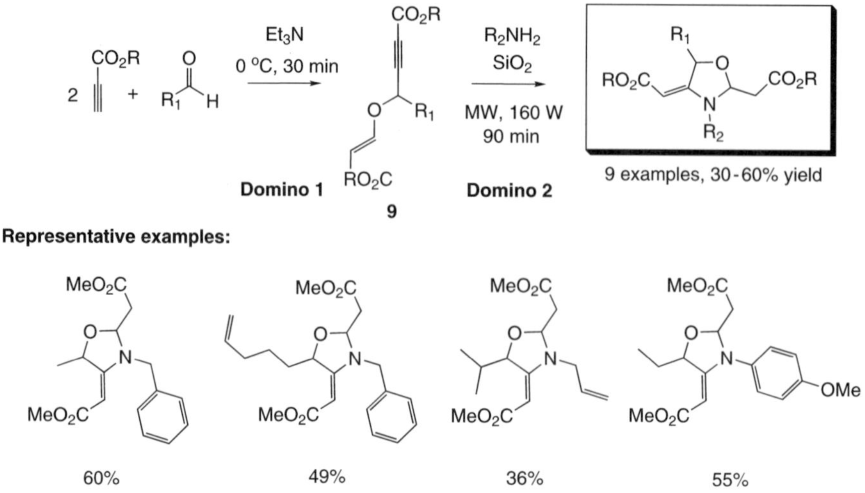

Representative examples:

60% 49% 36% 55%

Scheme 26 Domino synthesis of polysubstituted 1,3-oxazolidines

Benzoxazoles are routinely prepared in a two-step sequence comprising base-catalyzed bis-acylation of *ortho*-aminophenols followed by a Lewis-acid-assisted cyclization–dehydration reaction. Microwave flash heating of readily available acid chlorides and *ortho*-aminophenols in sealed reaction vessels delivered benzoxazoles in a one-pot process without aid of any additive such as base or Lewis acids [78]. Presumably, microwave heating (210 °C, 15 min) of *ortho*-aminophenols and acid chlorides produced monoacylated aminophenols and gaseous HCl as a byproduct. Because the reactions were performed in sealed tubes, gaseous HCl remained in the reaction media and catalyzed the concomitant cyclization–dehydration towards benzoxazoles. For comparison, conventional dioxane reflux at ambient pressure required 24 hours to give an 85% yield of benzoxazole **38**. Although toluene and dioxane were equally efficient as solvents, the latter rendered the isolation of the desired heterocycles more convenient. Thus, the reaction mixture was simply diluted with water and the precipitated benzoxazoles were collected by filtration. To demonstrate the suitability of the developed methodology (comprising rapid synthesis and simple workup) in medicinal and combinatorial chemistry, a representative 48-membered focused library of benzoxazoles was prepared (Scheme 27).

In a closely related publication, carboxylic acids were employed instead of acid chlorides in a microwave-assisted direct synthesis of 2-substituted benzoxazoles [79]. The reactions with 2-aminophenol were performed in a household microwave oven and worked well with aromatic, heteroaromatic, α,β-unsaturated and arylalkyl carboxylic acids (35–82% yields). Phthalic acid formed only mono-benzoxazoles, while the use of succinic acid led to a mixture of mono- and bis-benzoxazoles. Phthalic and succinic anhydrides could

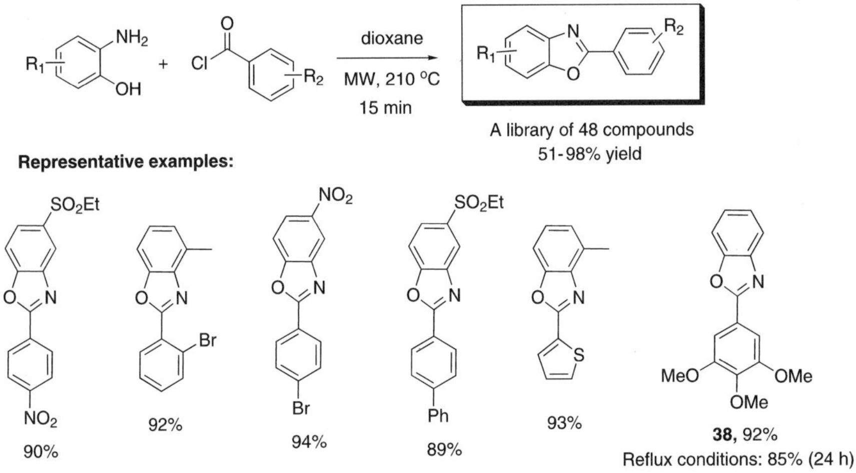

Scheme showing reaction conditions: dioxane, MW, 210 °C, 15 min. A library of 48 compounds, 51–98% yield.

Representative examples:

90% | 92% | 94% | 89% | 93% | **38**, 92% Reflux conditions: 85% (24 h)

Scheme 27 Microwave-assisted preparation of a library of benzoxazoles

be conveniently used in place of the acids. A related microwave-assisted cyclocondensation of 5-amino-4-hydroxy-3(2*H*)-pyridazinones with various carboxylic acids in the presence of polyphosphoric acid furnished oxazolopyridazinones [80].

2.7
Benzothiazoles

Benzothiazoles were obtained by a direct cyclocondensation of 2-aminothiophenol with a variety of carboxylic acids in the absence of any catalyst or dehydrating agent. The heterocycles were readily formed within 20 minutes in a household microwave oven [81]. Although direct comparison with the conventional thermal conditions was not made, reported literature precedents employed oil-bath heating of aminothiophenol with carboxylic acid at 220 °C for 4 hours in the presence of polyphosphoric acid [82] or $P_2O_5 - MeSO_3H$ (70 °C, 10 h) [83]. Consequently, the microwave methodology rendered clear advantages both in terms of reaction speed and milder conditions. A variety of carboxylic acids (aromatic, heteroaromatic, α,β-unsaturated, arylalkyl and cycloalkylcarboxylic acids) could be used and the reaction conditions were compatible with different functional groups such as chlorine, methoxy, phenoxy and thiophenoxy moieties. Bis-benzothiazoles could be obtained in the reaction with succinic and phthalic acids (Scheme 28).

Reduction of the reaction time was achieved by the use of microwave dielectric heating in synthesis of 2-cyanobenzothiazoles from anilines and 4,5-dichloro-1,2,3-dithiazolium chloride [84, 85]. $Mn(OAc)_3$ promoted the radical-mediated cyclization of aryl- and benzoyl-thioformanilides. The re-

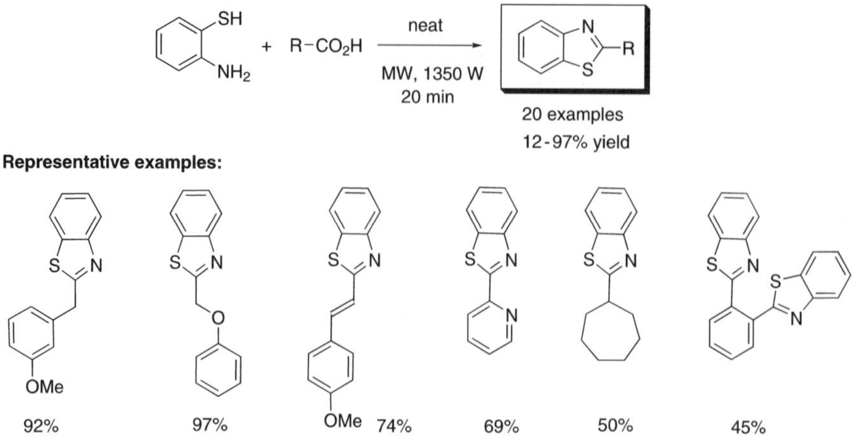
SH
NH2
+ R-CO2H
neat
MW, 1350 W
20 min
N
S
R
20 examples
12-97% yield
Representative examples:
OMe
92%
97%
OMe 74%
69%
50%
45%

Scheme 28 Cyclocondensation of 2-aminothiophenol with neat carboxylic acids

action required microwave heating in acetic acid at 110 °C for 6 minutes (household oven) to furnish a number of 2-substituted benzothiazoles [86]. The reactions in an oil-bath needed 6–10 hours to obtain comparable yields.

2.8 Triazoles

1,2,3-Triazoles are generally prepared by the 1,3-dipolar cycloaddition of an alkyne with an azide at elevated temperatures. Thus, reaction of organic azides with acetylenic amides was significant only after 12 h refluxing in toluene. As a contrast, microwave dielectric heating at 55–85 °C under solvent-free conditions furnished the corresponding disubstituted 1,2,3-triazoles within 30 minutes [87]. Heterocycles were formed as a mixture of two regioisomers and the major regioisomer was separated from

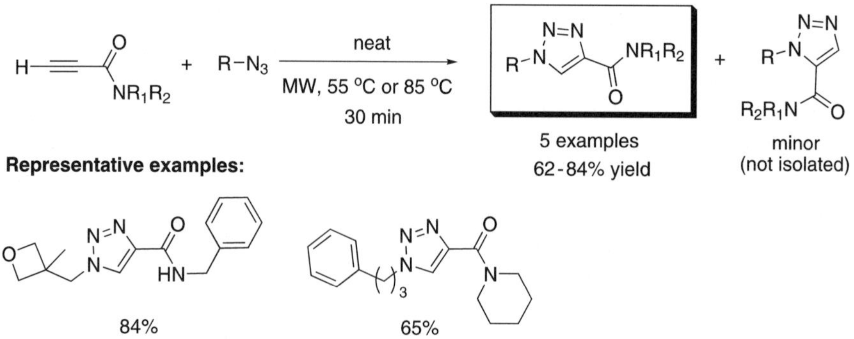
+ R-N3
neat
MW, 55 °C or 85 °C
30 min
5 examples
62-84% yield
minor
(not isolated)
Representative examples:
84%
65%

Scheme 29 1,3-dipolar cycloaddition of alkynes with an azide

the mixture by fractional recrystallization (Scheme 29). 1,2,4-Triazoles have also been synthesized in a condensation reaction of acid hydrazide, *S*-methyl isothioamide hydroiodide and NH$_4$OAc on the surface of silica gel under microwave irradiation [88].

2.9
Oxadiazoles, Thiadiazoles

Ley and co-workers developed a methodology for the preparation of 5-substituted-2-amino-1,3,4-oxadiazoles, their 2-aminosulfonylated analogues as well as the corresponding thiadiazole analogues [89]. Several features rendered the developed procedure especially useful for high-throughput automated synthesis of large compound libraries. First, the approach was diversity-oriented as it was based on a one-pot, three-component coupling of acylhydrazines, isocyanates (or isothiocyanates) and sulfonyl chlorides. Second, the methodology was designed to be divergent, allowing for selective formation of either 2-amino-1,3,4-oxadiazoles **39** or their 2-aminosulfonylated analogues **40** simply by choice of the appropriate base. Third, the attachment of a base to a polymer support rendered workup of the reaction mixture simple and convenient. Specifically, the polymer was removed by simple filtration and the product was isolated after passing through a silica cartridge. Finally, the use of microwave flash heating in combination with polymer-supported reagents resulted in rapid and clean transformations. Moreover, the use of a dedicated microwave synthesizer with an integrated liquid handling robot produced individual compounds for large compound libraries in an automated manner. Altogether, a 120-membered library of 5-substituted-2-amino-1,3,4-oxadiazoles and an impressive 850-membered library of distinct and isolated 2-aminosulfonylated analogues were prepared (Scheme 30).

Comprehensive screening of polymer-supported bases revealed that many bases could catalyze cyclodehydration to afford heterocycles **39**, but only especially strong bases such as PS-BEMP [for an earlier report on microwave-assisted cyclodehydration of 1,2-diacylhydrazines using polymer-supported phosphazene base (PS-BEMP) and TsCl see: [90]] catalyzed both cyclodehydration and subsequent sulfonamidation towards structures **40**. PS-DMAP was the most efficient base to promote the cyclodehydration en route to 2-amino-1,3,4-oxadiazoles **39**. To separate the desired heterocycles **39** from unreacted ureas, a difference in solubility and basicity between the starting materials and products **39** was exploited, using catch and release purification with a silica-bonded sulfonic acid sorbent. Workup and purification was facile also in the case of sulfonamides **40**, requiring only filtration through a short plug of silica and solvent evaporation.

A substrate-dependent transformation to either thiadiazoles **41** or oxadiazoles **40** occurred upon cyclodehydration of thiosemicarbazides and the selectivity of the reaction was found to be highly dependent on the electronic

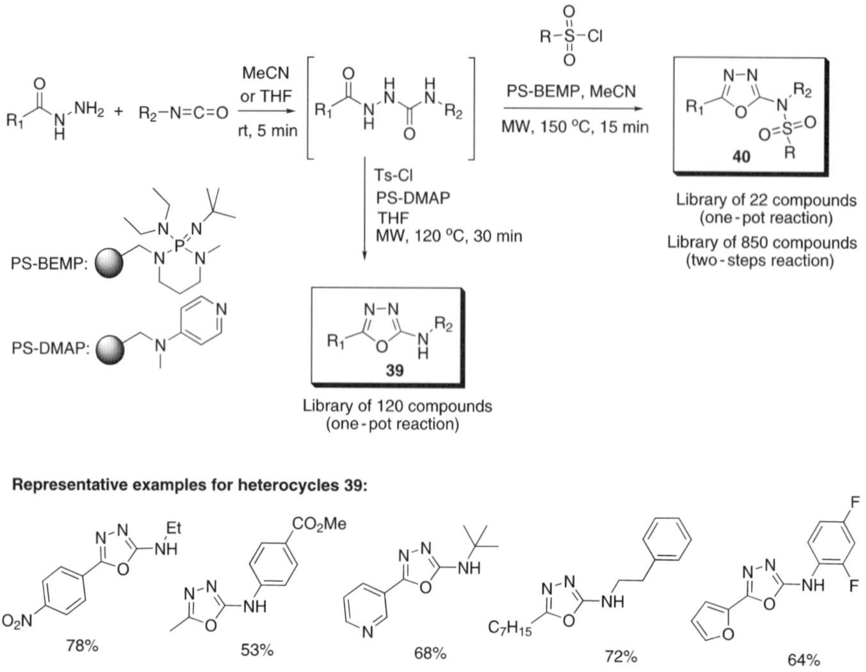

Scheme 30 Preparation of oxadiazole libraries using polymer-supported reagents

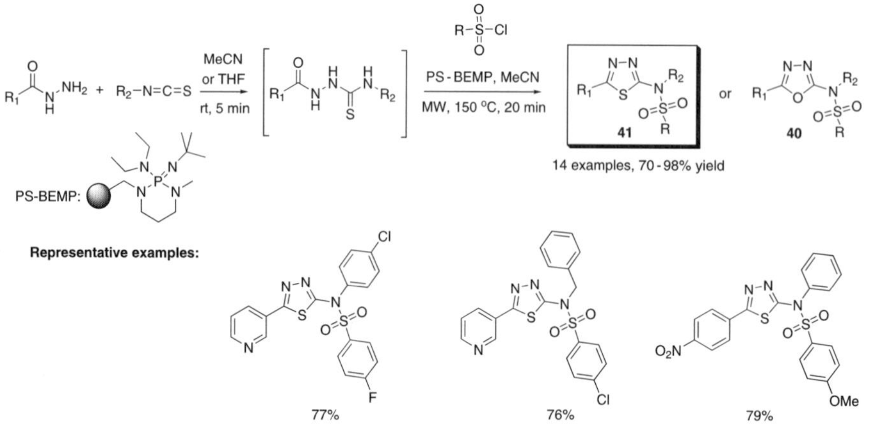

Scheme 31 Microwave-promoted cyclodehydration of thiosemicarbazides

characteristics of the R_1 and R_2 substituents [89]. Thus, electron-withdrawing R_1 substituents directed the formation of thiadiazoles **41**, while electron-donating R_1 groups or simple alkyl substituents in urea afforded oxadiazoles **40** in the cyclocondensation step (Scheme 31). Alternatively, thiadiazoles were prepared in a microwave-assisted thionation–cyclization sequence from 1,2-diacylhydrazines and Lawesson's reagent under solvent-free conditions [91].

A one-step synthesis of 1,3,4-oxadiazoles from readily available carboxylic acids and acid hydrazides was mediated by the polymer-supported reagent

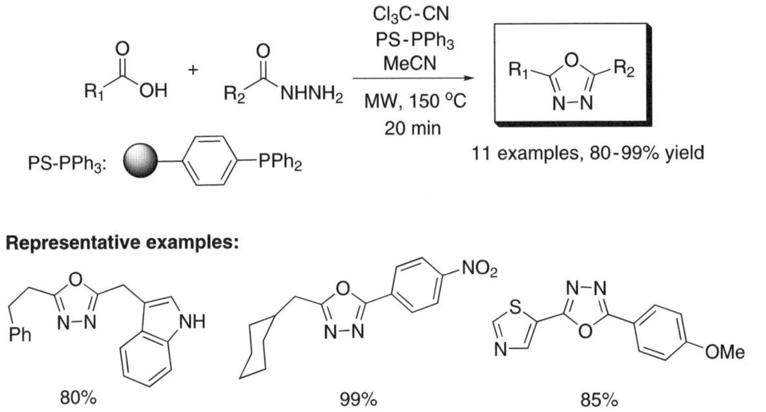

Scheme 32 One-step synthesis of 1,3,4-oxadiazoles under microwave irradiation

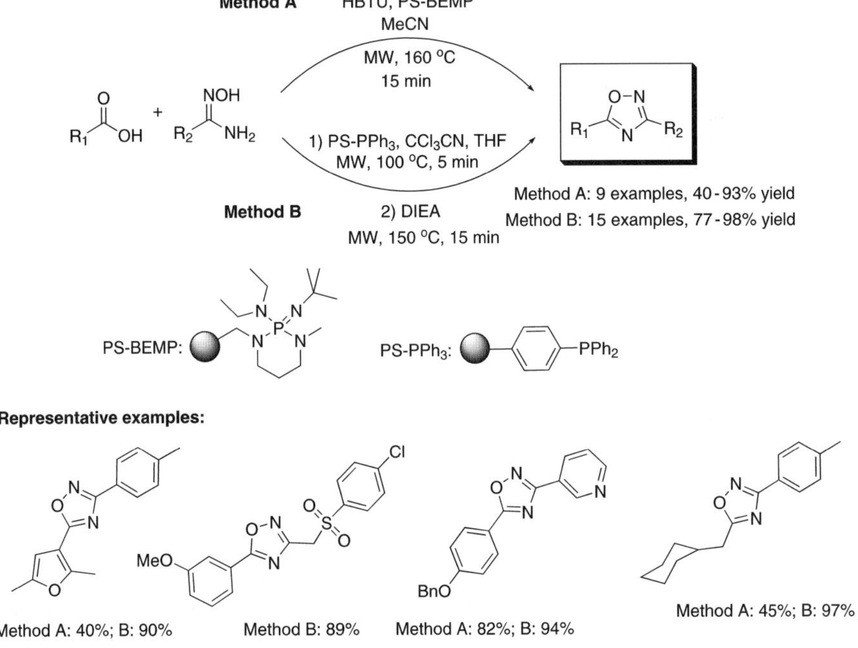

Scheme 33 Rapid generation of libraries of 1,2,4-oxadiazoles using solid-phase reagents

PS – PPh$_3$. A combination of PS – PPh$_3$ and CCl$_3$CN not only facilitated the formation of intermediate 1,2-diacylhydrazides by in situ conversion of carboxylic acids to acid chlorides, but also assisted subsequent cyclization to 1,3,4-oxadiazoles. The workup comprised simple filtration of the resin and evaporation of solvents, followed by flash chromatography. The cyclocondensation occurred within 20 min when heated in acetonitrile by microwaves in a sealed vial at 150 °C (68 °C above the boiling point) [92] (Scheme 32).

Two complementary methodologies were designed for rapid generation of libraries of 3,5-disubstituted 1,2,4-oxadiazoles from widely available carboxylic acids and amidoximes [93]. Both methods employed solid-phase reagents to simplify the purification process. Carboxylic acids were directly condensed with amidoximes in the presence of HBTU and an excess of PS-BEMP in acetonitrile (150 °C, 15 min). Alternatively, carboxylic acids were in situ converted to acid chlorides (with PS – PPh$_3$/CCl$_3$CN in THF) and subsequently reacted with amidoximes to furnish disubstituted oxadiazoles in good to excellent yields (Scheme 33).

2.10
Tetrazoles

The tetrazole functional group is of particular interest in medicinal chemistry since it is widely used as a bioisoster of the carboxyl group. Tetrazoles are usually prepared by the reaction of nitrile with various azide sources. Aryltetrazole boronate esters were prepared by the microwave-assisted reaction of the corresponding nitriles with TMS – N$_3$ in the presence of Bu$_2$SnO in 1,2-dimethoxyethane (DME) [94]. Characteristically, reflux for 22 hours in DME ($\sim$ 86 °C) in an oil bath was equal in terms of conversion to microwave irradiation for merely 20 min at 150 °C (99% yield of tetrazole **42**). Apparently, rapid heating by microwaves in conjunction with the use of a closed reaction vessel allowed the reaction to be performed well above the boiling point of the solvent, thus accelerating the cycloaddition. Indeed, a similar level of conversion (90% of tetrazole **42**) was observed in an oil bath utilizing the same conditions as the microwave-assisted reaction (150 °C, sealed tube, 20 min). These comparisons demonstrate that microwave heating in closed reaction vessels is a more effective and convenient method than conventional thermal techniques. The microwave-assisted cycloaddition could be readily scaled-up and run at higher concentrations in a sealed 80 mL microwave vessel (Scheme 34).

The sterically highly hindered di-*ortho*-substituted tetrazoles **43** were synthesized under similar conditions (TMS – N$_3$/Bu$_2$SnO in 1,4-dioxane, 140 °C) from the corresponding nitriles. Full conversion to the desired tetrazoles was not achieved even after prolonged reaction times (8 hours, 25–80% yield) and 15–63% of the starting nicotinonitriles were recovered [95] (Scheme 34).

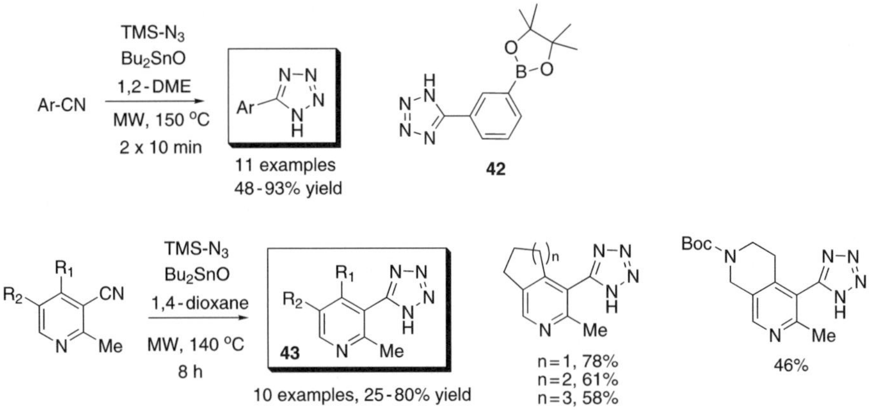

Scheme 34 Microwave-accelerated synthesis of tetrazoles

Tricyclic fused tetrazoles were prepared in a microwave-assisted tandem cycloaddition between TMS – N_3 and aromatic nitriles, followed by ring closure via an intramolecular N-allylation [96].

3
Six-membered Heterocycles

3.1
Pyridines, Dihydropyridines, Piperidines

Steroidal, alicyclic or aromatic annulated pyridines were prepared via a microwave-assisted, base-catalyzed Henry reaction of β-formyl enamides and nitromethane on an alumina support [97]. Highly substituted tri- and tetrasubstituted pyridines were synthesized in a Bohlmann–Rahtz reaction from ethyl β-aminocrotonate and various alkynones. The reaction involved a Michael addition–cyclodehydration sequence and was effected in a single synthetic step under microwave heating conditions [98]. An alternative approach towards polysubstituted pyridines was based on a reaction sequence involving an inverse electron-demand Diels–Alder reaction between various enamines **45** and 1,2,4-triazines **44** (Sect. 3.6), followed by loss of nitrogen and subsequent elimination–aromatization. Enamines **45** were formed in situ from various ketones and piperidine under one-pot microwave dielectric heating conditions [99]. Furthermore, a remarkable acceleration of the reaction speed (from hours and days to minutes) was observed in a microwave-assisted cycloaddition. Unsymmetrically substituted enamines **45** afforded mixtures of regioisomers (Scheme 35).

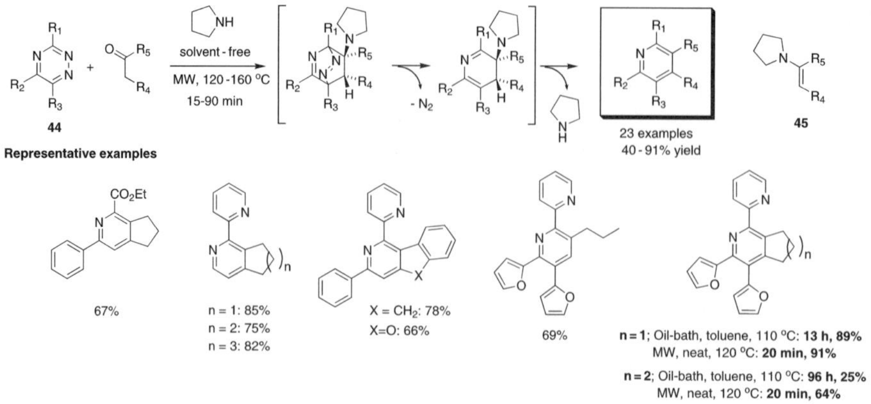

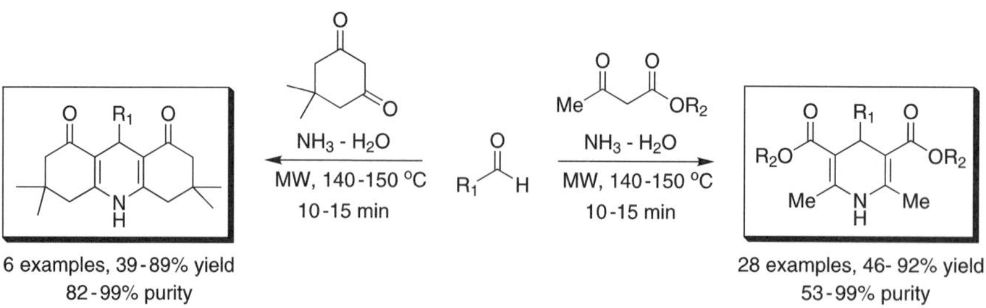

Scheme 35 Inverse electron-demand Diels–Alder reaction of enamines and 1,2,4-triazines

The Hantzsch synthesis of dihydropyridines represents a classical example of MCR, generating an array of diversely substituted heterocycles in a one-pot reaction procedure. Given that the reaction requires elevated temperatures and extended reaction times to proceed, acceleration of the process by microwave irradiation could be envisioned. Indeed, dielectric heating of aldehyde (aliphatic or aromatic) and 5 equivalents of β-ketoester in aqueous 25% NH_4OH (used both as reagent and solvent) at 140–150 °C for merely 10–15 min furnished 4-aryl-1,4-dihydropyridines in 51–92% yield after purification on a silica gel column [100]. The Hantzsch synthesis under reflux conditions ($\sim$ 100 °C) featured a remarkably longer time (12 hours) and lower yields (15–72%). To demonstrate the suitability of the procedure for the needs of combinatorial chemistry, a 24-membered library of 1,4-dihydropyridines (DHP) was prepared (Scheme 36).

Notwithstanding the reduced reaction times and improved yields, the need to use column chromatography to purify the target 1,4-DHP encumbers the application of the procedure for the fast preparation of screening libraries. To address the purification issue, various phase-separation techniques could be employed, such as solid-phase organic synthesis (SPOS).

Scheme 36 Microwave-promoted Hantzsch synthesis

However, the SPOS is usually associated with relatively long reaction times (due to the heterogeneous reaction conditions) as well as difficulties in monitoring the progress of the reaction. On the contrary, the use of soluble polymeric matrices allows the reaction to proceed under standard liquid phase conditions, while still keeping the advantages of purification using phase-separation techniques. For example, ionic liquids (IL), being immiscible both with a wide range of organic solvents and with water, can form a non-aqueous two-phase system. Consequently, the product isolation requires only extraction and washings. Moreover, after the first reactant is anchored to an IL phase, the excess reagents and by-products in subsequent reactions can be removed easily by solvent washing. Finally, the use of task-specific IL as a soluble polymeric matrix under the dielectric heating conditions intuitively seems to be especially beneficial, because ILs interact very efficiently with microwaves through the ionic conduction mechanism. Thus, condensation of IL-supported aldehyde **46**, β-ketoester **47** and NH_4OAc furnished IL-supported 1,4-dihydropyridines **48** (96–97% yield) after microwave heating at 120 °C for 10 min [101]. Product cleavage from the IL-support afforded carboxylic acids, esters or amides, depending on the cleavage conditions. Unsymmetrically substituted 1,4-DHP **51** were obtained in a similar way from IL-supported aldehyde **46**, dimedone **49** and aminocrotonate **50**, followed by transesterification (Scheme 37).

The use of 2,6-diaminopyrimidin-4-one **52** in a related solution phase cyclocondensation reaction with various aldehydes and 1,3-dicarbonyl compounds furnished a number of pyrido[2,3-d]pyrimidines **53** [102]. The $ZnBr_2$-catalyzed reactions proceeded much faster under microwave heating conditions than in an oil-bath, requiring only 20 min dielectric heating at 160 °C (vs. 3 days at 110 °C with oil-bath heating) to go to comple-

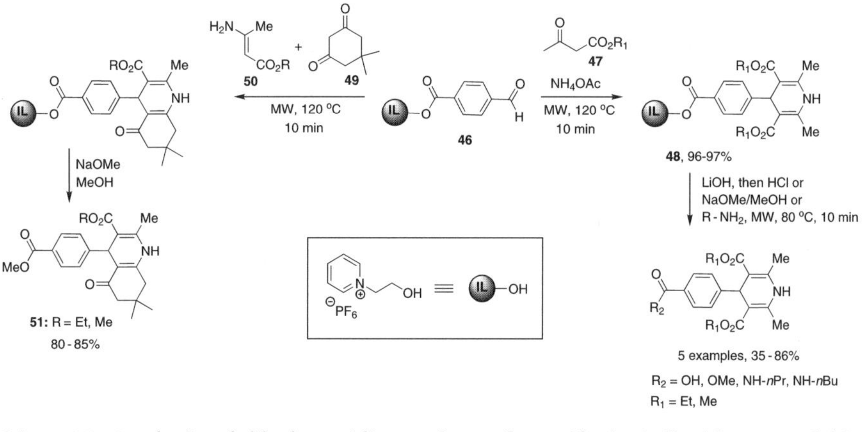

Scheme 37 Synthesis of dihydropyridines using task-specific ionic liquids as a soluble polymeric matrix

tion. The target heterocycles **53** were conveniently isolated by precipitation from the reaction mixture (Scheme 38). A number of related reports dealt with microwave-assisted preparation of fused pyrido[2,3-*d*]pyrimidines **54** and **55**, both under solvent-free conditions [103, 104] and on solid support [105]. *N*-hydroxylacridines were synthesized in another example of a microwave-assisted Hantzsch-type cyclocondensation of aryl aldoximes and dimedone [106].

An 18-membered library of highly substituted 2-pyridones has been prepared in a solution-phase one-pot, two-step MCR [107]. In the first step, preparation of enamines **57** from CH-active carbonyl compounds **56** and DMF–DMA was in all but two cases expedited by dielectric heating (100–170 °C, 5–10 min, solvent-free conditions). Subsequent cyclocondensation of enamines **57** with methylene-active nitrile **58** in the presence of a catalytic amount of piperidine occurred readily within 5 min upon microwave heating at 100 °C. It is noteworthy that the workup generally involved simple filtration of the target 2-pyridones, which precipitated from the reaction mixture upon cooling (Scheme 39). A series of *N*-aryl piperidines were obtained in a microwave-accelerated double *N*-alkylation of various anilines [12, 13] (see also synthesis of pyrrolidines, Sect. 2.1). Tetrahydropyridines were also prepared in a microwave-assisted intramolecular aza-Wittig reaction of chloro alkanes in the presence of NaN_3 and $(EtO)_3P$. The cyclization proceeded via in situ formation of alkyl azides [108].

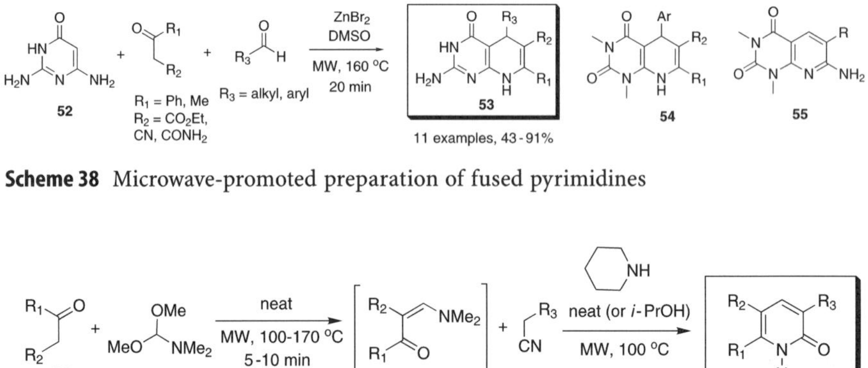

Scheme 38 Microwave-promoted preparation of fused pyrimidines

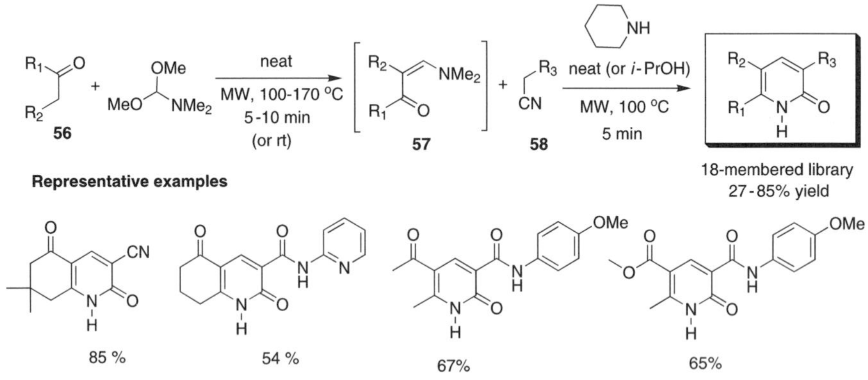

Scheme 39 Fast microwave-assisted synthesis of 2-pyridone libraries

3.2
Quinolines

The use of Sc(OTf)$_3$ as the catalyst facilitated the Skraup synthesis of 1,2-dihydroquinolines from anilines and a variety of dialkyl ketones at mild conditions (room temperature). Nevertheless, an elevated temperature was necessary if acetophenone was employed in the cyclocondensation with anilines [109]. Microwave dielectric heating at 150 °C for 50 min was sufficient to bring about the formation of the desired 1,2-dihydroisoquinolines (Scheme 40).

2-Aminoquinolines **62** have been prepared in a two-step, one-pot, three-component reaction of 2-azidobenzophenones, secondary amines and arylacetaldehydes [110]. The microwave-assisted reaction proceeded via the initial formation of enamines **59**. Subsequent addition of 2-azidobenzophenones **60** afforded the triazoline intermediates **61**, which underwent thermal rearrangement and cyclocondensation to furnish 2-aminoquinolines **62** (Scheme 41). Direct comparison with conventional thermal conditions demonstrated the superiority of microwave dielectric heating in terms of yields (73% vs. 31% of heterocycle **63** after 10 min at 180 °C). Furthermore, the formation of by-products due to decomposition of azide **60** was diminished in the microwave-assisted synthesis. Purification of the products was achieved using solid-phase extraction techniques.

Microwave irradiation considerably improved the reaction speed and yields of 2,4-disubstituted quinolines in a MCR of aldehydes, anilines and alkynes [111]. The cyclocondensation was catalyzed by montmorillonite clay doped with copper(I) bromide and was completed within 3–5 minutes (pulsed irradiation technique—1 min with 20 s off interval), when performed in a household microwave oven. Oil-bath heating at 80 °C for 3–6 hours was necessary to achieve comparable yields of quinolines (71–90%) (Scheme 42).

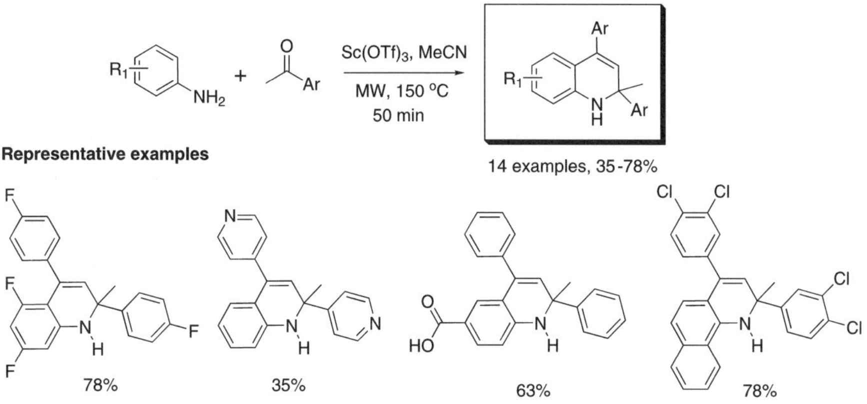

Scheme 40 Microwave-promoted Skraup synthesis of 1,2-dihydroquinolines

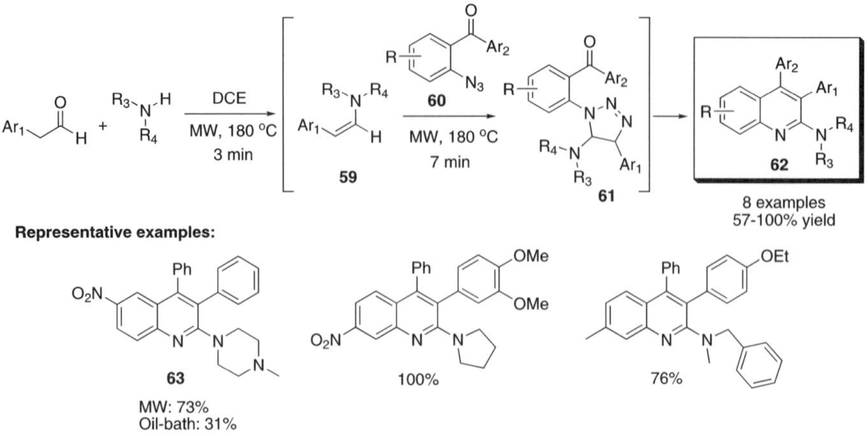

Representative examples:

8 examples
57–100% yield

63

MW: 73%
Oil-bath: 31%

100%

76%

Scheme 41 Rapid synthesis of aminoquinolines under microwave irradiation

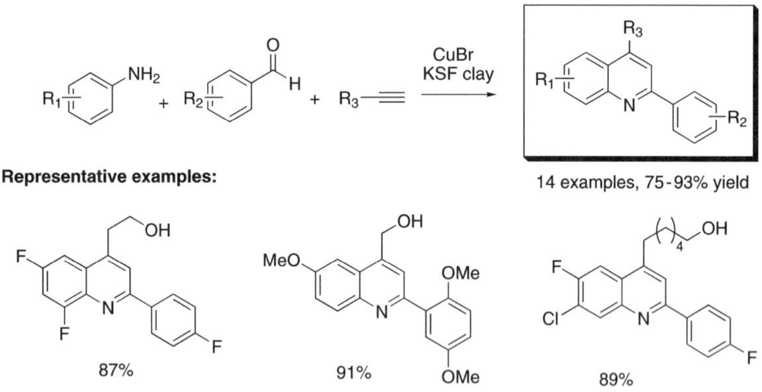

Representative examples:

14 examples, 75–93% yield

87%

91%

89%

Scheme 42 Solvent-free one-pot synthesis of 2,4-disubstituted quinolines

The Friedländer annulation is one of the most straightforward approaches towards poly-substituted quinolines. Thus, a 22-membered library of quinolines was synthesized in a TsOH-catalyzed cyclocondensation–dehydration of 2-aminoaryl ketones and 2-aminoarylaldehydes with ketones in a household microwave oven (with power control) under solvent-free conditions [112]. It was observed that the Friedländer reaction occurred readily also in an oil-bath (at 100 °C). To compare the conventional and dielectric heating conditions precisely, a purpose-built monomode microwave system with temperature control was employed instead of the household oven. The experiments at 100 °C under otherwise identical conditions demonstrated that the dielectric heating protocol was only slightly faster. Products were isolated by a simple precipitation–neutralization sequence (in the case of solid products) or neutralization–extraction for oily or low melting point products (Scheme 43).

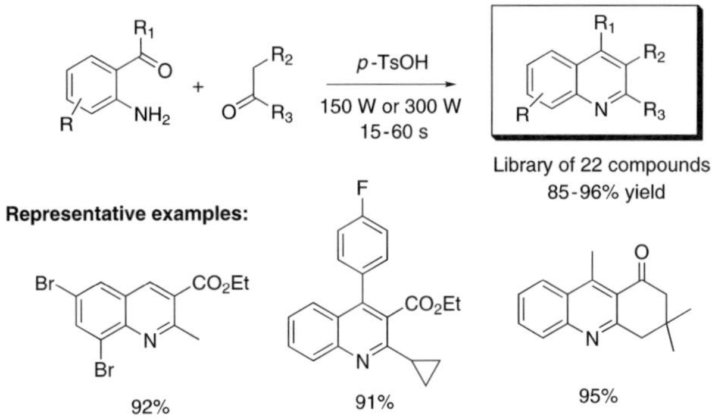

Scheme 43 Preparation of poly-substituted qinolines in the Friedländer annulation

Pyrido-fused tetrahydroquinolines were assembled in a Sc(OTf)$_3$ catalyzed aza-Diels–Alder MCR of 1,4-dihydropyridines, anilines and aldehydes. Although the cyclocondensation occurred within 12 hours at room temperature, the possibility to reduce the reaction time to just 5 min was demonstrated by microwave dielectric heating at 80 °C [113].

3.3
Pyrimidines

Dihydropyrimidinones are routinely synthesized in the Biginelli three component cyclocondensation reaction between CH-acidic carbonyl compounds, aldehydes and ureas (or thioureas) under strongly acidic conditions. Several improved protocols employed Lewis acids instead of traditional mineral acids with microwave heating. Thus, recent reports utilized montmorillonite KSF clay under solventless conditions [microwave heating in a household oven for 5–17 min (1200 W) vs. 6 h oil-bath at 110 °C] [114], NBS in N,N-dimethylacetamide (3–6 min, 600 W) [115] and catalytic amounts (10 mol %) of iodine adsorbed on neutral alumina (1 min, 90 °C) [116]. Alternatively, lanthanide catalysts (Yb(OTf)$_3$ or LaCl$_3$) were used by Kappe and Stadler in an automated sequential microwave-assisted synthesis of a 48-membered dihydropyrimidine library via the Biginelli reaction [117]. Yb(OTf)$_3$ catalyzed the cyclocondensation with ureas, while LaCl$_3$ was the superior catalyst in the case of thioureas. Microwave flash heating at 120 °C in pressurized vials, well above the boiling point of the solvents ($bp_{(EtOH)} = 78$ °C and $bp_{(AcOH)} = 117$–118 °C) brought down the reaction time from 4–12 hours (under reflux conditions) to 10–20 min. This rendered the sequential generation of a 48-membered library feasible within 12 hours. The library was generated in a fully automated and unattended mode using stock solutions of starting materials (CH-acidic compounds **64** and aldehydes **65**) and the liquid handling tools of the microwave synthesizer. Consequently, the solvents

(EtOH – AcOH 1 : 3) were chosen to ensure the complete dissolution of the starting components. Workup and isolation of the products became a critical issue in the context of the high speed of the methodology. The proper choice of the above-mentioned solvents made the workup simple and efficient as the formed products many times precipitated directly from the reaction mixture after completion of the reaction (Scheme 44).

An alternative solution to the workup issue relied on the attachment of CH-acidic compounds **64** to a soluble polymer support (PEG-4000). The approach improved the yields of the dihydropyrimidinones **66** by the use of a 2-fold excess of other components—urea and aldehyde in the microwave-assisted solvent-free cyclocondensation [118]. Another single-step approach towards 4,5-disubstituted pyrimidines was based on cyclocondensation of a variety of aromatic, heterocyclic and aliphatic ketones, formamide and HMDS as the ammonium source [119]. The high temperature (215 °C) required to effect the formation of pyrimidines was secured by microwave dielectric heating in sealed vessels (Scheme 45).

A small library of di- and trisubstituted pyrimidines was prepared by condensation of amidines and guanidines with a range of alkynones. The reaction could be performed under conventional conditions (reflux in acetonitrile, ca. 82 °C), albeit 2 hours was required for the reaction to go to completion. Microwave dielectric heating in sealed vessels at 120 °C (ca. 38 °C above the boiling point of acetonitrile) diminished the reaction time to 40 min [120, 121] (Scheme 46).

Pyrido[2,3-*d*]pyrimidines **70** were synthesized in a one-pot three-component cyclocondensation of α,β-unsaturated esters **67**, CH-active nitriles **68** and amidines **69** [122]. While reflux in THF or MeOH for 24 hours was required

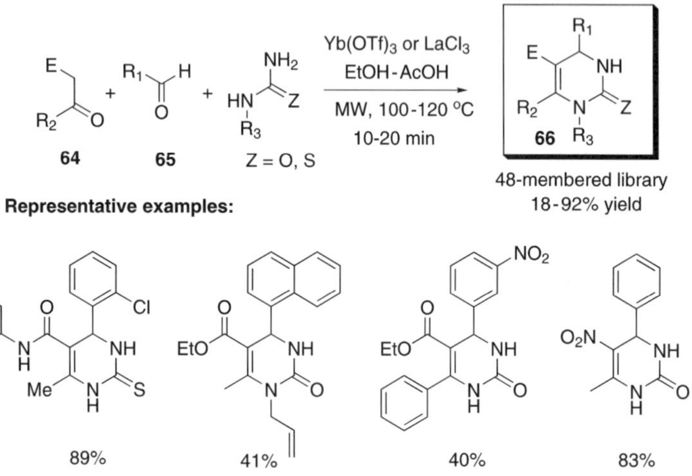

Scheme 44 Microwave-promoted automated synthesis of dihydropyrimidine library

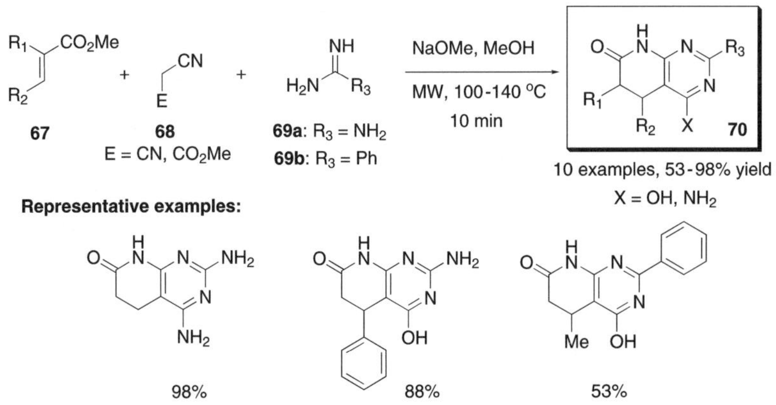

Scheme 45 A single-step synthesis of pyrimidines

Scheme 46 Rapid condensation of amidines and guanidines with alkynones

Scheme 47 Three-component synthesis of pyrido[2,3-*d*]pyrimidines

to complete the cyclocondensation in an oil-bath, full conversion was effected after merely 10 min by microwave heating in sealed vessels at 100–140 °C. Notably, pyrido[2,3-*d*]pyrimidines **70** were conveniently isolated from the reaction mixture by simple filtration when guanidine **69a** ($R_3 = NH_2$) was used in the cyclocondensation. The amidine **69b** ($R_3 = Ph$) derived target heterocycles **70** required purification by flash chromatography (Scheme 47).

3.4
Quinazolines

The most common synthetic method towards quinazolin-4-ones is the Niementowski reaction, a cyclocondensation of anthranilic acid with formamide which requires high temperatures (130–150 °C) and long reaction times (6 hours). It is noteworthy that a remarkable reduction of the reaction time (20 min) was achieved under microwave heating conditions (150 °C) [123]. Moreover, microwave-accelerated reactions were cleaner and afforded higher yields than those under conventional thermal conditions (Scheme 48).

Liu and co-workers have developed an efficient three-component, one-pot, two-step synthesis of 3*H*-quinazolin-4-ones from readily available carboxylic acids and amines [124] (Scheme 49). The versatility of the methodology is remarkable. Simple variation of the starting materials allows not only for the decoration of the quinazolin-4-one core with diverse substituents, but also

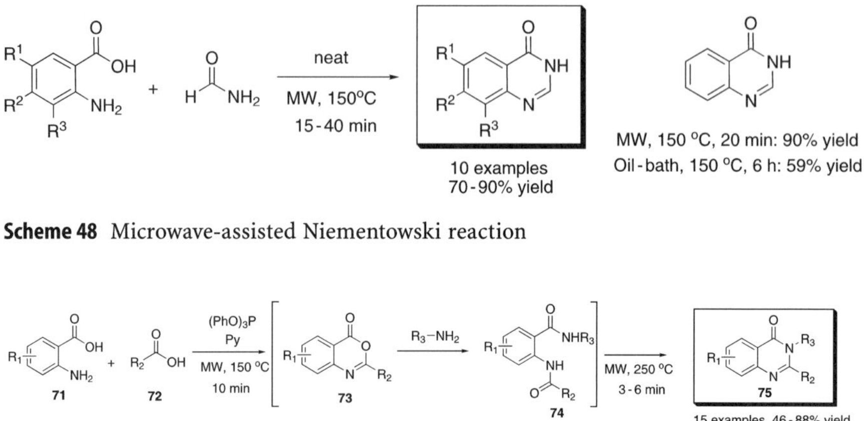

Scheme 48 Microwave-assisted Niementowski reaction

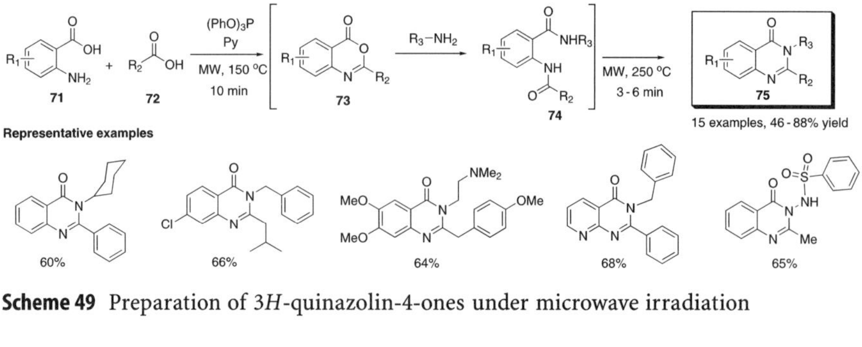

Scheme 49 Preparation of 3*H*-quinazolin-4-ones under microwave irradiation

for the construction of a variety of highly complex quinazolinone-derived heterocyclic systems such as pyrazino[2,1-*b*]quinazolines [125], pyrrolo[2,1-*b*]quinazolines [126] and quinazolinobenzodiazepines [127]. Thus, a variety of pharmacologically relevant complex heterocyclic systems and biologically active natural products could be synthesized rapidly and in an automated manner. Along with acceleration of the reaction speed, the use of microwave dielectric heating induced transformations that were difficult to effect in an oil-bath. The microwave-promoted cyclocondensation of anthranilic acid **71** with a carboxylic acid **72** in the presence of (PhO)₃P, followed by dehydration, afforded intermediate benzoxazinones **73**. Subsequent reactions with amines under conventional thermal conditions (reflux in pyridine) were sluggish, resulting in moderate yields of the desired heterocycle **75** (< 50%), which was accompanied by the intermediate diamide **74** and multiple side products (Scheme 49). On the contrary, dielectric heating at 150 °C effected clean conversion to the desired quinazolinone **75**, thus providing access to hitherto unavailable substitution patterns. Acid chlorides could be employed instead of carboxylic acids **72** and sulfonyl hydrazide performed as well as aliphatic and aromatic amines (Scheme 49).

The methodology was successfully extended to a one-pot total synthesis of complex heterocyclic systems such as pyrazino[2,1-*b*]quinazolines **79**, encountered in nature as alkaloids **80–82** (Scheme 50) [125]. To assemble the pyrazino[2,1-*b*]quinazoline core, *N*-Boc protected amino acid **76** was employed instead of carboxylic acid **72** (Scheme 49) in the synthesis of the corresponding intermediate benzoxazinones **77**. The subsequent reaction with an amine moiety of another amino acid ester **78** was accompanied by concomitant cleavage of the *N*-Boc protecting group and diketopiperazine-like cyclization (for the one-pot deprotection–cyclization reaction of *N*-Boc dipeptide esters to afford 2,5-piperazinedione under microwave dielectric heating, see: [128]) to afford the target heterocycle **79**. Hence, the total

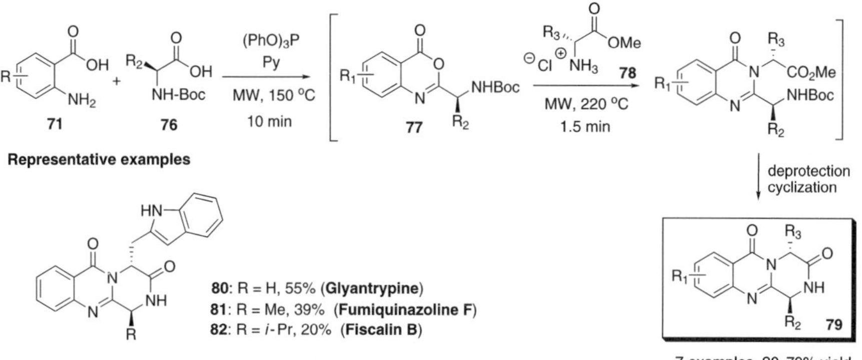

Scheme 50 One-pot total synthesis of pyrazino[2,1-*b*]quinazolines

synthesis of alkaloids **80–82** was effected in a one-pot reaction comprising a three-component, four-step sequential process. Although partial epimerization of both stereogenic centers took place under the microwave heating conditions, simple recrystallization of quinazolines **80–82** increased the optical purity to > 99% ee. The wide chemistry scope of the methodology renders it suitable for preparation of natural product-templated libraries (Scheme 50).

An alternative protocol towards pyrazino[2,1-*b*]quinazolines **79** relied on cyclocondensation of diketopiperazine-derived lactim ethers with anthranilic acid. Microwave dielectric heating in a domestic oven (600 W irradiation power) furnished heterocycles **79** within 3–5 min, while the corresponding reaction in an oil-bath required 2 hours heating at 120–140 °C [129]. The use of *N*-protected ω-amino acids **83** in the microwave-assisted reaction with anthranilic acid **71** furnished pyrrolo[2,1-*b*]quinazolines **85** via transannular cyclization of the intermediate cyclic diamide **84** [126]. Subsequent in situ condensation with a variety of aldehydes furnished isaindigotone **86** and analogues, possessing cytotoxic activity (Scheme 51).

A microwave-assisted domino reaction of two equivalents of anthranilic acid **71** and *N*-Boc amino acid **87** to furnish quinazolinobenzodiazepinones **88** was elaborated [127] (Scheme 52). The development of the domino process was possible because the homo-coupling of anthranilic acid was presumably slower than the desired reaction between acids **71** and **87**. The use of microwave dielectric heating was beneficial as the formation of the seven-membered benzodiazepine ring required a higher temperature (230 °C) and a longer reaction time (20 min) compared to the six-membered analogues **79**

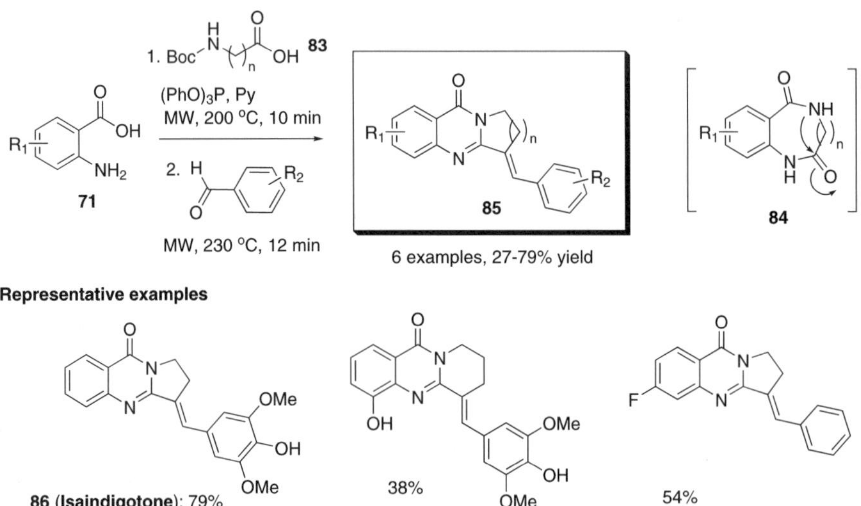

Scheme 51 Microwave-promoted preparation of pyrrolo[2,1-*b*]quinazolines

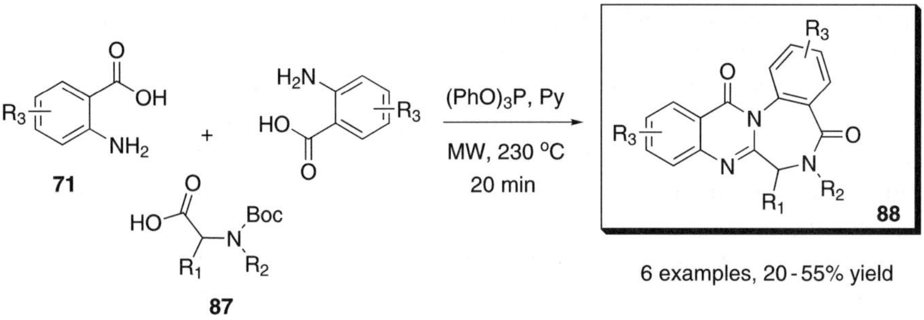

Scheme 52 Domino synthesis of fused benzodiazepinones

and **85** ($n = 2$). The desired heterocycles were purified by preparative HPLC (Scheme 52).

3.5
Piperazines, Pyrazines

A 20-membered library of piperazines **93** has been prepared in a parallel format utilizing microwave-assisted direct annulation of primary amines with resin-bound bis-mesylate **89**. Diazaspiro[5.5]undecane scaffolds **94** and **95** were obtained in a similar way employing the corresponding polymer-supported reagents **90** and **91** [130] (Scheme 53). The use of resin-bound bis-mesylates **89–91** allowed the utilization of a large excess of amine (10 equivalents) to bring the annulation to completion. The excess amine was subsequently removed by simple filtration and washing. Furthermore, the attachment of bis-mesylate **89** to a polymer support prevented possible contamination of the target heterocycles **93–95** with products from oligomerization side-reactions. It is worthwhile to note that solid-phase synthesis has also several drawbacks due to the heterogeneous nature of the process, such as relatively long reaction times, causing the thermal degradation of the polymer support. The use of microwave dielectric heating not only secured the high temperature (160 °C) necessary to drive the annulation to completion, but also reduced the reaction time, thus avoiding the degradation of the polymer backbone (it has been demonstrated that resins can withstand microwave irradiation even at 200 °C for short reaction times such as 20–30 min [131]). As the resin itself is a rather poor microwave absorber, sufficiently strong absorbtion of microwave energy was achieved by the use of a polar solvent—NMP. Moreover, the high boiling point (202 °C) of NMP allowed the annulation to be performed in open vessels. Subsequently, libraries of heterocycles **93–95** were prepared in a 96-well plate in a specialized microwave synthesizer designed for parallel synthesis. The workup of the annulation reaction (washing and filtration) and the cleavage from the resin was performed in an automated manner. The whole process exemplifies the automated parallel synthesis of a small focused library (Scheme 53).

Lindsley and co-workers developed a general procedure towards the collection of diverse heterocyclic scaffolds from common 1,2-diketone intermediates **96**. Substituted quinoxalines **97**, fused pyrazolo[4,5-*g*]quinoxalines **98** and imidazolo[3,4-*g*]quinoxalines **99** as well as pyrido[2,3-*b*]pyrazines **100** and thieno[3,4-*b*]pyrazines **101** have been prepared in excellent yields [132] (Scheme 54), employing optimized reaction conditions (microwave heating of equimolar mixtures of 1,2-diketone **96** and diamine components at 160 °C for 5 min in 9 : 1 MeOH – AcOH). The use of microwave irradiation resulted in reduced reaction times (5 min vs. 2–12 hours), improved yields as well as the suppressed formation of polymeric species; a characteristic of traditional

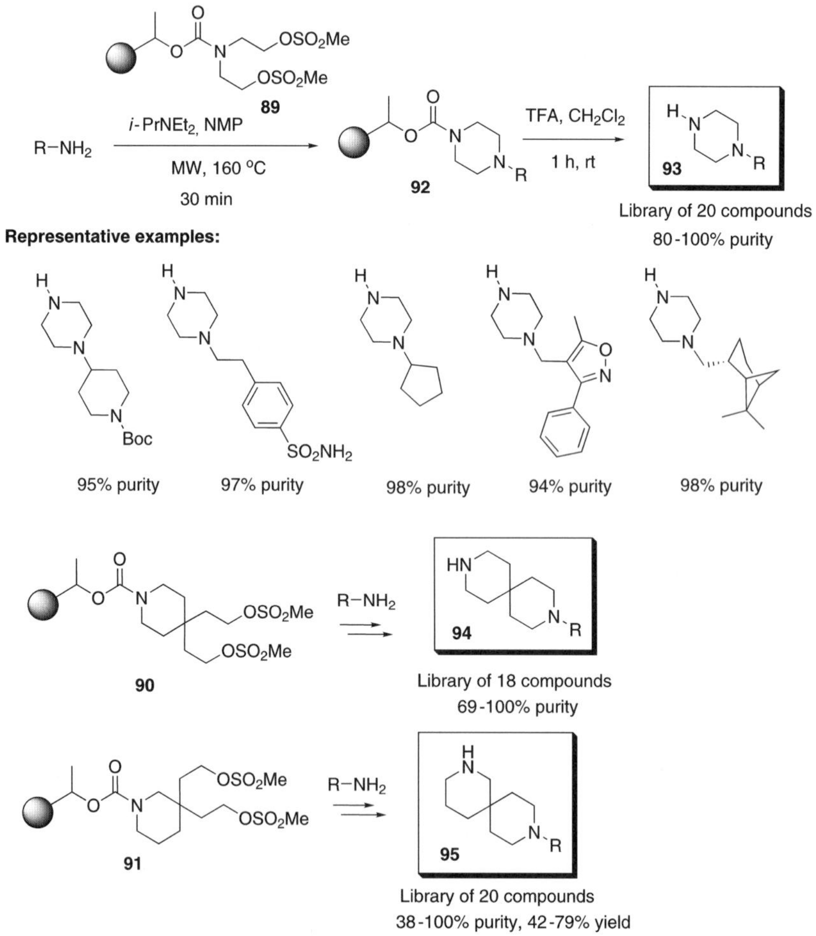

Scheme 53 Annulation of primary amines with resin-bound bismesylate reagents

thermal conditions. Thus, in contrast to heating in an oil-bath (at 50–70 °C) and even to the room temperature reactions, microwave irradiation at 160 °C for 5 min afforded thieno[3,4-b]pyrazine **101** (R$_1$ = Ph) in 72% yield with no detectable polymerization side-product (Scheme 54). The developed methodology was employed for the rapid preparation of two 200- and 50-membered focused libraries of 2,3-diaryl quinoxalines during lead generation and optimization of allosteric AKT protein kinase inhibitors [133]. Mono-substituted quinoxalines could be accessible also from α-hydroxyketones and symmetrical diamines in microwave-assisted solventless reactions provided that catalytic amounts (1 mol%) of an in situ oxidant (activated MnO$_2$) were employed [134].

Microwave dielectric heating accelerated Ugi four-component condensations of N-Boc aniline, acid, aldehyde and isonitrile, leading to quinoxalinones **105** and benzimidazoles [135]. To simplify the purification of the Ugi condensation products, fluorous phase organic synthesis was employed as an alternative to other, traditional phase separation techniques, for example, polystyrene resins. Thus, attachment of a fluorous N-Boc protecting group ("fluorous tag") **103** to a starting diamine **102** delivered the Ugi four-component condensation product **104**, which was soluble in fluorocarbon solvents while being immiscible with water, as well as many common organic solvents. Subsequently, the Ugi product **104** containing the F-Boc group was conveniently separated from the reaction mixture by fluorous solid-phase extraction (FSPE). Cleavage of the F-Boc protecting group was accompanied with concomitant cyclization to furnish the substituted quinoxalinones **105** (and benzimidazoles). The target heterocycles were obtained in moderate to excellent purity after FSPE (Scheme 55) [135]. Related microwave-assisted Ugi three-component MCR furnished 2,5-diketopiperazines [136]. Alternatively, 2,5-piperazines were prepared in a microwave-accelerated cyclocon-

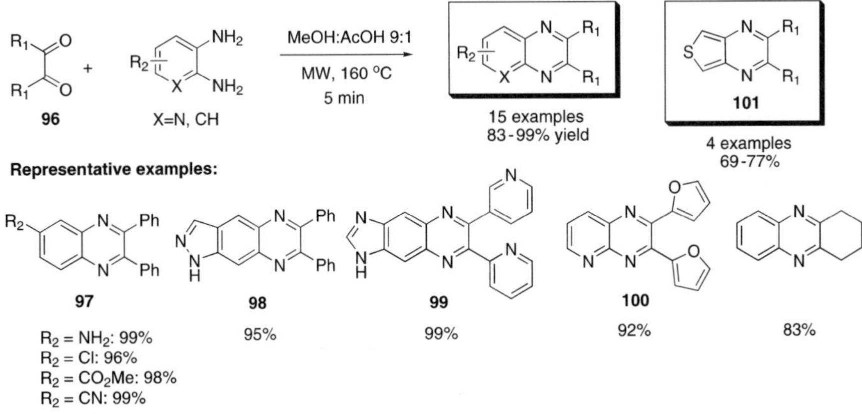

Scheme 54 Preparation of diverse fused pyrazines

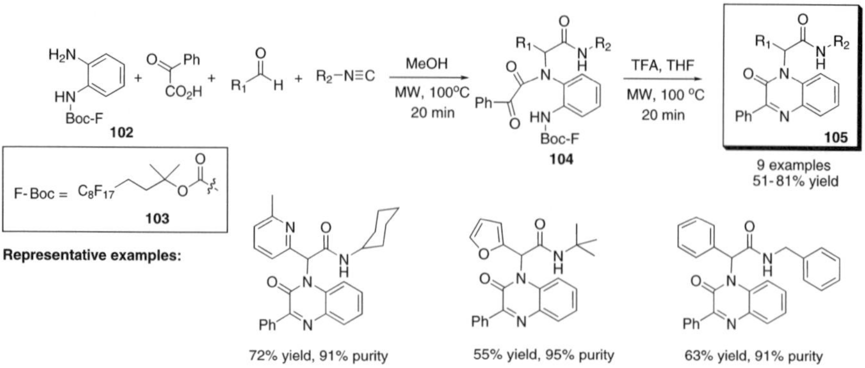

Scheme 55 Microwave-assisted fluorous phase Ugi four-component condensation

densation of Boc-protected dipeptide esters [128]. The use of microwave irradiation accelerated the synthesis of benzopiperazinones [137] as well as chiral quinoxalinones [138] on a soluble PEG polymer-support (Sect. 2.4).

3.6
Triazines

The 1,2,4-triazine core is a synthetically important scaffold because it could be readily transformed into a range of different heterocyclic systems such as pyridines (Sect. 3.1) via intramolecular Diels–Alder reactions with acetylenes. 1,2,4-Triazines have been synthesized by the condensation of 1,2-diketones with acid hydrazides in the presence of NH_4OH in acetic acid for up to 24 h at reflux temperature. Microwave dielectric heating in closed vessels allowed the reaction to be performed at 180 °C (60 °C above the boiling point of acetic acid). As a result, the reaction time was reduced to merely 5 minutes. Subsequently, a 48-membered library of 1,2,4-triazines was generated from diverse acyl hydrazides and α-diketones [139]. Two thirds of the desired heterocycles precipitated from the reaction mixture upon cooling with > 75% purity, while the remaining part of the library was purified by preparative LCMS (Scheme 56).

The "specific microwave effects" were presumably responsible for the observed 40-fold acceleration in microwave-assisted synthesis of 1,3,5-triazines **107** compared to conventional thermal conditions. Thus, microwave heating of benzonitrile and dicyandiamide **106** in an ionic liquid ([bmim]PF_6) in the presence of powdered KOH at 130 °C for just 12 min afforded 2,4-diaminotriazine **107** in 87% yield. Under otherwise identical conditions the reaction in a pre-heated oil-bath (130 °C temperature) took 8 hours to afford the target heterocycle **107** in 79% yield [140] (Scheme 57).

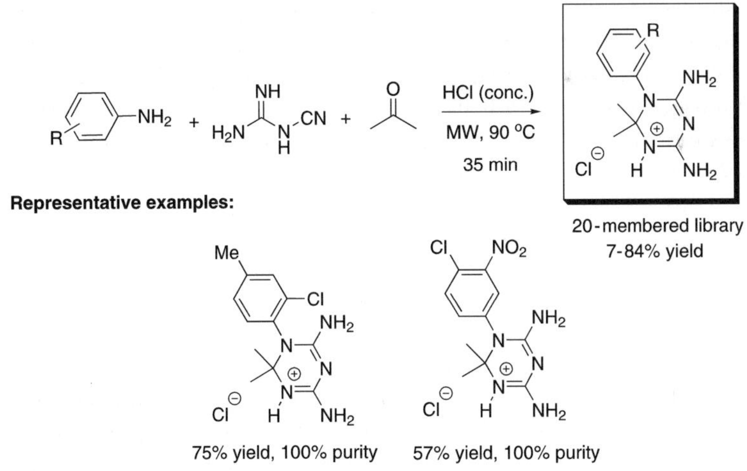

Scheme 56 Synthesis of a library of 1,2,4-triazines

Scheme 57 Microwave-accelerated synthesis of 1,3,5-triazines in ionic liquid

Scheme 58 Parallel synthesis of 1,3,5-triazine library

A 20-membered library of 1,3,5-triazines was prepared in a parallel format employing a one-pot, three-component condensation of anilines, dicyandiamide and acetone. The reaction time was reduced from 22 hours (reflux in acetone at $\sim 60\,°C$) to 35 min by using microwave dielectric heating at $90\,°C$ in a sealed tube [141]. It is noteworthy that the workup consisted simply of cooling the reaction mixture to $4\,°C$ and a subsequent filtration of the precipitated product (Scheme 58).

3.7
Benzopyrones (Coumarines, Chromones)

Specific microwave effects were proposed to accelerate the synthesis of chromones via cyclocondensation of phloroglucinol with β-ketoesters under solvent-free conditions [142]. The reaction presumably proceeded via formation of an intermediate β-diketone **112** in a [4 + 2] cycloaddition reaction of α-oxo ketene **109** with phloroglucinol **110**, followed by a Fries rearrangement. As the magnitude of heating by microwaves depends on the dielectric properties of the molecules, the greater the polarity of reacting species, the more efficient is the absorbance of the microwave energy. Consequently, for solvent-free reactions, where the polarity of the system is increased along the reaction coordinate (from the ground state to the transition state), enhancement of reactivity would be anticipated [143]. Microwave dielectric heating presumably facilitated both the [4 + 2] cycloaddition step and the subsequent Fries rearrangement of intermediate β-keto-ester **111**, because development of charges in the transition states during the reaction path could be envisioned (Scheme 59). Indeed, the yields under microwave irradiation were 4-fold higher than those in a conventionally heated reaction under otherwise identical conditions (reaction time 10 min, equal speed of heating from $25\,°C$ to $125\,°C$). Accordingly, a series of chromones **113** was synthesized within 3–12 min by microwave heating of neat phloroglucinol and β-ketoesters at $240\,°C$. The workup involved a simple acid-base extraction (Scheme 59).

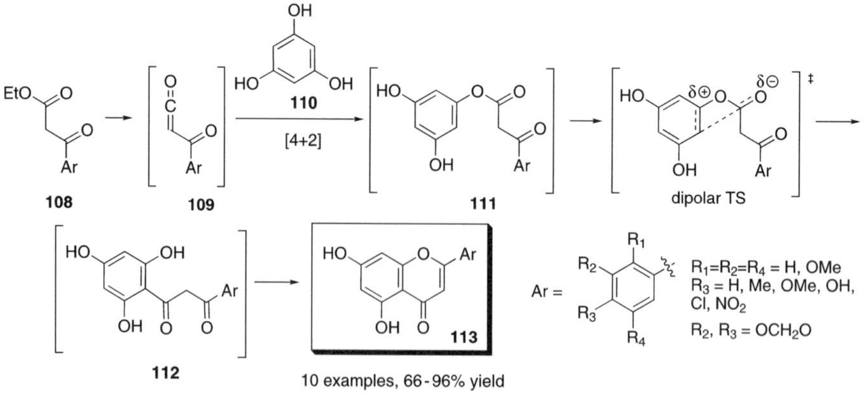

Scheme 59 Microwave-promoted synthesis of chromones

Alternatively, *ortho*-hydroxy β-diketones (such as **112**, Scheme 59) have been obtained in a microwave-assisted Baker–Venkataraman rearrangement of *O*-benzoylated *ortho*-hydroxyacetophenones in the presence of base (pyridine/KOH) [144]. Subsequent dehydrative cyclization of β-diketones under dielectric heating conditions was catalyzed by H_2SO_4 or $CuCl_2$ [145]. The closely related solvent-free cyclocondensation of β-ketoesters with phenoles (the Pechmann reaction) furnished coumarins in the presence of various acid catalysts such as montmorillonite K10 [147]. Graphite was employed as a chemically inert support to increase the efficiency of microwave absorbtion as it has been shown that graphite couples strongly with microwaves and efficiently transfers thermal energy to the absorbed reagents [147].

4
Conclusions

The speed and efficiency of dielectric heating renders the microwave technology particularly suitable for the rapid and automated production of heterocyclic libraries. Furthermore, the proper choice of microwave processing techniques (solvent-free, solid- or polymer-supported conditions, etc.) can simplify the workup and avoid laborious and time consuming purification of target products. Utilization of specific microwave effects (selective heating of reacting species) imparts an unusual reactivity that many times cannot be achieved by conventional heating.

Acknowledgements We thank Professor E. Lukevics, Dr. J. Fotins and Dr. R. Zemribo for proof-reading of the manuscript and artist L. Putnina for designing the graphical abstract.

References

1. Kappe CO, Dallinger D (2006) Nature Rev: Drug Disc 5:51
2. Tietze LF (1996) Chem Rev 96:115
3. Mingos DMP, Baghurst DR (1991) Chem Soc Rev 20:1
4. Bagnell L, Bliese M, Cablewski T, Strauss CR, Tsanaktsidis J (1997) Aust J Chem 50:921
5. Bagnell L, Cablewski T, Strauss CR, Trainor RW (1996) J Org Chem 61:7355
6. Hamelin J, Bazureau J-P, Texier-Boullet F (2002) Microwaves in Heterocyclic Chemistry. In: Loupy A (ed) Microwaves in Organic Synthesis. Wiley, Weinheim, p 253
7. Besson T, Brain CT (2004) Heterocyclic Chemistry Using Microwave-Assisted Approaches. In: Lidström P, Tierney JP (eds) Microwave-Assisted Organic Synthesis. Blackwell, Oxford, p 44
8. Xu Y, Guo Q-X (2004) Heterocycles 63:903
9. Kappe CO, Stadler A (2005) Microwaves in Organic and Medicinal Chemistry. Wiley, Weinheim

10. Molteni V, Ellis DA (2005) Curr Org Synth 2:333
11. Westman J (2004) Speed and Efficiency in the Production of Diverse Structures: Microwave-Assisted Multicomponent Reactions. In: Lidström P, Tierney JP (eds) Microwave-Assisted Organic Synthesis. Blackwell, Oxford, p 102
12. Ju Y, Varma RS (2005) Org Lett 7:2409
13. Ju Y, Varma RS (2006) J Org Chem 71:135
14. Mayo KG, Nearhoof EH, Kiddle JJ (2002) Org Lett 4:1567
15. Grigg R, Martin W, Morrisa J, Sridharana V (2003) Tetrahedron Lett 44:4899
16. Balan D, Adolfsson H (2004) Tetrahedron Lett 45:3089
17. Declerck V, Ribière P, Martinez J, Lamaty F (2004) J Org Chem 69:8372
18. Perreux L, Loupy A (2002) Nonthermal Effects of Microwaves in Organic Synthesis. In: Loupy A (ed) Microwaves in Organic Synthesis. Wiley, Weinheim, p 61
19. Kuhnert N (2002) Angew Chem Int Ed 41:1863
20. Strauss CR (2002) Angew Chem Int Ed 41:3589
21. Garbacia S, Desai B, Lavastre O, Kappe CO (2003) J Org Chem 68:9136
22. Minetto G, Raveglia LF, Taddei M (2004) Org Lett 6:389
23. Rao HSP, Jothilingam S, Scheeren HW (2004) Tetrahedron 60:1625
24. Tejedor D, González-Cruz D, Garcia-Tellado F, Marrero-Tellado JJ, Rodriguez ML (2004) J Am Chem Soc 126:8390
25. Bashiardes G, Safir I, Mohamed AS, Barbot F, Laduranty J (2003) Org Lett 5:4915
26. Perreux L, Loupy A (2001) Tetrahedron 57:9199
27. Joule JA, Mills K (2000) Heterocyclic Chemistry. Blackwell Science, Oxford
28. Siu J, Baxendale IR, Ley SV (2004) Org Biomol Chem 2:160
29. Lipińska TM (2004) Tetrahedron Lett 45:8831
30. Lipińska TM, Czarnocki SJ (2006) Org Lett 8:367
31. Lachance N, April M, Joly M-A (2005) Synthesis 2571
32. Poondra RR, Turner NJ (2005) Org Lett 7:863
33. Akamatsu H, Fukase K, Kusumoto S (2004) Synlett, p 1049
34. Bora U, Saikia A, Boruah RC (2003) Org Lett 5:435
35. Barchin BM, Cuadro AM, Alvarez-Builla J (2002) Synlett, p 343
36. Appukkuttan P, Van der Eycken E, Dehaen W (2005) Synlett, p 127
37. Bouchard J, Wakim S, Leclerc M (2004) J Org Chem 69:5705
38. Hoener APF, Henkel B, Gauvin J-C (2003) Synlett, p 63
39. Wolkenberg SE, Wisnoski DD, Leister WH, Wang Y, Zhao Z, Lindsley CW (2004) Org Lett 6:1453
40. Xu Y, Wan L-F, Salehi H, Deng W, Guo Q-X (2004) Heterocycles 63:1613
41. Sparks RB, Combs AP (2004) Org Lett 6:2473
42. Balalaie S, Hashemi MM, Akhbari M (2003) Tetrahedron Lett 44:1709
43. Coleman CM, MacElroy JMD, Gallagher JF, O'Shea DF (2002) J Comb Chem 4:87
44. Durand-Reville T, Gobbi LB, Gray BL, Ley SV, Scott JS (2002) Org Lett 4:3847
45. De Luca L, Giacomelli G, Porcheddu A (2005) J Comb Chem 7:905
46. Henkel B (2004) Tetrahedron Lett 45:2219
47. Kim YJ, Varma RS (2004) Tetrahedron Lett 45:7205
48. Boufatah N, Gellis A, Maldonado J, Vanelle P (2004) Tetrahedron 60:9131
49. VanVliet DS, Gillespie P, Scicinski JJ (2005) Tetrahedron Lett 46:6741
50. Perumal S, Mariappan S, Selvaraj S (2004) ARKIVOC viii, p 46
51. Brain CT, Steer JT (2003) J Org Chem 68:6814
52. Ireland SM, Tye H, Whittaker M (2003) Tetrahedron Lett 44:4369
53. Schwerkoske J, Masquelin T, Perun T, Hulme C (2005) Tetrahedron Lett 46:8355
54. Lu Y, Zhang W (2004) QSAR Combi Sci 23:827

55. Bendale PM, Sun C-M (2002) J Comb Chem 4:359
56. Lin M-J, Sun C-M (2004) Synlett, p 663
57. Su Y-S, Lin M-J, Sun M-C (2004) Tetrahedron Lett 46:177
58. Su Y-S, Sun M-C (2005) Synlett, p 1243
59. Lee M-J, Sun C-M (2004) Tetrahedron Lett 45:437
60. Lin M-J, Sun C-M (2003) Tetrahedron Lett 44:8739
61. Yeh W-B, Lin M-J, Lee M-J, Sun C-M (2003) Molecular Diversity 7:185
62. Paul S, Gupta M, Gupta R, Loupy A (2002) Synthesis, p 75
63. Muccioli GG, Poupaert JH, Wouters J, Norberg B, Poppitz W, Scriba GKE, Lambert DM (2003) Tetrahedron 59:1301
64. Molteni V, Hamilton MM, Mao L, Crane CM, Termin AP, Wilson DM (2002) Synthesis, p 1669
65. Westman J, Lundin R, Stalberg J, Ostbye M, Franzen A, Hurynowicz A (2002) Comb Chem High Throughput Screen 5:565
66. Giacomelli G, Porcheddu A, Salaris M, Taddei M (2003) Eur J Org Chem, p 537
67. Westman J, Lundin R (2003) Synthesis, p 1025
68. De Luca L, Giacomelli G, Porcheddu A, Salaris M, Taddei M (2003) J Comb Chem 5:465
69. Giacomelli G, De Luca L, Porcheddu A (2003) Tetrahedron 59:5437
70. Chiacchio U, Rescifina A, Saita MG, Iannazzo D, Romeo G, Mates JA, Tejero T, Merino P (2005) J Org Chem 70:8991
71. Vasquez TE, Nixey T, Chenera B, Gore V, Bartberger MD, Sun Y, Hulme C (2003) Molecular Diversity 7:161
72. Pabba C, Wang H-J, Mulligan SR, Chen Z-J, Stark TM, Gregg BT (2005) Tetrahedron Lett 46:7553
73. Klapars A, Huang X, Buchwald SL (2002) J Am Chem Soc 124:7421
74. Tejedor D, Santos-Expósito A, González-Cruz D, Marrero-Tellado JJ, Garcia-Tellado F (2005) J Org Chem 70:1042
75. Lee JC, Choi HJ, Lee YC (2003) Tetrahedron Lett 44:123
76. Paul S, Nanda P, Gupta R, Loupy A (2004) Tetrahedron Lett 45:425
77. Trost BM, Dogra K, Franzini M (2004) J Am Chem Soc 126:1944
78. Pottorf RS, Chadha NK, Katkevics M, Ozola V, Suna E, Ghane H, Regberg T, Player MR (2003) Tetrahedron Lett 44:175
79. Kumar R, Selvam C, Kaur G, Chakraborti AK (2005) Synlett, p 1401
80. Frolov EB, Lakner FJ, Khvat AV, Ivachtchenko AV (2004) Tetrahedron Lett 45:4693
81. Chakraborti AK, Selvam C, Kaur G, Bhagat S (2004) Synlett, p 851
82. Phoon CW, Ng PY, Ting AE, Yeo SL, Sim MM (2001) Bioorg Med Chem Lett 11:1647
83. Kashiyama E, Hutchinson I, Chua MS, Stinson SF, Phillips LR, Kaur G, Sausville EA, Bradshaw TD, Westwell AD, Stevens MFG (1999) J Med Chem 42:4172
84. Frère S, Thiéry V, Bailly C, Besson T (2003) Tetrahedron 59:773
85. Frère S, Thiéry V, Besson T (2003) Synth Commun 33:3789
86. Mu X-J, Zou J-P, Zeng R-S, Wu J-C (2005) Tetrahedron Lett 46:4345
87. Katritzky AR, Singh SK (2002) J Org Chem 67:9077
88. Rostamizadeh S, Tajik H, Yazdanfarahi S (2003) Synth Commun 33:113
89. Baxendale IR, Ley SV, Martinelli M (2005) Tetrahedron 61:5323
90. Brain CT, Brunton SA (2001) Synlett, p 382
91. Kiryanov AA, Sampson P, Seed AJ (2001) J Org Chem 66:7925
92. Wang Y, Sauer DR, Djuric SW (2006) Tetrahedron Lett 47:105
93. Wang Y, Miller RL, Sauer DR, Djuric SW (2005) Org Lett 7:925
94. Schulz MJ, Coats SJ, Hlasta DJ (2004) Org Lett 6:3265

95. Bliznets IV, Vasil'ev AA, Shorshnev SV, Stepanov AE, Lukyanov SM (2004) Tetrahedron Lett 45:2571
96. Ek F, Manner S, Wistrand L-G, Frejd T (2004) J Org Chem 69:1346
97. Chetia A, Longchar M, Lekhok KC, Boruah RC (2004) Synlett, p 1309
98. Bagley MC, Lunn R, Xiong X (2002) Tetrahedron Lett 43:8331
99. Sainz YF, Raw SA, Taylor RJK (2005) J Org Chem 70:10086
100. Öhberg L, Westman J (2001) Synlett, p 1296
101. Legeay J-C, Van den Eynde JJ, Bazureau JP (2005) Tetrahedron 61:12386
102. Bagley MC, Singh N (2002) Synlett, p 1718
103. Devi I, Kumar BSD, Bhuyan PJ (2003) Tetrahedron Lett 44:8307
104. Devi I, Bhuyan PJ (2004) Synlett, p 283
105. Agarwal A, Chauhan PMS (2005) Tetrahedron Lett 46:1345
106. Tu S, Miao C, Gao Y, Fang F, Zhuang Q, Feng Y, Shia D (2004) Synlett, p 255
107. Gorobets NY, Yousefi BH, Belaj F, Kappe CO (2004) Tetrahedron 60:8633
108. Singh PND, Klima RF, Muthukrishnan S, Murthy RS, Sankaranarayanan J, Stahlecker HM, Patel B, Gudmundsdóttir AD (2005) Tetrahedron Lett 46:4213
109. Theoclitou M, Robinson LA (2002) Tetrahedron Lett 43:3907
110. Wilson NS, Sarko CR, Roth GP (2002) Tetrahedron Lett 43:581
111. Yadav JS, Reddy VS, Rao RS, Naveenkumar V, Nagaiah K (2003) Synthesis, p 1610
112. Jia C-S, Zhang Z, Tu S-J, Wang G-W (2006) Org Biomol Chem 4:104
113. Carranco I, Díaz JL, Jiménez O, Vendrell M, Albericio F, Royo M, Lavilla R (2005) J Comb Chem 7:33
114. Mitra AK, Banerjee K (2003) Synlett, p 1509
115. Hazarkhani H, Karimi B (2004) Synthesis, p 1239
116. Saxena I, Borah DC, Sarma JC (2005) Tetrahedron Lett 46:1159
117. Stadler A, Kappe CO (2001) J Comb Chem 3:624
118. Xia M, Wang Y (2003) Synthesis, p 262
119. Tyagarajan S, Chakravarty P (2005) Tetrahedron Lett 46:7889
120. Bagley MC, Hughes DD, Lubinu MC, Merritt EA, Taylor PH, Tomkinson NCO (2004) QSAR Combi Sci 23:859
121. Bagley MC, Hughes DD, Taylor PH (2003) Synlett, p 259
122. Mont N, Teixido J, Borell JI, Kappe CO (2003) Tetrahedron Lett 44:5385
123. Alexandre F-R, Berecibar A, Besson T (2002) Tetrahedron Lett 43:3911
124. Liu J-F, Lee J, Dalton AM, Bi G, Yu L, Baldino CM, McElory E, Brown M (2005) Tetrahedron Lett 46:1241
125. Liu JF, Ye P, Zhang B-L, Bi G, Sargent K, Yu L, Yohannes D, Baldino CM (2005) J Org Chem 70:6339
126. Liu J-F, Ye P, Sprague K, Sargent K, Yohannes D, Baldino CM, Wilson CJ, Ng S-C (2005) Org Lett 7:3363
127. Liu J-F, Ye P, Zhang B, Bi G, Sargent K, Yu L, Yohannes D, Baldino CM (2005) J Org Chem 70:10488
128. López-Cobenas A, Cledera P, Sánchez JD, Pérez-Contreras R, López-Alvarado P, Ramos MT, Avendaño C, Menéndez JC (2005) Synlett, p 1158
129. Cledera P, Sánchez JD, Caballero E, Avendaño C, Ramos MT, Menéndez JC (2004) Synlett, p 803
130. Macleod C, Martinez-Teipel BI, Barker WM, Dolle RE (2006) J Comb Chem 8:132
131. Pérez R, Beryozkina T, Zbruyev OI, Haas W, Kappe CO (2002) J Comb Chem 4:501
132. Zhao Z, Wisnoski DD, Wolkenberg SE, Leister WH, Wang Y, Lindsley CW (2004) Tetrahedron Lett 45:4873

133. Lindsley CW, Zhao Z, Leister WH, Robinson RG, Barnett SF, Deborah D-J, Raymond EJ, Hartman GD, Huff JR, Huber HE, Duggan ME (2005) Bioorg Med Chem Lett 15:761
134. Kim SY, Park KH, Chung YK (2005) Chem Commun, p 1321
135. Zhang W, Tempest P (2004) Tetrahedron Lett 45:6757
136. Cho S, Keum G, Kang SB, Han S-Y, Kim Y (2003) Molecular Diversity 6:283
137. Chang W-J, Yeh W-B, Sun C-M (2003) Synlett, p 1688
138. Tung C-L, Sun C-M (2004) Tetrahedron Lett 45:1159
139. Zhao Z, Leister WH, Strauss KA, Wisnoski DD, Lindsley CW (2003) Tetrahedron Lett 44:1123
140. Peng Y, Song G (2004) Tetrahedron Lett 45:5313
141. Lee H-K, Rana TM (2004) J Comb Chem 6:504
142. Seijas JA, Vázquez-Tato MP, Carballido-Reboredo R (2005) J Org Chem 70:2855
143. Perreux L, Loupy A (2002) Nonthermal Effects of Microwaves in Organic Synthesis. In: Loupy A (ed) Microwaves in Organic Synthesis. Wiley, Weinheim, p 61
144. Tsukayama M, Kawamura Y, Ishizuka T, Hayashi S, Torii F (2003) Heterocycles 60:2725
145. Kabalka GW, Mereddy AR (2005) Tetrahedron Lett 46:6315
146. Besson T, Thiéry V, Frère S (2001) Tetrahedron Lett 42:2791
147. Laporterie A, Marquié J, Dubac J (2002) Microwave-assisted Reactions on Graphite. In: Loupy A (ed) Microwaves in Organic Synthesis. Wiley, Weinheim, p 219

Top Curr Chem (2006) 266: 103–144
DOI 10.1007/128_046
© Springer-Verlag Berlin Heidelberg 2006
Published online: 10 June 2006

Microwave-Assisted and Metal-Catalyzed Coupling Reactions

P. Nilsson[1] · K. Olofsson[1] · M. Larhed[2] (✉)

[1]Biolipox AB, Berzelius Väg 3, SE-17165 Solna, Sweden

[2]Department of Medicinal Chemistry, Organic Pharmaceutical Chemistry,
Uppsala University, Box 574, SE-75123 Uppsala, Sweden
mats@orgfarm.uu.se

Abstract The review highlights microwave-accelerated metal-catalyzed transformations of aryl and vinyl halides (or pseudo halides), providing a selection of carbonylations, Heck- and Sonogashira reactions, nucleophilic substitutions and cross-couplings. Examples are chosen in such a way as to illustrate the large diversity of microwave applications within this class of coupling reactions, including solution-phase, solid-phase and fluorous chemistry. In most cases, the reactions were performed on a small scale using sealed vessels and irradiation times of less than one hour.

Keywords Homogeneous catalysis · Cross-coupling · Microwave · Heck · Sonogashira · Buchwald–Hartwig

Abbreviations

AIBN	2,2′-azo-bis-isobutyronitrile
binap	2,2′-bis(diphenylphosphino)-1,1′-binaphthyl
Cy	cyclohexyl
dba	dibenzylideneacetone
DCM	dichloromethane
DMA	dimethylacetamide
DIEA	diisopropylethylamine
dppe	1,3-bis(diphenylphosphino)ethane
dppf	1,1′-bis(diphenylphosphino)ferrocene
GC	gas chromatography
MW	microwave irradiation
n-Bu	normal-butyl
NMP	1-methyl-2-pyrrolidone
palladacycle	trans-di(μ-acetato)-bis[o-(di-o-tolyl-phosphino)benzyl]dipalladium(II)
PEG	polyethylene glycol
rac	racemic
TBAF	tetrabutylammonium fluoride
t-Bu	tert-butyl
TON	turnover number
triflate	trifluoromethanesulfonyl

1
Introduction

The outcome of an organic reaction is completely dependent on the delivery of energy and the timing of its introduction into the reaction system. If applied carelessly, heat can destroy the most robust organic reactions. Thus, it is crucial to choose an appropriate energy source in organic chemistry [1]. Among the sources available, microwave irradiation has huge potential to provide controlled energy directly to the molecules of interest. Accordingly, since the mid-1980s, a vast number of research reports have demonstrated that accelerated chemical rates can be achieved when reactions are heated with microwave irradiation instead of using traditional sources of heat [2–6].

With the onset of the twenty-first century, dedicated microwave devices intended for controlled small-scale synthesis in sealed vessels have become commercially available. This equipment has facilitated easy programming and documentation of time, temperature and irradiation power during reactions, while at the same time providing high reaction control and straightforward operation. Furthermore, modern microwave heating reactors are highly efficient energy sources because of their ability to rapidly heat the bulk of the reaction mixture directly, avoiding wall effects and the associated decomposition of sensitive catalysts [7–10]. One environmentally friendly feature of such systems is their reduced energy consumption compared with the classical heating of small-scale reactions [11]. Rapid heating to high temperatures in sealed vessels with microwave heating also permits the use of a small amount of low-boiling solvent, allowing easy work-up and reduction of solvent waste.

The development of fast, reliable and convenient chemical processes is important for the whole field of organic chemistry. In fact, in many cases even a slight drop in chemical yield is acceptable when compensated for by large gains in reaction time or improved ease of handling [12–15]. These features apply in particular to iterative reaction optimization in commercial situations where the reaction time is an important parameter [16]. In particular, the pharmaceutical industry is continuously on the look-out for new rapid synthetic methods and strategies for the discovery and development of new drugs.

Homogeneous transition metal catalysis has revolutionized organic chemistry over the past 40 years [17]. This group of reactions, which comprises a plethora of versatile carbon–carbon and carbon–heteroatom bond-forming processes and asymmetric transformations, constitutes an ever-growing part of the arsenal of synthetic tools available to the modern organic chemist. Indeed, these days metal catalysis is present in nearly all novel target synthesis carried out at the laboratory scale. In the last few decades much effort has been devoted to extending the scope of palladium-, copper- and nickel-catalyzed reactions proceeding via aryl- or vinylmetal intermediates [18–20]. Metal-mediated transformations have proven to be especially valuable for introducing substituents onto aromatic core structures, and they perform equally well in both inter- and intramolecular applications. Furthermore, in homogeneous catalysis there are currently many different metal sources and ligands that can be used to fine-tune the activity of catalytic complexes and the selectivity of reactions. An important advantage is that hydroxyl, carboxylic acid and amine groups usually do not need protection. However, the long reaction times frequently required for complete conversion have unfortunately (with some exceptions) limited the exploitation of the full potential of homogeneous catalysis in fine-chemical, high-throughput and medicinal synthesis [8, 21]. Rapid and reliable microwave methods are therefore highly desirable.

Microwave-heated organic reactions can sometimes be smoothly conducted in open vessels, but often it is of interest to work with closed systems, especially if superheating (which often results in reduced reaction times) is desired [3, 5]. The use of disposable septum-sealed vessels designed for straightforward pressurized processing and automation is essential in this case for both safety and productivity reasons. Further, if "flash-heated" metal-catalyzed reactions can be combined with modern high-speed purification techniques, the large repertoire of smooth transformations available will make this methodology highly attractive for the fast creation of compound libraries. The overall goal should be to fundamentally change and improve the way that metal-catalyzed chemistry is being performed today.

The advancements already achieved, in both academia and in industry, in the field of microwave-assisted metal catalysis are still remarkable in many ways. Five years ago it was still possible to nurse an ambition to cover all of the literature in this area within the scope of a book chapter [22]. In 2006 this would be extremely difficult, as the number of publications has increased significantly together with the range of transformations investigated. Thus, microwave heating is in fact no longer viewed as a back-up method, but to an increasing extent as the preferred heating method [4].

This review attempts to provide an overview of microwave-promoted metal-catalyzed transformations of aryl and vinyl halides (or pseudo halides), providing a personal selection of both pioneering and very recently published work. Covered areas include carbonylative transformations, Heck and Sonogashira reactions, nucleophilic substitutions and cross-couplings. Because of the diversity of the microwave systems used, the reader should consult the original references for detailed descriptions of settings and instrumentation.

2
Palladium-Catalyzed Carbonylations

Carbon monoxide is clearly a useful chemical building block [23]. There are however some definite disadvantages with the use of toxic and explosive gaseous reagents in a laboratory environment. Not only has the administration of CO to the reaction medium got to be safe, but the safety of the transport and storage of pressurized carbon monoxide containers must be considered. This reduces the utility of carbon monoxide in general, particularly in small-scale synthetic applications. A convenient way of avoiding many of these problems would be to exploit a liquid or solid reagent with the ability to release carbon monoxide in situ during a reaction. Thus, the number of reports of successful applications of condensed CO-sources in organic chemistry is increasing at a steady pace [24].

Palladium-catalyzed carbonylation reactions with aryl halides are powerful methods of generating aromatic amides, hydrazides, esters and carboxylic acids [25]. We have previously reported the exploitation of $Mo(CO)_6$ as a robust carbon monoxide-releasing reagent in palladium-catalyzed carbonylation reactions [26–29]. This stable and inexpensive solid delivers a fixed amount of carbon monoxide upon heating or by the addition of a competing molybdenum coordinating ligand (for example DBU). This allows for direct liberation of carbon monoxide in the reaction mixture without the need for external devices.

One of our initial forays into $Mo(CO)_6$-promoted carbonylations included the investigation of intermolecular reactions using amines as nucleophiles to form secondary and tertiary aromatic amides from aryl bromides and iodides [27]. Subsequent work using the activating preligand t-Bu$_3$PHBF$_4$ has allowed for the extension of this chemistry into examples using aryl chlorides as substrates (Scheme 1) [30].

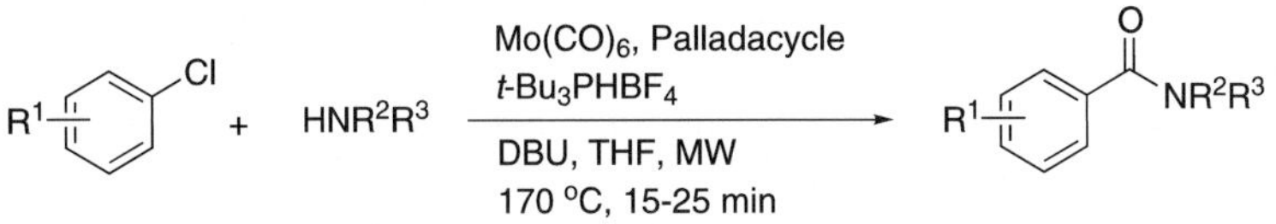

$R^1 = CF_3, CO_2Me, Me, OMe$
Amines = Bn, n-Bu, t-Bu, piperidine, aniline

Yield 51-91%
20 examples

Scheme 1 Microwave-promoted aminocarbonylations of aryl chlorides using $Mo(CO)_6$ as a solid carbon monoxide source

The preparation of primary benzamides by aminocarbonylation of aryl halides with ammonia as the nucleophile is recognized as being considerably more troublesome than performing the corresponding processes with organic amines, since the reaction requires two toxic gaseous reactants, ammonia and carbon monoxide [31]. In addition, the nucleophilicity of ammonia is limited and the Pd(II) coordination strength is high. Stimulated by reports on $Mo(CO)_6$-mediated reductive cleavage of $N-O$ bonds [32], hydroxylamine hydrochloride was investigated as a convenient solid ammonia equivalent [33]. Employing this in situ ammonia-carbonylation approach, primary benzamides were prepared from both aryl iodides and bromides in a straightforward manner (Scheme 2). The methodology was further applied to the synthesis of a complex HIV-1 protease inhibitor.

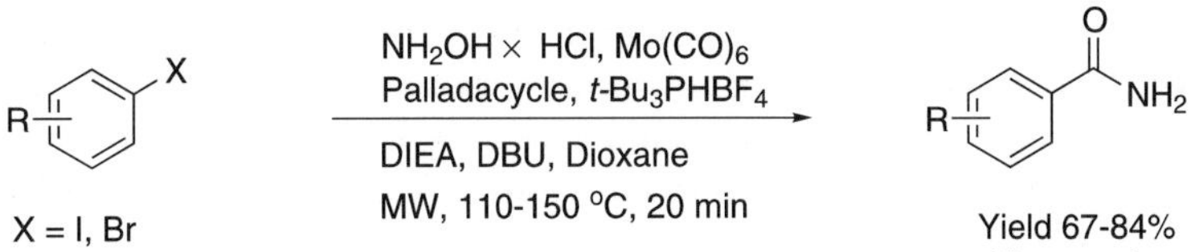

$X = I, Br$
$R = CF_3, F, Me, OMe$

Yield 67-84%
16 examples

Scheme 2 Preparation of primary benzamides from aryl iodides and bromides by in situ carbonylation

Despite the numerous reports dealing with combinatorial diversification of the dihydropyrimidinone (DHPM) template, transition metal-catalyzed derivatizations have been poorly explored. In a collaborative effort between the groups of C.O. Kappe and M. Larhed, high-speed transition metal-catalyzed functionalizations of this important scaffold were investigated [34].

Three different 4-(bromophenyl)-substituted DHPMs were considered as suitable starting materials for amino- and alkoxycarbonylations. Good-to-excellent yields were obtained in the aminocarbonylations of *meta-* and *para-*bromo compounds with *n*-butylamine, aniline, benzylamine and morpholine as nucleophiles (Scheme 3). Carbonylations of the related *ortho*-bromo substrate were less productive. In addition, the study was expanded to include a number of alkoxycarbonylations.

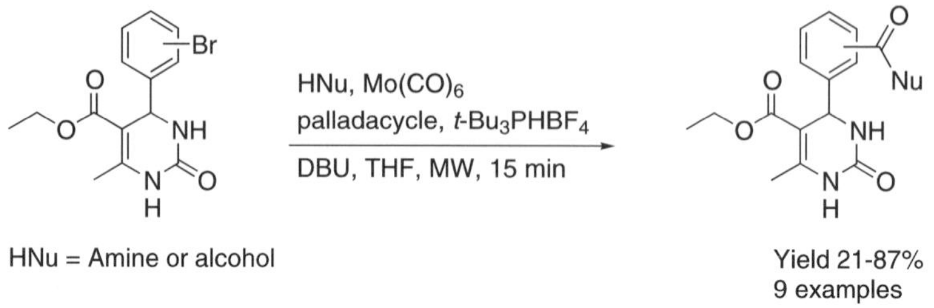

Scheme 3   Palladium-catalyzed amino- and alkoxycarbonylations of 4-(bromophenyl)-DHPMs using Mo(CO)$_6$ as the CO source

Among the different solvent alternatives in organic chemistry, water is extremely cheap and nontoxic [35–37]. In addition to these two general advantages, several positive aspects are expected when using water as the reaction medium for microwave superheated protocols [9, 10]. Firstly, water is rapidly heated by microwave irradiation to high reaction temperatures, where water acts as a less polar pseudo-organic solvent. Secondly, precise control of the reaction temperature is easily achieved because of the very high heat capacity of water, and thirdly, the lack of flammable properties also make the use of water safe in pressurized exothermic reactions. In addition, there is a need to implement more sustainable methods, not only for large-scale production, but also for lab-scale medicinal chemistry research. In an effort to develop carbonylation reactions in neat water, the potential of carrying out palladium-catalyzed Mo(CO)$_6$-promoted aminocarbonylations was examined [38, 39]. Rewardingly, aryl iodides, bromides and chlorides were all rapidly transformed into aromatic amides under microwave conditions (Scheme 4). In spite of the use

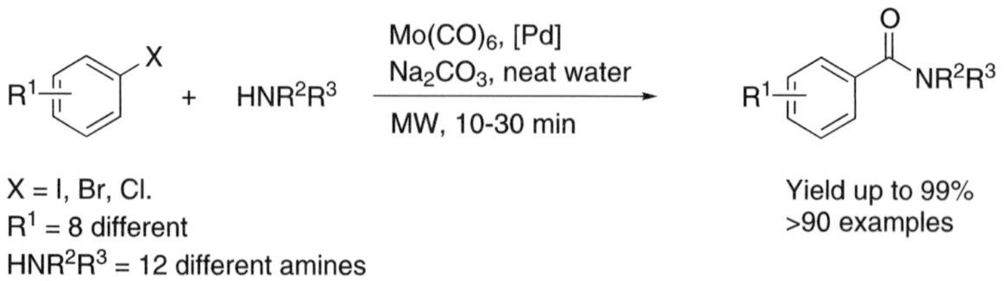

Scheme 4   Fast aminocarbonylation of aryl iodides, bromides and chlorides in neat water

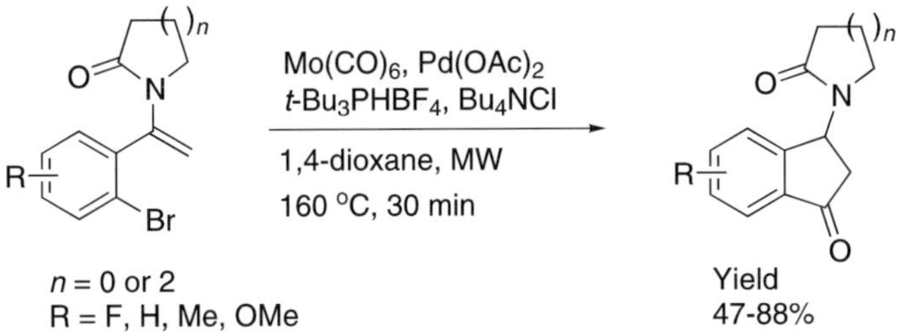

$n = 0$ or 2
$R = F, H, Me, OMe$

Yield
47–88%

Scheme 5 Rapid generation of 3-acylaminoindanones by in situ carboannulation

of water as a solvent, aminocarbonylation strongly dominated over hydroxy-carbonylation, providing good yields of both secondary and tertiary model amides.

Recently, $Mo(CO)_6$ was used as the CO-source in the palladium-catalyzed generation of 3-acylaminoindanones [40]. These target structures were prepared from o-bromoaryl-substituted enamides after 30 min of microwave heating (Scheme 5).

3
The Heck Reaction

In the late 1960s, the research teams of Moritani–Fujiwara and Heck independently discovered the palladium-mediated arylation and alkenylation of alkenes [41]. The stoichiometric version of this vinylic substitution reaction was further refined into a catalytic system, mainly by Richard Heck and his group [41]. In the last thirty-five years, this selective palladium-catalyzed transformation, generally known as the Heck reaction, has been extensively explored and used in several diverse areas such as the preparation of natural products, pharmaceuticals, agrochemicals and dyes, because of the mildness and high selectivity of the reaction [42, 43]. The Heck arylation in Scheme 6 was reported in 1996 and constitutes the first example of microwave-promoted, palladium-catalyzed C – C bond formation [44]. The power of the microwave flash heating methodology is amply manifested by the short reaction times and good yields of these couplings. The reactions were conducted in a single-mode cavity in septum-sealed Pyrex vessels with-

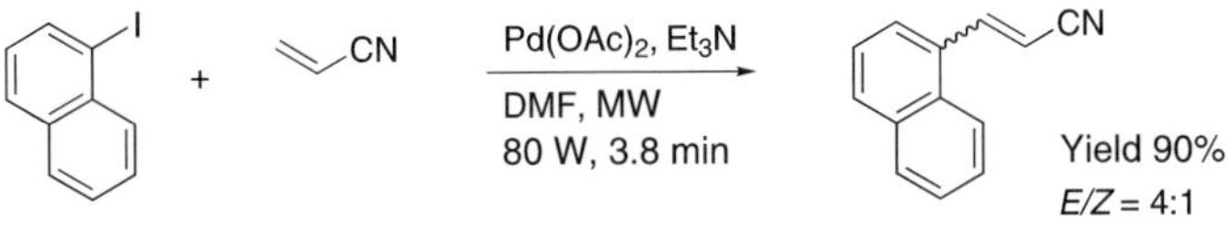

Yield 90%
$E/Z = 4:1$

Scheme 6 Heck coupling with 1-iodonaphthalene and acrylonitrile

out temperature control. The reaction in Scheme 6 (and in seven additional Heck coupling examples) was originally carried out with classic heating in the absence of solvent. To enhance the yields and reduce the reaction times, 0.5 mL of DMF was added to increase the polarity and the dielectric loss tangent [1] of the reaction mixture. This small modification of the original reaction conditions allowed the isolation of the products in high purity after very short reaction times (2.8–4.8 min).

Compared to the dramatic increase in microwave heating in cross-coupling and metal-mediated nucleophilic substitutions, the number of published microwave-assisted Heck reactions is limited. Thus, only seven additional examples emphasizing different concepts will be presented in this section.

A dimeric 4-hydroxyacetophenone oxime-derived palladacycle was used by Nájera et al. as a very efficient precatalyst for selective and microwave-assisted arylations of acrolein diethyl acetal, furnishing 3-arylpropanoate products [45]. In the presence of Cy_2NMe and tetrabutylammonium bromide (TBAB), higher reaction rates and slightly improved yields were realized under microwave conditions (Scheme 7).

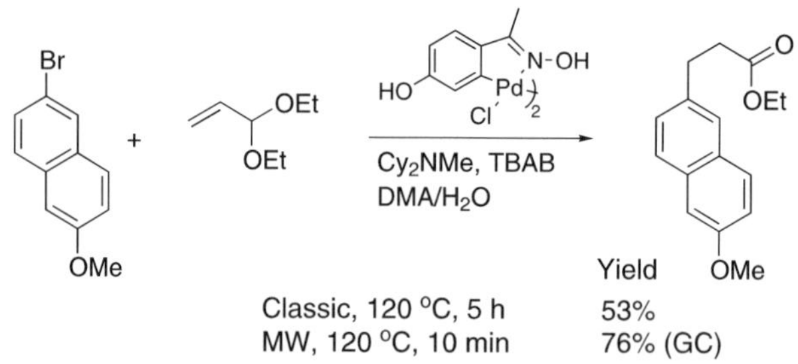

Scheme 7 Heck synthesis of an ethyl 3-arylpropanoate

Microwave-heated Heck reactions in water using ultralow palladium catalyst concentrations have been performed by Arvela and Leadbeater [46]. Different catalyst concentrations were investigated using a commercially available 1000 ppm palladium solution as the catalyst source. Impressively, useful Heck arylations were performed with palladium concentrations as low as 500 ppb (Scheme 8).

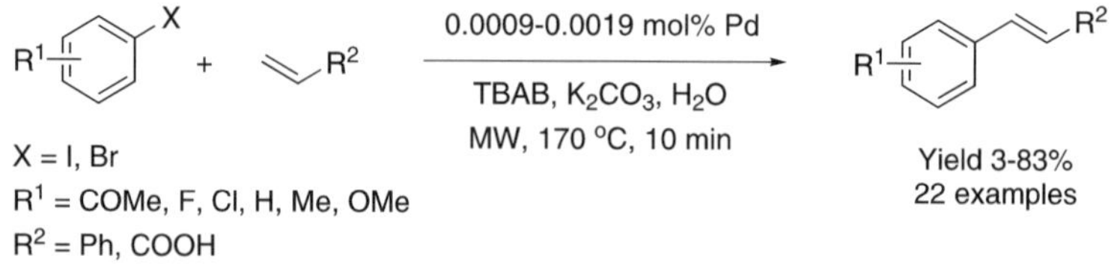

Scheme 8 Microwave-promoted Heck reactions using ultralow palladium concentrations

The use of ionic liquids in combination with microwave heating has great benefits, as the high boiling point and low vapor pressure of ionic liquids are combined with a propensity to interact strongly with microwave fields. In 2002, 1-butyl-3-methylimidazolium hexafluorophosphate (bmimPF$_6$) was therefore evaluated as a solvent for the Heck reaction under microwave conditions. Terminal monoarylations of electron-poor butyl acrylate were carried out under high-density irradiation, affording good-to-excellent yields using old-fashioned palladium chloride as the precatalyst [47]. In addition, the authors demonstrated that the catalyst was immobilized in the ionic liquid, allowing recycling of the "ionic catalyst phase" in five consecutive Heck reactions. Finally, a β,β-diphenylacrylate was synthesized in 45 min by a straightforward double Heck arylation using an excess of phenyl bromide and the sterically hindered 1,2,2,6,6-pentamethyl-piperidine (PMP) as the base (Scheme 9).

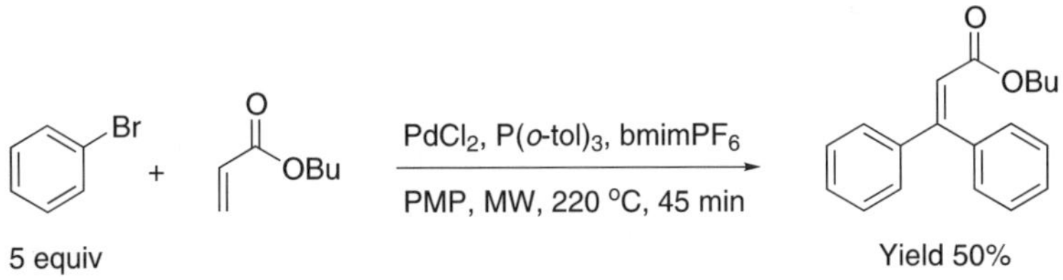

Scheme 9 Double Heck arylation with an ionic liquid as solvent

The recent advances in using relatively cheap and easily available chloroarenes in palladium(0)-catalysis instead of bromo- or iodoarenes is arguably one of the most exciting developments in chemistry today [48]. A paper from 2003 dealt with Heck-couplings performed with both activated and deactivated chloroarenes in ionic liquid-doped 1,4-dioxane [49]. The coupling between butyl acrylate and 4-chloroacetophenone needed 30 min at 180 °C when high-density microwave heating was applied, while standard heating at the same temperature required 1 hour and delivered a reduced yield (Scheme 10).

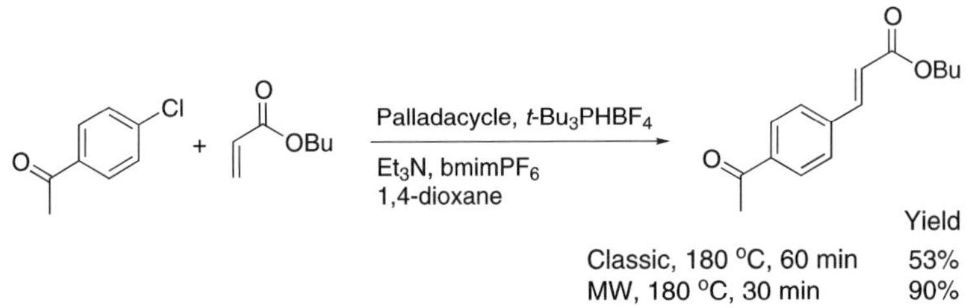

Scheme 10 Heck coupling of an aryl chloride

Using enol triflates, Heck vinylation of electron-rich olefins generally provides branched 1,3-butadiene products [41]. However, by incorporating a palladium(II)-coordinating tertiary amino group into a vinyl ether, an efficient vinylpalladium presentation and full terminal selectivity were realized, affording linear 1-alkoxy-1,3-butadienes [50]. These highly regioselective vinylations were completed in less than 30 min under single-mode microwave irradiation, as compared to overnight reactions with conventional heating. However, slightly lower E/Z-stereoselectivities and chemical yields were in many cases obtained in the high-temperature microwave-mediated couplings compared to the corresponding traditional reactions (Scheme 11).

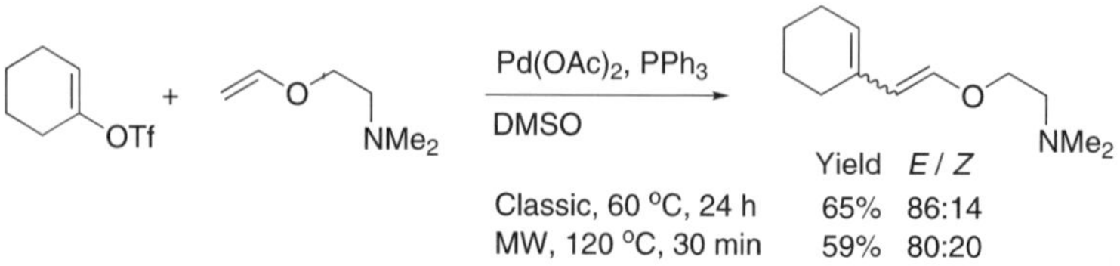

Scheme 11 Chelation-controlled terminal Heck vinylation of an enol ether

A nonclassical substrate for the Heck reaction is 2,3-epoxycyclohexanone. The reactivity of this molecule under Heck coupling conditions is most likely attributed to its in situ isomerization to 1,2-cyclohexanedione. The 1,2-diketone subsequently reacts with aryl bromides as an olefin via the enol tautomer. Thus, within 5 to 30 min of directed microwave heating of the aqueous PEG mixture, up to 13 different C3-arylations were conducted using less than 0.05 mol % palladium acetate and no phosphine ligand (Scheme 12) [51].

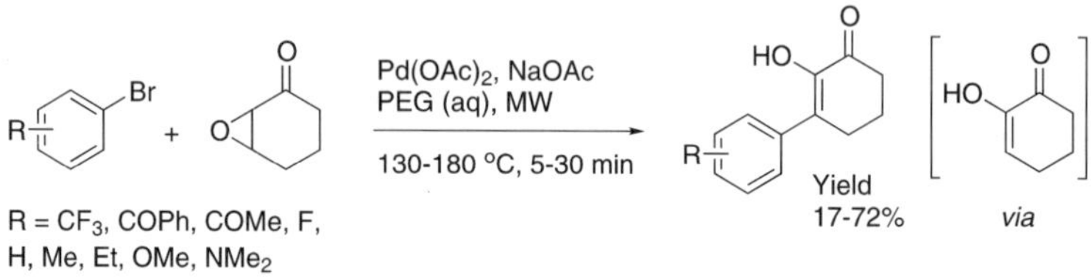

Scheme 12 A one-pot isomerization–arylation of 2,3-epoxycyclohexanone

The useful and selective reactivity of arylboronic acids makes them favorite building blocks for many modern organic chemistry applications. Arylboronic acids also serve as highly useful arylpalladium precursors in palladium(II)-catalyzed oxidative Heck reactions. Andappan et al. developed a microwave-accelerated protocol for oxidative Heck couplings using copper(II) acetate as the palladium reoxidant [52]. The method proved to be

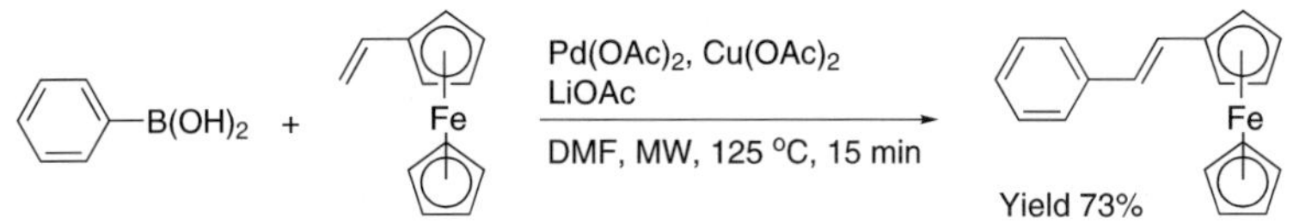

Scheme 13 A microwave-assisted oxidative Heck reaction

highly robust with both different electron-poor olefins and a diverse set of arylboronic acids. One example is presented in Scheme 13.

4
The Sonogashira Coupling

The Sonogashira reaction, using copper and palladium catalysis, is a general and reliable protocol used to cross-couple terminal alkynes with aryl/vinyl halides, forming aryl alkynes in an easy manner [53]. In 2000 Kabalka and co-workers discovered a solvent-free protocol that could be used to couple aryl iodides with alkyl and phenylacetylenes in a domestic oven, utilizing a palladium catalyst on alumina support [54]. In 2001 Erdélyi and Gogoll published a detailed investigation concerning the influence of mono-mode microwave irradiation on the efficiency and productivity of the Sonogashira coupling employing several different arylmetal precursors (Scheme 14) [55].

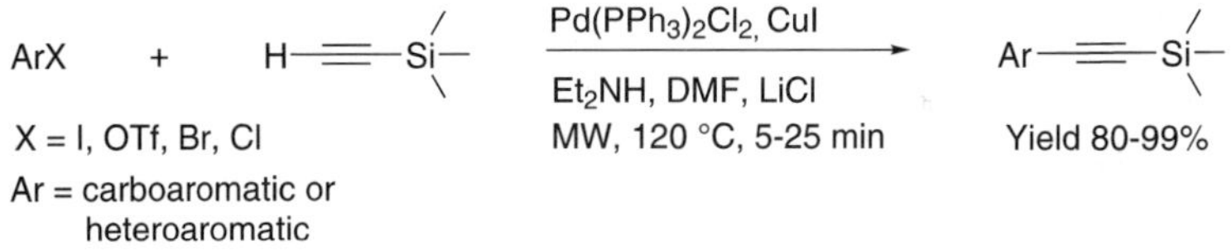

Scheme 14 Aryl coupling of trimethylsilylacetylene via a palladium-catalyzed Sonogashira protocol

Modern reports of microwave-promoted Sonogashira reactions involve the usage of solvent-free [56] conditions and attachment of the aryl halide onto a solid support [57–60] (e.g., polystyrene and PEG 4000). Notably, new catalyst systems have been introduced encompassing nickel [61] and copper [62] catalysis and even "transition metal-free" transformations [63]. The report published by the group of Wang using nickel catalysis is intriguing, since it demonstrates the use of 1,1-dibromostyrene precursors, liberating aryl acetylenes in situ (Scheme 15). The authors indicate that the addition of copper greatly accelerates the reaction, causing it to be completed after only 3 minutes of heating in a domestic oven.

Performing several reaction steps in a one-pot procedure enhances throughput, especially in combination with rapid microwave heating. A smooth se-

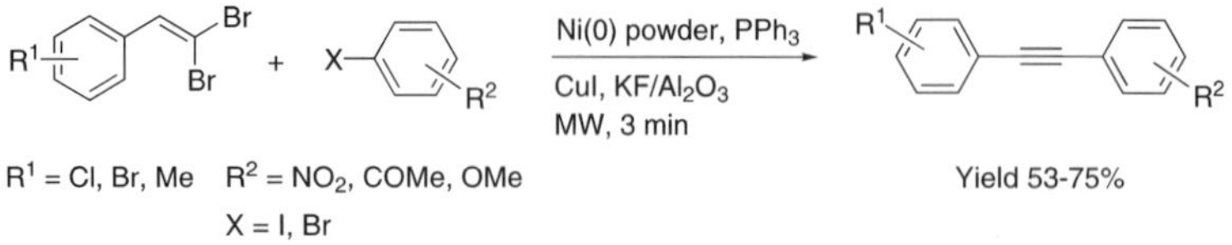

Scheme 15 Nickel-catalyzed solvent-free Sonogashira coupling

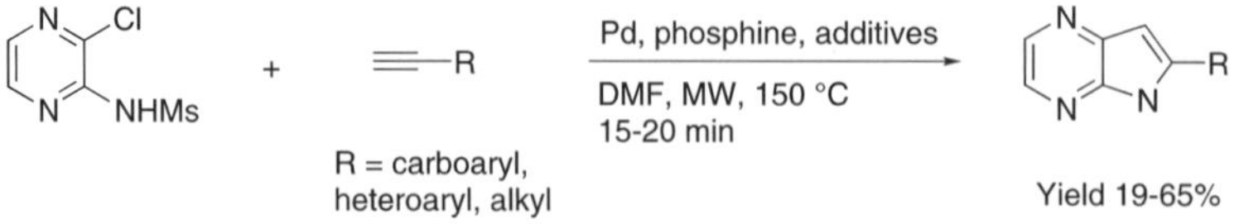

Scheme 16 A microwave-promoted three-step one-pot cyclization

quential three-step and one-pot Sonogashira coupling-heteroannulation and deprotection procedure has been exemplified by Hopkins and Collar employing microwave irradiation (Scheme 16) [64].

The option to perform Sonogashira and similar reactions while avoiding the use of transition-metals is indisputably very attractive, mainly due to cost and environmental issues. Erik van der Eycken and his group have exemplified the omission of transition metals in the coupling of phenyl acetylene with different aryl halides, employing microwave heating at 175 °C with a carbonate base and TBAB in water [63]. The reaction times spanned between 5–25 minutes, and good-to-excellent yields were obtained. The crude reaction mixture was analyzed using atomic absorption spectrophotometry to rule out any latent contamination of significant transition metals beyond 1 ppm. A similar "transition metal-free" water-based methodology using poly(ethyleneglycol) instead of TBAB as the phase-transfer reagent has also been reported by Leadbeater [65]. Later, the same group discovered that Suzuki couplings can proceed with a palladium concentration as low as 0.5 ppb, indicating that a very small amount of catalyst can, in favorable cases, still be productive due to very high catalyst turnover numbers (TON) [66].

5
Cyano Couplings

The direct synthesis of aryl- or alkyl nitriles from cyanide and organohalide precursors is revered in synthetic chemistry, as the nitriles represent a flexible functionality that can easily be converted into (for example) carboxylic acids, esters, amides, amidines, amines and various heterocycles [67], such as thiazoles, oxazolidones, triazoles and tetrazoles [68]. The tetrazole group is becoming increasingly popular in medicinal chemistry as a bioisosteric replacement for the carboxyl group.

A development of the palladium-catalyzed cyanation of aryl bromides using zinc cyanide was published in the mid-1990s [69]. Normally, the transformation from halide to nitrile called for at least 5 hours of heating, and the subsequent cycloaddition to the tetrazole is known to involve even longer reaction times. Hallberg has described a single-mode microwave methodology for the palladium-catalyzed synthesis of both aryl and vinyl nitriles with zinc cyanide, starting from the corresponding bromides [70]. The processing times were short, and full conversions were achieved (Scheme 17).

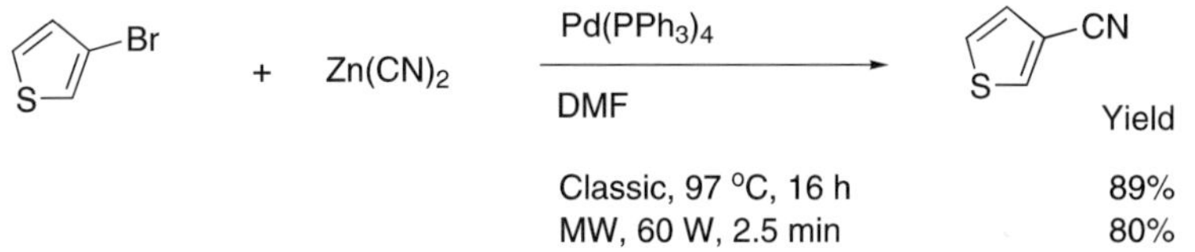

	Classic, 97 °C, 16 h	89%
	MW, 60 W, 2.5 min	80%

Scheme 17 Palladium-catalyzed cyanation of a heteroaryl bromide into a heteroaryl nitrile

Furthermore, cyanation followed by a subsequent cycloaddition, forming a tetrazole, was implemented as a rapid one-pot procedure on TentaGel-support, as depicted in Scheme 18. Only an insignificant decay of the solid support was experienced under the high-temperature conditions applied.

In the last few years numerous reports have been published in the field of microwave-promoted aryl halide cyanation, utilizing nickel [71], palladium [72, 73] and copper [74, 75] catalysis. Even water [75] and ionic liquids [76] have proven useful as solvents in these processes. Srivastava and Collibee have exemplified a swift and dynamic procedure using polymer-supported triphenyl phosphine to enable easy subsequent removal through filtration [72]. As shown in Scheme 19, both bromides and iodides could be activated using palladium catalysis in DMF. Even without optimization of the individual reaction times, the overall process time involving simple filtration and extraction for compound isolation appears to be short.

Aryl triflates, produced from the corresponding hydroxy arene using triflic anhydride or triflic imide with microwave heating [77] (or by other means)

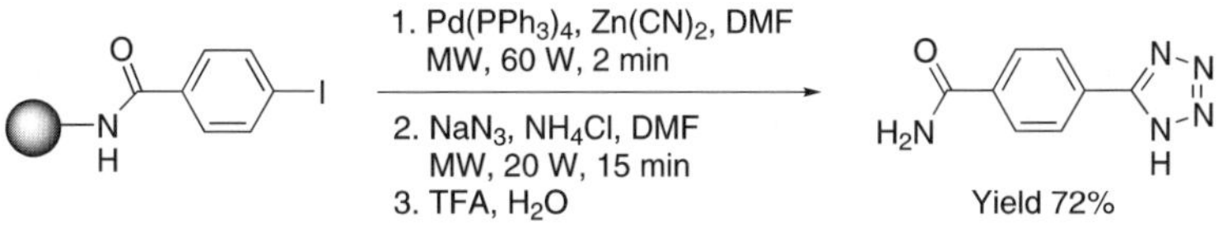

Scheme 18 Three-step one-pot procedure for tetrazole synthesis on a polymer support

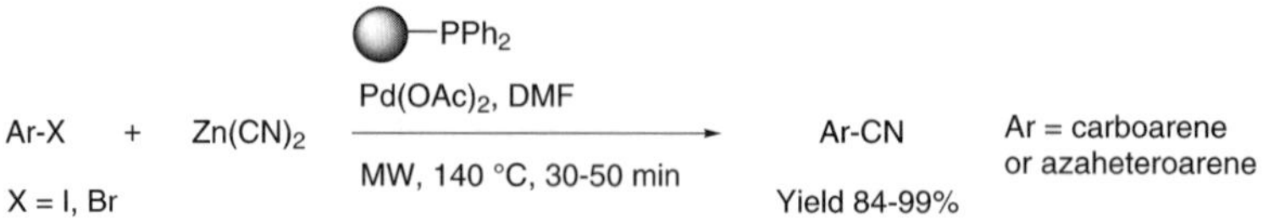

$$\text{Ar-X} \quad + \quad \text{Zn(CN)}_2 \quad \xrightarrow[\text{MW, 140 °C, 30-50 min}]{\substack{\bullet\text{-PPh}_2 \\ \text{Pd(OAc)}_2,\ \text{DMF}}} \quad \text{Ar-CN}$$

X = I, Br   Yield 84–99%

Ar = carboarene or azaheteroarene

Scheme 19 Palladium-catalyzed aryl cyanation using a triaryl phosphine on polymer-support

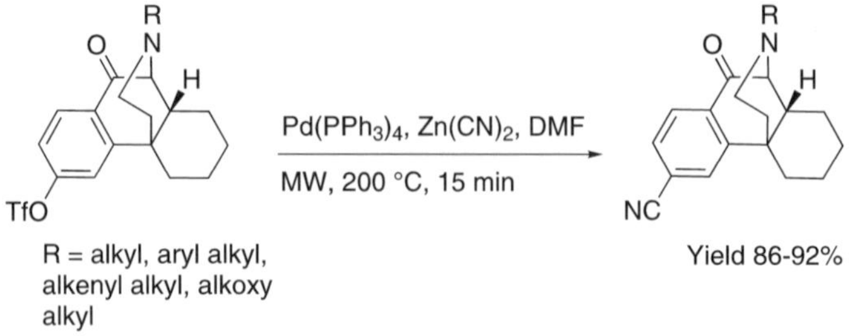

R = alkyl, aryl alkyl, alkenyl alkyl, alkoxy alkyl

Yield 86–92%

Scheme 20 Effective cyanation using aryl triflates and palladium catalysis

are important halide substitutes, but they can on occasion show poor reactivity and stability. Nonetheless, Zhang and Neumeyer reported on the palladium-catalyzed activation of triflates in nitrile couplings for the preparation of diverse κ-opioid receptor ligands [73]. A reaction time of 15 minutes at 200 °C was suitable for complete exchange to take place, producing yields of 86–92% in sealed reaction vessels (Scheme 20).

6
Aryl–Nitrogen Couplings

The simultaneous development by the groups of Hartwig and Buchwald in 1994 of aryl amination chemistry helped trigger a burst of interest surrounding catalyzed C – N couplings in general, and aryl–nitrogen bond formation in particular [78]. Catalytic aryl amine couplings are generally sluggish processes, especially when copper catalysis is used. Accordingly, to improve the kinetics a multitude of microwave-enhanced methods have been implemented in connection to the development of catalytic protocols. An overview of the transformations in this section is presented in Scheme 21.

A domestic oven and open containers were used in the work by Sharifi, who performed couplings between aryl bromides and alkyl amines (yields: 32–86%) [79]. The palladium precatalyst $Pd[P(o-\text{tolyl})_3]_2Cl_2$ was found to be the most efficent catalyst when toluene was used as solvent and sodium *tert*-butoxide as base. About the same time, a similar approach to arylating alkyl and aryl amines was reported by Hallberg [80]. Mono-modal microwave

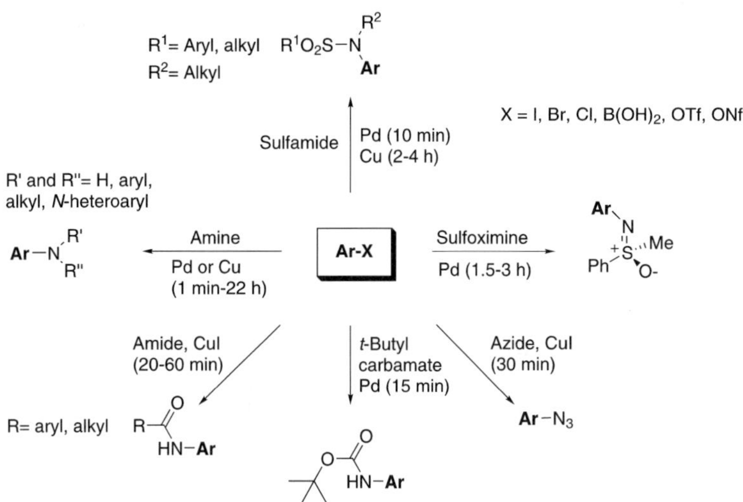

Scheme 21 Microwave-heated copper- and palladium-catalyzed *N*-arylations

irradiation of closed reaction vessels allowed the process to be completed within an impressive 4 minutes in DMF, duplicating the standard protocol introduced by Hartwig and Buchwald. In a recent publication, Beifuss and coworkers employed aniline arylation as a chemoselective key step in the formation of the desired phenazine unit. High-temperature microwave conditions permitted an improved reaction rate without compromising the high yield (Scheme 22).

In a medicinal chemistry program aiming to discover p38 MAP kinase inhibitors, the group of Skjaerbaeck introduced microwaves in order to achieve aryl decoration of the aniline backbone [81]. A systematic optimization of the catalyst, solvent, base and reaction time/temperature delivered a general procedure for the expeditious production of the desired aryl aminobenzophenones within 3–15 minutes, as demonstrated in Scheme 23.

Aryl nonaflates were successfully employed as thermally stable arylating agents in a study by the group of Buchwald, giving fast reactions ranging

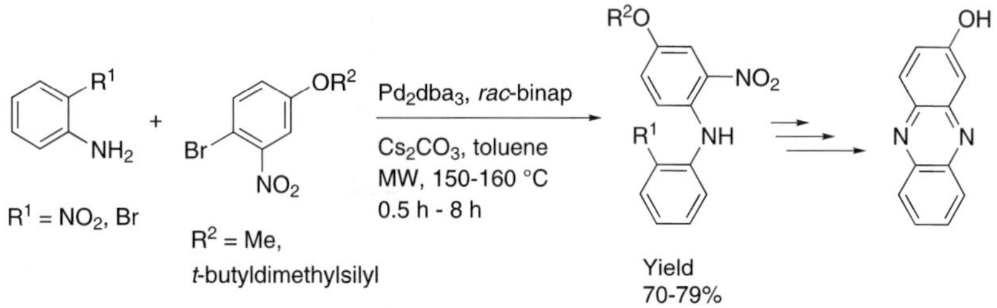

Scheme 22 Microwave-promoted diarylamine formation

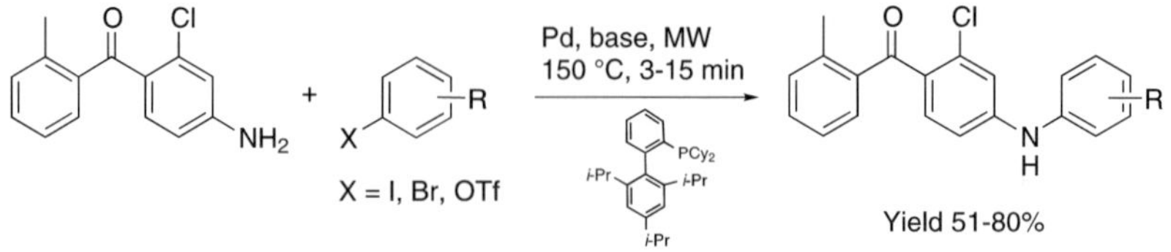

Pd, base, MW
150 °C, 3-15 min

X = I, Br, OTf

Yield 51-80%

Scheme 23 Construction of p38 MAP kinase inhibitors utilizing rapid microwave chemistry

from 1–45 min. The coupling of aromatic amines using palladium, phosphine ligands and microwave heating furnished up to 99% yield in favorable cases [82]. In addition, amination of azaheteroaryl bromides and chlorides has also been reported to be efficiently executed within 10 minutes, using standard reaction conditions and microwave heating [83, 84]. Brain and Steer reported on a synthesis of benzimidazoles via an intramolecular cyclization process using an amidine moiety as the N-nucleophile [85]. The reaction was fast and efficient and combined with a "catch and release" strategy, featuring capture of the benzimidazole on an acidic resin in the purification stage, meaning that a very interesting procedure for quick compound production was obtained (Scheme 24).

Weigand and Pelka have examined microwave-enhanced aminations of aryl chlorides and bromides using amine resins as the nitrogen nucleophile [86]. The normally very sluggish reaction (18 h, reflux) using polystyrene Rink resin and electron-poor chlorides and bromides could be finalized within 15 minutes by employing a sealed vessel protocol with DME/t-BuOH 1 : 1 as solvent at 130 °C. This high-speed technique gave yields equally as high as the much slower method with oil bath heating.

Aryl chlorides are more reluctant to participate in amination than most other aryl halides/pseudohalides. To tackle this problem, Caddick et al. examined the effect of palladium-N-heterocyclic carbenes as catalysts in rapid microwave-promoted reactions [87]. *Para*-tolyl and -anisyl chloride were reacted with aromatic and aliphatic amines in mostly good yields within 6 minutes of heating at 160 °C. Reactions using anisyl, tolyl or phenyl chlorides and aliphatic amines have also been reported by Maes et al. using a phosphine ligand and a strong base, which creates the desired products after 10 minutes of heating at 110–200 °C [88].

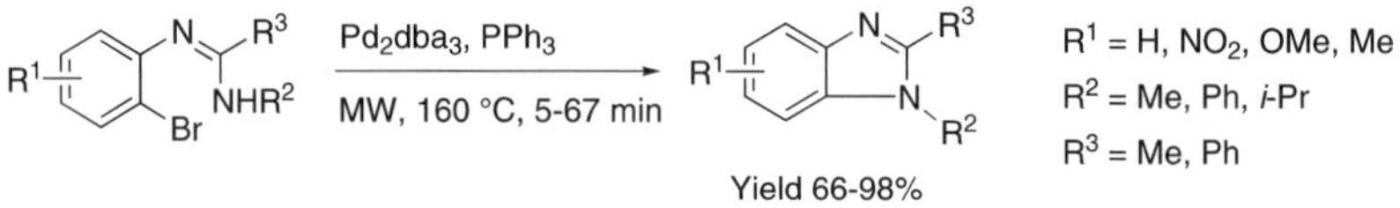

Pd$_2$dba$_3$, PPh$_3$
MW, 160 °C, 5-67 min

Yield 66-98%

R^1 = H, NO$_2$, OMe, Me
R^2 = Me, Ph, i-Pr
R^3 = Me, Ph

Scheme 24 Palladium-catalyzed intramolecular amidination generating substituted benzimidazoles

The development of large-scale microwave applications (gram-to- kilogram scale) is under steady progress, ultimately aiming to circumvent the associated physical constraints imposed by the larger volumes used (i.e., microwave penetration depth, stirring, hotspot formation) [89–91]. Very recently Maes reported on a palladium-catalyzed aryl amination application performed on the 20 mmol scale using 10–100 mL vessels and four different models of dedicated microwave synthesizers [92]. All four studied reactors gave, with small model-dependent variations, the expected outcome of the Buchwald–Hartwig couplings, as illustrated in Scheme 25. It was however found that a unique heating profile was obtained for every model investigated, depending on the choice of solvent and vessel volume at constant magnetron power output [92].

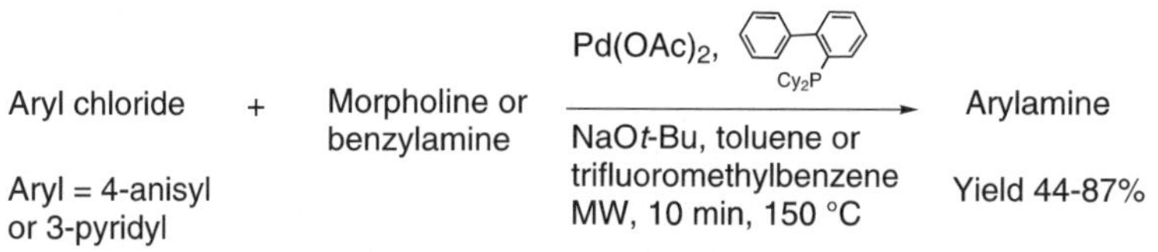

Scheme 25 Batch-scale aryl aminations under microwave irradiation

The Ullmann coupling, being a copper-catalyzed *N*-arylation, is known to be more sluggish than the corresponding palladium-catalyzed transformation. Even so, Wu and coworkers succeeded to accelerate reaction times down to only one hour with preserved chemoselectivity (Scheme 26) [93]. A set of aromatic azaheterocycles produced yields of 49–91% after 1–22 hours of microwave heating.

As early as 1999 Combs published a microwave protocol for *N*-arylation of imidazoles, pyrazoles and 1,2,3-triazoles linked to a solid support [94]. In this report, *p*-tolyl boronic acid was used as the arylating agent under Cu(II) catalysis. The reaction mixture was heated in a domestic oven for three 10-second periods with manual stirring in-between, delivering yield of 55–64% of the cleaved product. The closely associated Goldberg reaction has also been reported by Lange et al. to benefit from microwave heating at high substrate concentrations in the synthesis of *N*-aryl-2-piperazinones (Scheme 27) [95].

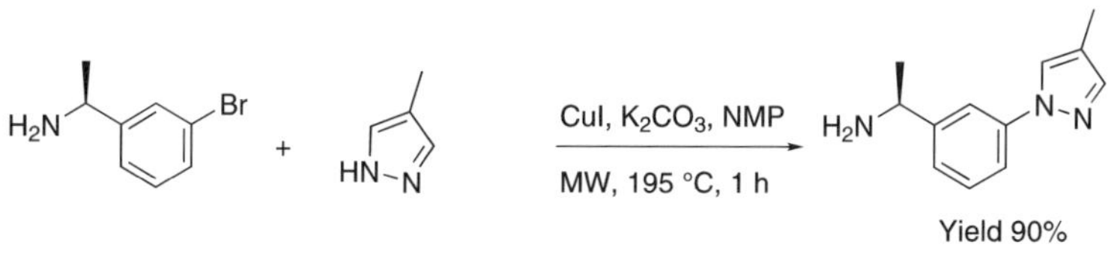

Scheme 26 Microwave-accelerated Ullmann coupling

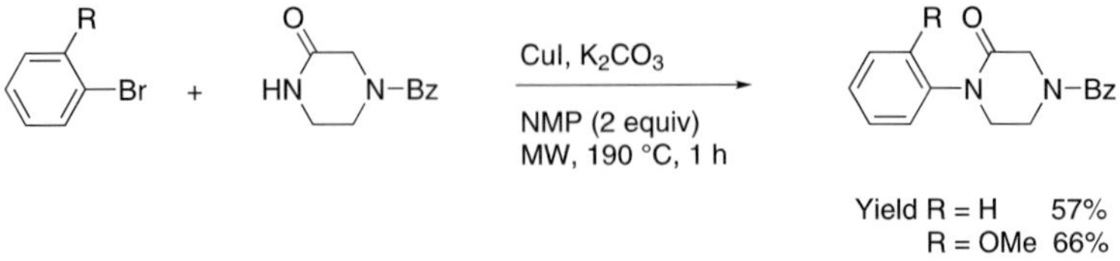

Scheme 27 Copper-catalyzed Goldberg arylation of protected 2-piperazinone

Dihydropyrimidones are highly valued as templates for the development of pharmaceuticals, and accordingly a method for the efficient *N*-arylation of the urea functionality has been reported by the groups of Larhed and Kappe, as described in Scheme 28 [34]. Regrettably, Biginelli condensations do not allow—as highlighted by the authors—the direct synthesis of *N*3-arylated dihydropyrimidone, thus illustrating the importance of this straightforward protocol for the chemical diversification of the heterocyclic template.

Interestingly, in the same publication, palladium-catalyzed carbamoylation and amidations were also carried out on the aforementioned template using rapid microwave heating in good to excellent yields (62% and 88–72%).

Turner and Poondra have described an efficient intramolecular one-pot and two-step Goldberg aryl amidation that gives access to useful *N*-substituted oxindoles (Scheme 29) [96].

A similar theme was used by Pabba in the synthesis of indazoles via a one-pot and two-step condensation–arylation sequence [97]. Here Cu(I) catalysis was used to assemble aryl hydrazines with 2-halobenzaldehydes or 2-halo-acetophenones to deliver the target molecules in high yields after two short

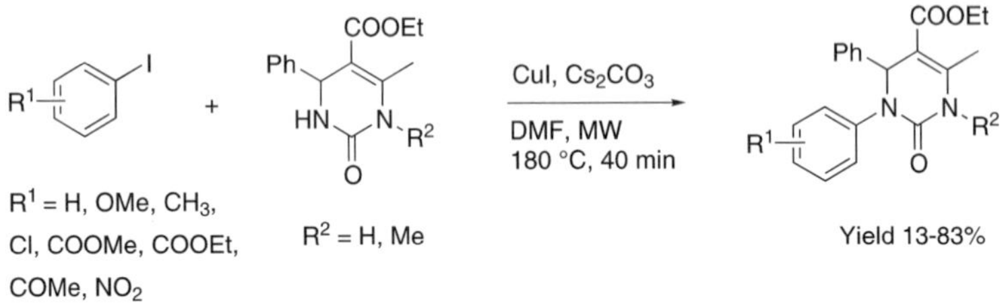

Scheme 28 Copper-catalyzed *N*3-arylation of dihydropyrimidones

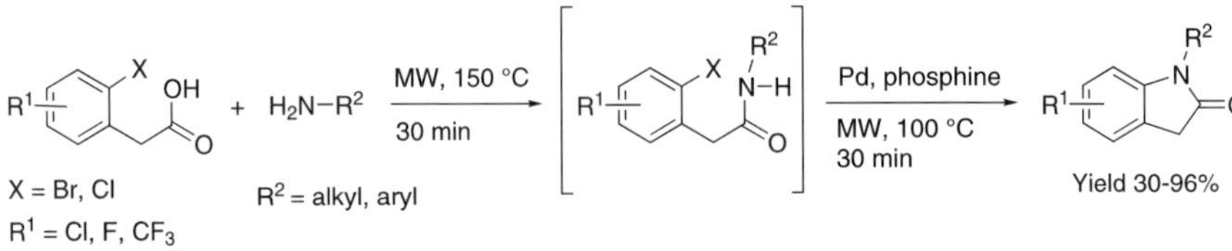

Scheme 29 A two-step preparation of substituted oxindoles via transient amide formation

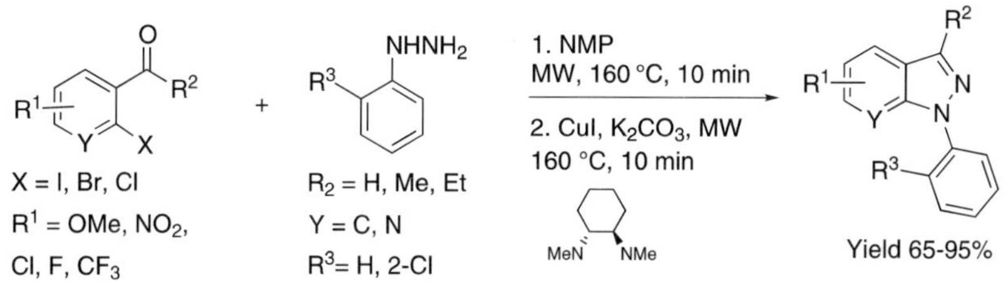

Scheme 30 Convenient two-step manufacture of indazole analogs

microwave irradiation periods (Scheme 30). Regular heating was reported to require longer reaction times.

The N-arylsulfonamide functionality recurs constantly in medicinal chemistry due to its privileged pharmacological profile. The prospect of attaching aryl groups directly to the sulfonamide via an N-aryl coupling is often advantageous since the alternatives, for example the reaction between sulfonyl chloride and arylamine, are commonly troubled by the poor nucleophilicity of the arylamines. Wu and He recently published a usable microwave protocol for copper-catalyzed N-arylations, employing aryl iodides and bromides [98]. Under sealed vessel conditions at 195 °C, moderate-to-good yields (54–90%) were reached within 2–4 h using NMP as the solvent and potassium carbonate as base. A related methodology was published by Cao and colleagues, where they used palladium catalysis together with aryl chlorides to arylate a set of alkyl and aryl sulfonamides (Scheme 31) [99].

The group of Harmata has explored a route that can be used to effectively couple aryl chlorides with methylphenylsulfoximines [100]. Using palladium acetate and *rac*-binap with a large excess of aryl chlorides as coupling partners and cesium carbonate as the base, yields of 10–94% were attained after one or two 1.5-hour irradiation periods at 135 °C. Switching to an aryl triflate, and using a surplus of the sulfoximines (five equivalents) furnished an impressive 94% yield (Scheme 32).

Aryl azides constitute central precursors for heterocyclic compound generation (e.g., "click" chemistry applications) [101]. Interestingly, a safe, convenient and microwave-compatible aryl azidonation process has been dis-

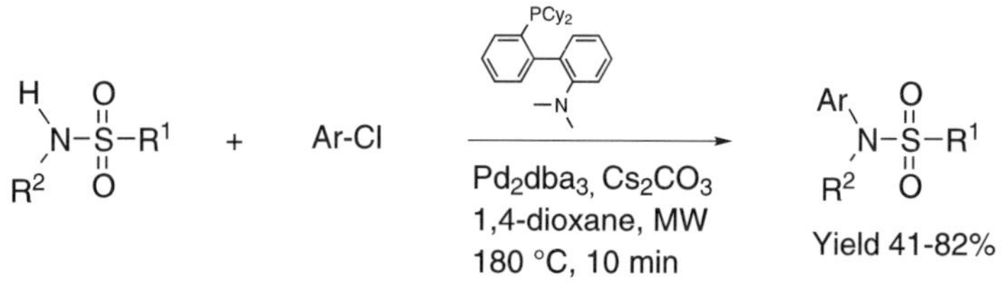

Scheme 31 Palladium-catalyzed N-arylation of sulfonamides using aryl chloride

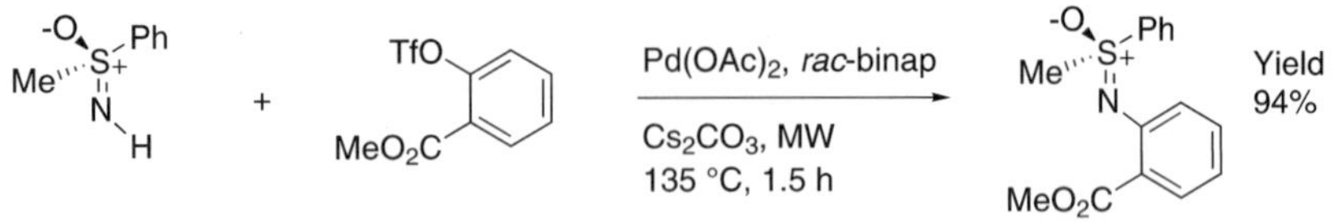

Scheme 32 Palladium-catalyzed sulfoximine N-arylation using an aryl triflate

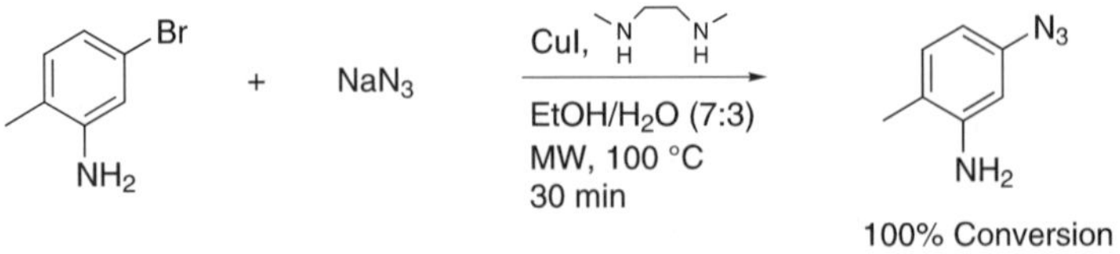

Scheme 33 Copper(I)/diamine-catalyzed aryl azide synthesis

closed [102]. It was shown that with an accurate choice of ligand and solvent system, full conversion could be reached within 30 min by employing copper catalysis in the azidonation of an aryl bromide (Scheme 33).

7
Aryl–Oxygen Bond Formation

Nowadays, the coupling of organohydroxyls with aryl halides, using copper or palladium as catalysts, is a routine method of synthesizing aryl ethers. The copper-catalyzed Ullmann aryl ether synthesis is attractive since the copper is cheap and relatively nontoxic. The reaction is also compatible with several functional groups. However, in contrast to the palladium-catalyzed equivalent, Ullmann reactions require larger amounts of catalyst and typically show slower kinetics. Stockland Jr and coworkers have developed a method that uses organosoluble copper clusters to avoid these issues [103]. A set of alkyl aryl ethers was prepared using only 0.4 mol % of copper clusters to scrutinize conventional and microwave heating (Scheme 34). The yields achieved through 11 hours of classic heating at 110 °C were similar to those obtained using the microwave, but when performed under an air atmosphere the yields plummeted for both heating methods.

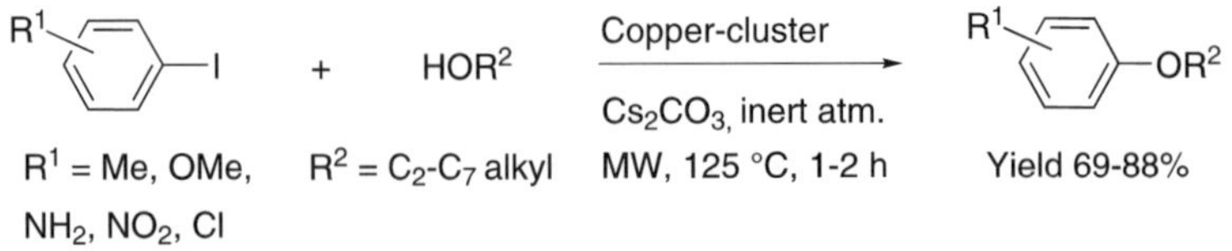

Scheme 34 Copper-catalyzed arylation of aliphatic alcohols

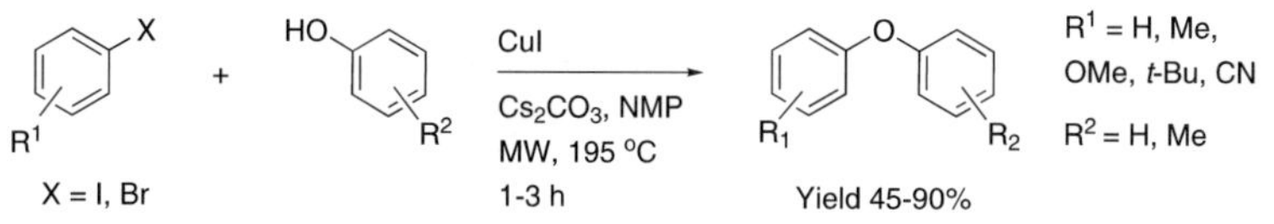

Scheme 35 Copper(I)-catalyzed arylation of phenols

He and Wu have developed a method for the arylation of phenols using aryl iodides and bromides [104]. The coupling between 4-*t*-butyl-iodobenzene and phenol was conducted both thermally and with directed microwave heating, affording 74% and 90% yields respectively at the same reaction temperature (195 °C). Unfortunately, the authors report the protocol to be mismatched with the less expensive aryl chlorides (Scheme 35).

8
Aryl–Phosphorus Coupling

A Teflon autoclave was used in the transformation of aryl iodides to aryl phosphonates, useful precursors to aryl phosphonic acids, in a study made by Villemin [105]. With a domestic microwave oven, the reaction times were successfully shortened compared to those from classic heating from 10 hours to 4–22 minutes. Aryl iodides exhibited good reactivity while bromides gave lower yields and triflates very slow reactions (Scheme 36). It is interesting to note that the reactions were implemented with short reaction times in the nonpolar solvent toluene, which is essentially microwave-transparent [5].

In a separate study, Villemin reported on the coupling between triethyl phosphite and aryl halides. The process was only effectively catalyzed by nickel and palladium amongst the transition metals examined (Ni, Pd, Co, Fe, Cu) [106]. Using sealed vessels and an inert atmosphere, the reactions were finalized within 5 minutes when a maximum temperature of about 200 °C was used (Scheme 37).

Kappe and Stadler have invented a microwave protocol enabling quick access to triaryl phosphines via coupling of diphenylphosphine with aryl halides and triflates [107]. Because of the value of phosphine ligands in assorted transition metal-catalyzed reactions, convenient routes for their pro-

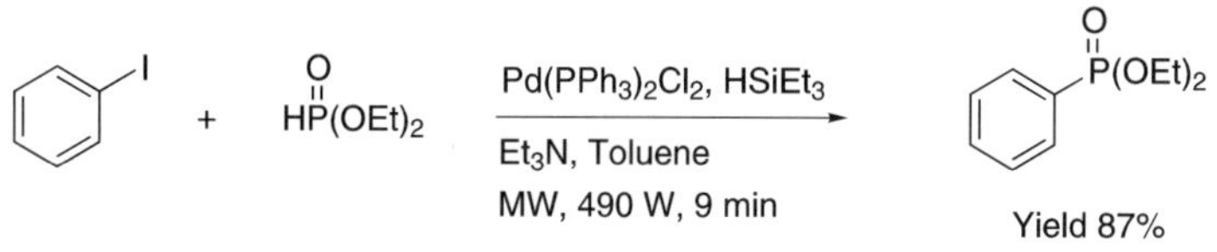

Scheme 36 Fast palladium-catalyzed phosphonation using diethylphosphite

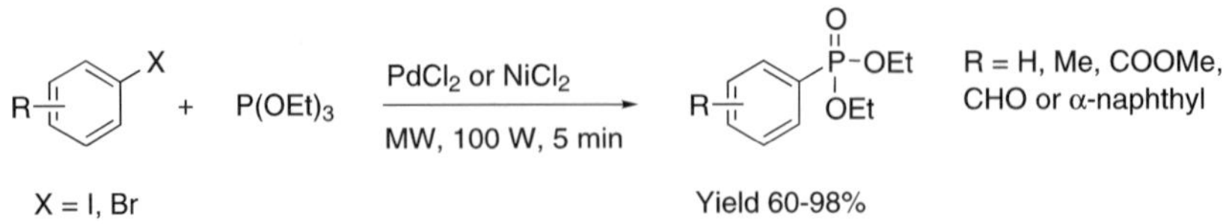

Scheme 37 Palladium- and nickel-catalyzed synthesis of aryl phosphonates

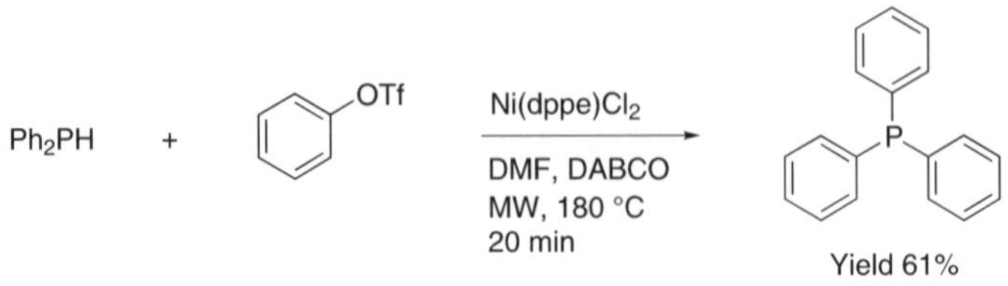

Scheme 38 Nickel-catalyzed synthesis of triphenylphosphine

duction are important. Both homogeneous palladium and nickel, together with heterogeneous palladium catalysts, were evaluated. The more atypical substrate phenyl triflate could also be coupled using nickel catalysis in an expeditious manner (see above, Scheme 38). Reactions with other aryl halides proceeded in 26–85% yields after 3–30 minutes of microwave heating at 180–200 °C.

9
Aryl–Sulfur Bond Formation

Thiation of aromatic compounds is utilized both for the preparation of aryl sulfides and for the generation of sulfur heteroarenes. The diaryl thioether motif is found in the structure of several approved pharmaceuticals, such as antihistamines. In a recent letter, Wu and He disclosed that Cu(I) catalysis is very efficient for the coupling of thiophenols with aryl iodides (Scheme 39) [108]. The procedure is very similiar to the diaryl ether synthesis under microwave irradiation reported by the same authors.

Kappe and Lengar have elucidated the versatility of microwaves in a sulfur phenylation of a thiourea [109]. Under pressurized conditions and microwave

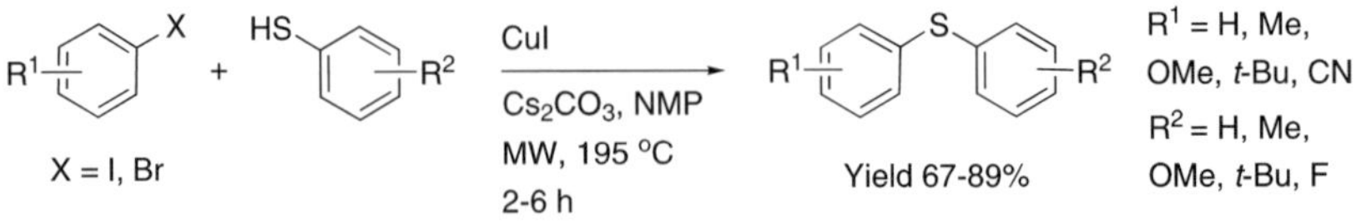

Scheme 39 Formation of a diaryl sulfide linkage with copper catalysis

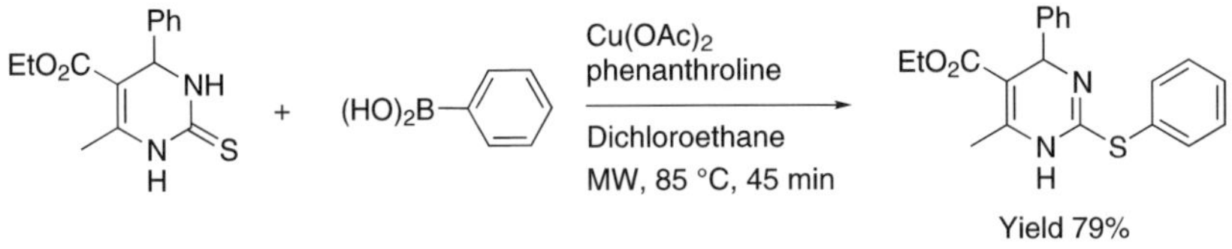

Scheme 40 Sulfur phenylation using phenyl boronic acid under stoichiometric Cu(II) conditions

irradiation, the reactions could be settled within an hour, as presented in Scheme 40. In comparison, the corresponding standard reaction run at room temperature with dichloromethane takes four days to reach completion with a similar yield (72%).

The group of Besson has exemplified the preparation of sulfur-containing aromatic heterocycles via an intramolecular aryl sulfur coupling in order to establish a benzothiazol substructure during a multistep synthesis [110]. In a previous report, the same group investigated the scope and limitation of this key-step transformation, as presented below in Scheme 41 [111]. All reactions were duplicated using conventional heating (oil bath) at reflux temperature and produced similar yields after 45–60 minutes.

Sulfonylation of arenes is normally carried out using sulfonyl chloride and a stoichiometric amount of Lewis or Brönstedt acids as the catalyst. Dubac and coworkers discovered a practical method using microwave high-temperature conditions that demanded only 5–10 mol% of $FeCl_3$ (in relation to the sulfonyl chloride) for full conversion [112]. Arenes encompassing alkyl-benzenes, anisole and halobenzenes were sulfonylated using different aryl-sulfonyl chlorides. A representative example is depicted in Scheme 42. The sulfonylations generally proceeded with good *para*-regioselectivity, with the electron-rich bromoanisole substrate being an exception.

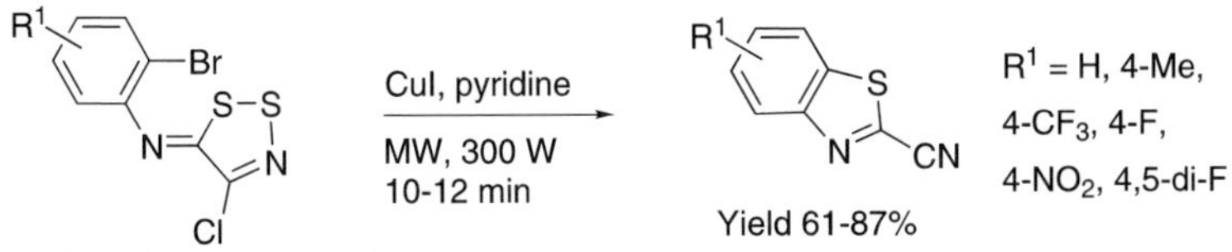

Scheme 41 Benzothiazole synthesis via copper-catalyzed cyclization–elimination

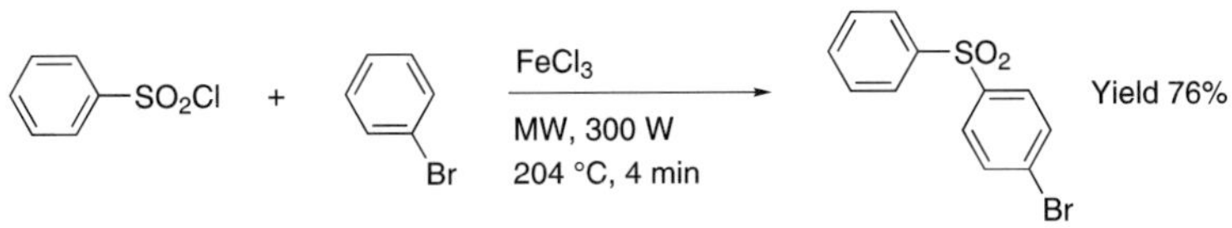

Scheme 42 Iron(III) chloride-catalyzed regio- and chemoselective sulfonylation

10
Aryl Halide Substitution Reactions

Halogen substitutions in aryl halides are useful in transition metal-catalyzed processes where the halide is taken advantage of as a leaving group in order to form the vital arylmetal complex in the catalytic cycle. Because of the intrinsic reactivity differences among the halides a halogen exchange reaction can alleviate problems, depending on the chosen catalytic system [113]. In medicinal chemistry, incorporation of fluorine or chlorine in metabolically exposed positions is a widespread practice, as is the introduction of ^{124}I, ^{76}Br and ^{18}F nuclides as radiotracers in positron emission tomography (PET) imaging studies [114]. Due to the high cost of radionuclides and the rapidity of decay in many cases it is obligatory to introduce the radioactive nucleus at the end-point of the synthesis in order to secure satisfying radiochemical yields [115]. Accordingly, minimizing the reaction time is crucial when using short-lived nuclides, making the application of high-speed microwave chemistry highly desirable. Leadbeater and his research group have reported on microwave-enhanced nickel-catalyzed halogen substitution procedures (Scheme 43) [116]. Yields were generally good to excellent, and a comparison with oil bath heating was also made using a closed vessel at 170 °C for four hours, which revealed equal productivity.

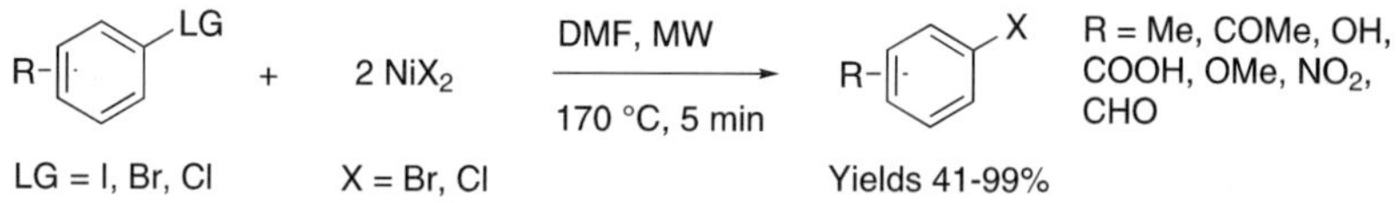

R–LG + 2 NiX₂ $\xrightarrow{\text{DMF, MW}}$ R–X

DMF, MW, 170 °C, 5 min

R = Me, COMe, OH, COOH, OMe, NO₂, CHO

LG = I, Br, Cl   X = Br, Cl   Yields 41-99%

Scheme 43 Nickel-catalyzed halide substitution reactions under microwave irradiation

11
Cross-Coupling Reactions

11.1
The Suzuki–Miyaura Reaction

In 1979 Suzuki discovered that cross-coupling reactions of organoboron compounds proceeded in the presence of ordinary bases, such as hydroxide or alkoxide ions [117]. The conditions proved to be generally applicable and the Suzuki reaction is arguably the most versatile among the cross-coupling reactions today. For example, the reaction has attracted the interest of several research teams involved in high-throughput chemistry, since a wide variety of boronic acids are commercially available [118].

The first microwave-promoted Suzuki couplings were published in 1996 (Scheme 44). Phenyl boronic acid was coupled with 4-methylphenyl bromide

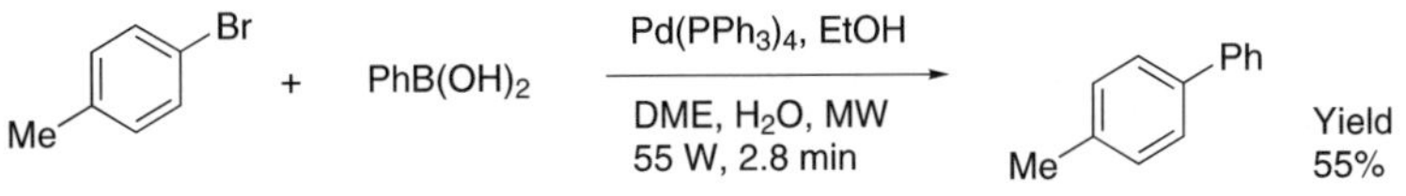

Scheme 44 Suzuki coupling of phenyl boronic acid with *p*-tolyl bromide

to give a fair yield of product after a reaction time of less than four minutes under single-mode irradiation. The same reaction had previously been conducted under classic conditions with a reported reaction time of four hours [44].

In 2005, very similar reactions were shown to proceed smoothly in continuous flow reactors (Scheme 45). The yield of couplings with aryl bromides and iodides were overall high, although the authors noted that it was not clear exactly how long a sample was irradiated due to uncertainties regarding the focus of the irradiation over the capillary column [119].

The energy efficiency of microwave-heated Suzuki–Miyaura reactions has also been investigated [11].

The development of catalytic systems using neat water as solvent is of high importance to industrial and environmentally friendly applications. In this respect, water is perhaps the ultimate solvent because of its lack of toxicity and ready availability. Leadbeater has published several papers where the Suzuki–Miyaura reaction has been optimized for aqueous conditions [9, 120]. Aryl bromides and iodides were coupled and the corresponding products isolated in good yields with an attractive ligandless protocol. Some reactions gave increased yields with the addition of tetrabutylammonium bromide (TBAB) [121]. Recently, an application for a scaled-up Suzuki–Miyaura synthesis in water using an automated batch stop-flow apparatus was also published (Scheme 46) [89].

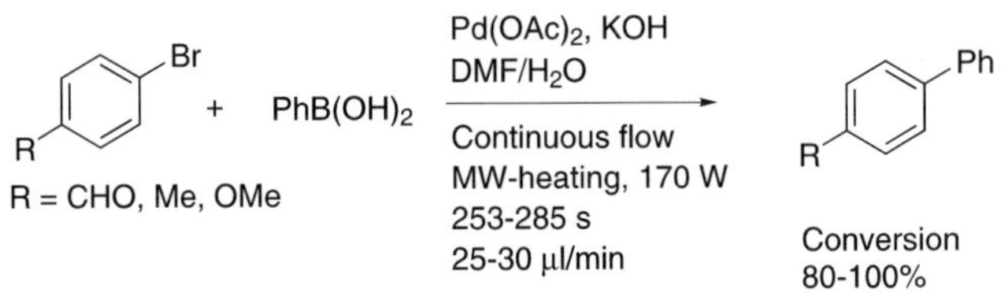

Scheme 45 Microwave-heated continuous flow reactions

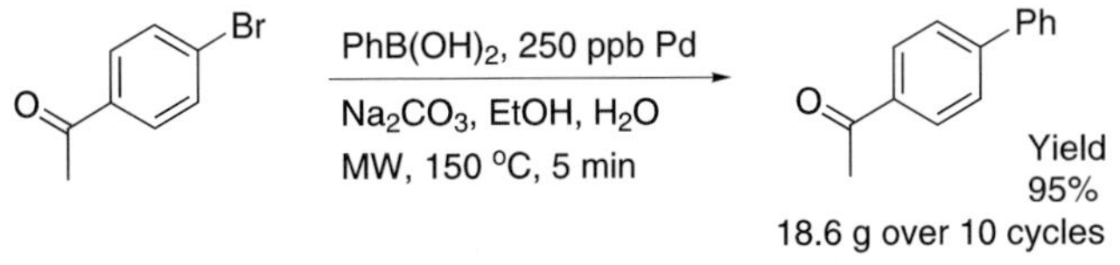

Scheme 46 Scale-up of a Suzuki–Miyaura coupling in water

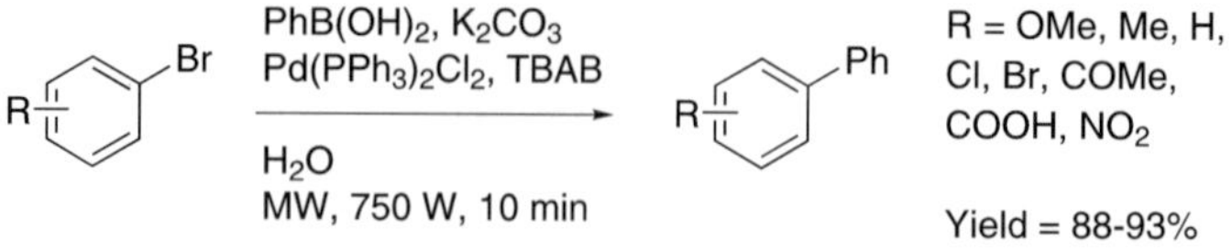

Scheme 47 Suzuki–Miyaura couplings in water

Aryl bromides were used by another team in the coupling of phenylboronic acid in water, using easily available starting materials and an improved domestic microwave oven (Scheme 47) [122].

The development of microwave-heated reactions using aryl chlorides has attracted the interest of several research groups. Transition metal-catalyzed reactions with aryl chlorides were elusive for a long time and were generally only successful with very high reaction temperatures and special reaction conditions. Lately, new catalytic systems, most notably those presented by Fu [48], have spurred the development of several schemes for the microwave-assisted activation of aryl chlorides.

Efficient Suzuki–Miyaura reactions using electron-rich aryl chlorides were reported with the $Pd(OAc)_2/PCy_3$ catalytic combination. The reaction conditions allowed several bases to be used but the authors chose the inexpensive potassium phosphate (Scheme 48) [123].

N-Heterocyclic carbene ligands were used with good results in the coupling of sluggish electron-rich aryl chlorides with phenylboronic acid. These ligands permitted a reaction with a comparatively low reaction temperature and a short reaction time (Scheme 49) [124].

Recently, a paper was published describing Suzuki–Miyaura coupling in DMF and water with aryl chlorides using the air- and moisture-stable

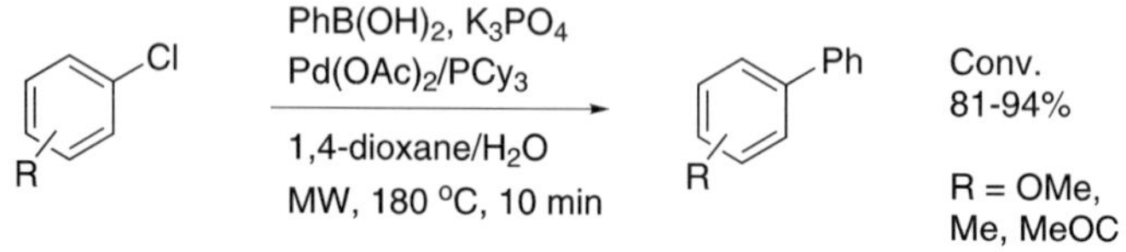

Scheme 48 Suzuki–Miyaura reaction with an aryl chloride

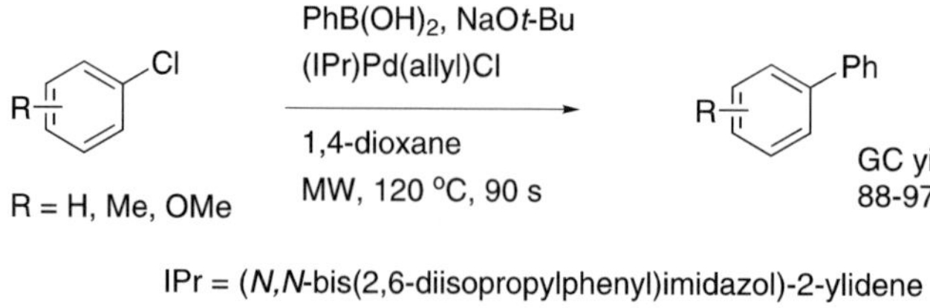

Scheme 49 Reactions with aryl chlorides using a carbene ligand

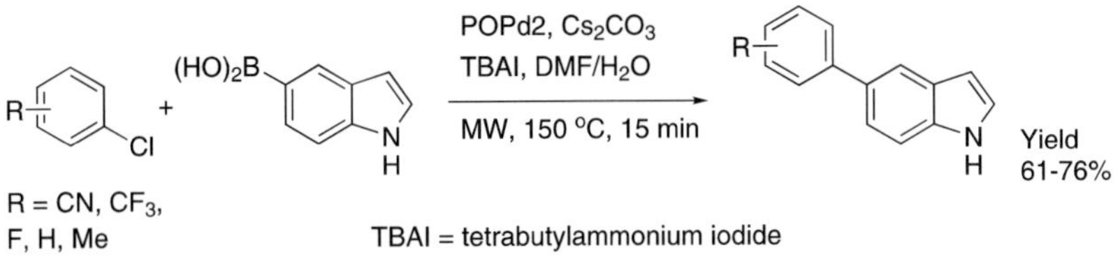

Scheme 50 Suzuki–Miyaura couplings with the POPd2 catalyst

dihydrogen di-μ-chlorodichlorobis(di-*tert*-butylphosphinito-κ P)dipalladate (POPd2) catalyst (Scheme 50) [125].

Water was also used in the coupling of aryl chlorides with the electron-neutral phenylboronic acid. Here Leadbeater and Arvela took advantage of simultaneous cooling of the reaction vessel by compressed air while the bulk of the reaction mixture was heated by microwaves. This allowed the successful coupling of both electron-rich and -poor aryl chlorides in moderate-to-excellent yields. The rationale was that careful control over the reaction temperature is needed to prevent the destruction of the aryl chloride in the reaction mixture. A significant advantage could be seen with the simultaneous cooling technique, especially in the case of electron-rich and -neutral aryl chlorides as compared to reactions without cooling [126]. Suzuki–Miyaura couplings in water using di(2-pyridyl)methylamine-palladium dichloride complexes [127] and Pd(PPh$_3$)$_2$Cl$_2$ [128] have also been reported.

Suzuki couplings performed with PEG as a nontoxic reaction medium have been published by Varma (Scheme 51) [129].

Interestingly, the Suzuki reaction was shown to proceed smoothly also on polymeric support as early as ten years ago, and high yields of a variety of products were reported under these reaction conditions (Scheme 52) [130].

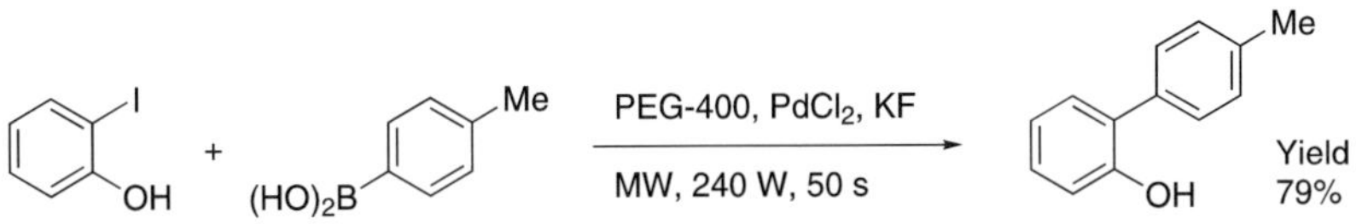

Scheme 51 Suzuki coupling in PEG as a nontoxic reaction medium

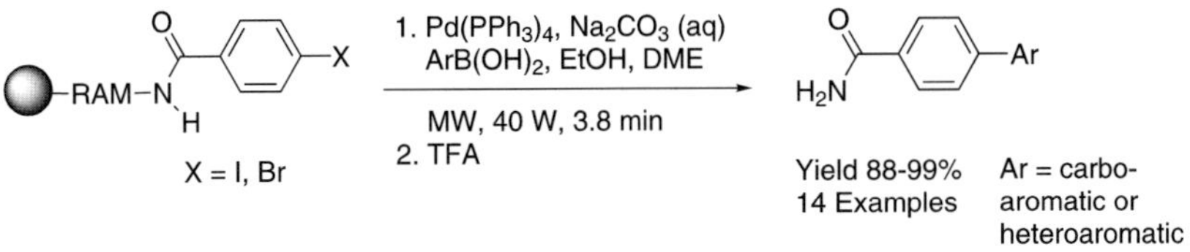

Scheme 52 Suzuki couplings on polymer support

4-Bromo- and 4-iodobenzoic acid linked to Rink-amide TentaGel gave a conversion of more than 99% within four minutes. The impressive yields suggested a high potential for the use of microwave-assisted reactions on polymeric resins [131].

For applications in high-speed synthesis, one interesting paper investigated different polyethylene-supported palladium catalysts (FibreCat) and their efficiency in the Suzuki–Miyaura reaction. The supported catalysts have attracted some interest due to the possibilities of recycling and their convenience at the work-up stage, where the catalyst can be easily filtered off. The more electron-rich and more reactive systems generally gave higher conversions and shorter reaction times. The reactions could all be conducted under ambient atmosphere and performed better than standard homogeneous systems as measured by the purity of the products. The reactions made with supported palladium were, when the conversion was quantitative, pure enough to be collected simply by solid-phase extraction over Si-carbonate. Aryl iodides, aryl bromides, aryl triflates and electron-poor aryl chlorides all gave excellent yields, while electron-rich aryl chlorides gave moderate yields (Scheme 53) [132].

Clean reactions and high yields were reported in the Suzuki–Miyaura reaction with polystyrene-supported palladium using an improved commercial microwave oven and reflux conditions. The polystyrene-based reagents are described as a borderline class of catalysts that retain the advantages of homogeneous reaction systems while possessing the ease of recovery and work-up of heterogeneous catalysts [133]. In this method, the polystyrene-supported Pd is prepared rapidly by ultrasound treatment and subsequently used in the reactions. The catalysts were insensitive to heat and there was no need for an inert atmosphere. Initial attempts to perform the reaction in pure water failed and a mixture of toluene and water was used throughout. Electron-rich and -poor aryl bromides were all reactive, and a comparison with conventional oil bath heating delivered comparable yields, although with longer reaction times (Scheme 54) [133].

A recent addition to this field is the polymer-supported di(2-pyridyl)-methylamine–palladium dichloride complex covalently attached to a styrene-alt-maleic acid anhydride copolymer. Turnover numbers as high as 10^5 were reported and a couple of microwave-heated Suzuki–Miyaura reactions could be performed in neat water [134]. 2-Pyridinealdoxime-based Pd(II)-

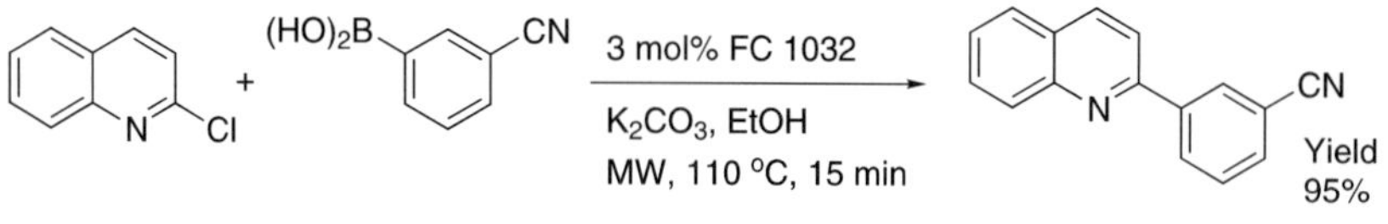

Scheme 53 Suzuki–Miyaura coupling with an alkyl phosphine-supported catalyst (FC 1032)

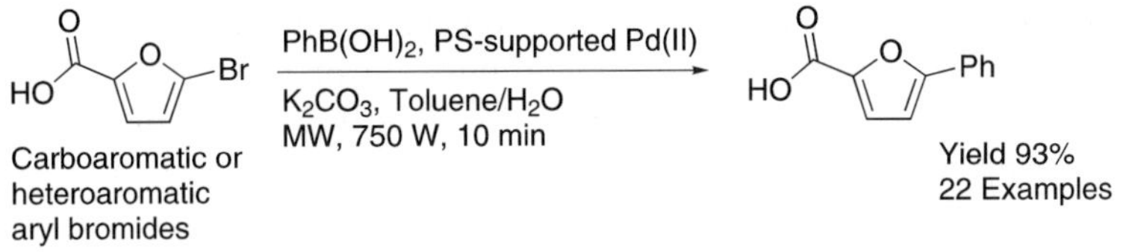

Scheme 54 Suzuki–Miyaura couplings with polystyrene-supported catalyst

complexes covalently anchored to a glass/polymer composite material were used with good effect in a number of Suzuki–Miyaura couplings under microwave irradiation using water as solvent. A range of different aryl chlorides and bromides were evaluated under both thermal and microwave conditions. The example given (Scheme 55) is one of the reactions where microwave heating was found to give substantially improved yields as compared to standard, thermal heating [135].

An alternative to solid-supported catalysts are catalysts that are insoluble themselves [136]. A pyridine-aldoxime ligand was presented and evaluated in the Suzuki–Miyaura reaction using water as a solvent. Using an Irori kan to contain the polymeric catalyst, the reaction could be repeated 14 times without any noticeable reduction in efficiency. The optimized reaction conditions were then used to create a small library of approximately 30 biaryl compounds using aryl iodides, bromides, triflates as well as an activated chloride (Scheme 56) [136].

A modern development and variation of solid support is fluorous chemistry [137, 138]. Fluorous chemistry is an emerging field that takes advantage of the unique physical and solubility properties of perfluorinated organic compounds. Many recent publications have underlined the special properties

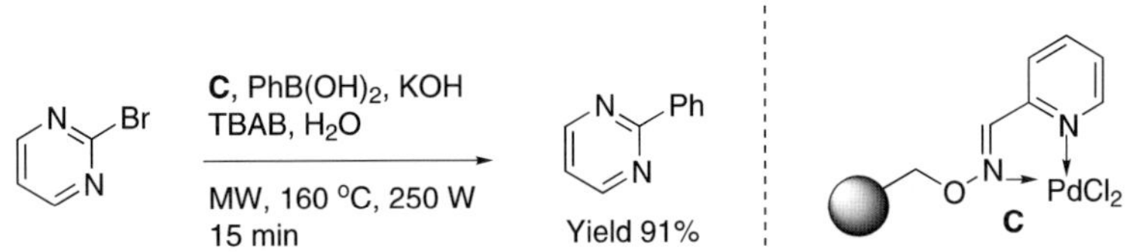

Scheme 55 Suzuki–Miyaura coupling with glass/polymer-supported catalyst

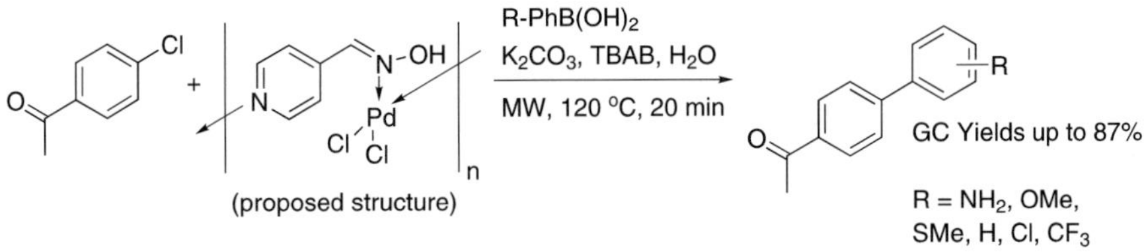

Scheme 56 Suzuki–Miyaura couplings with an insoluble pyridine-aldoxime catalyst

of fluorous chemistry, where the attractive features of solution-phase chemistry are combined with the convenient work-up of solid-phase chemistry, without the disadvantages of the latter. Zhang used perfluorooctylsulfonates as a coupling partner in the Suzuki–Miyaura reaction, where the perfluorooctylsulfonate group functioned as the leaving group (a pseudo triflate) in the coupling while it also had a high enough degree of fluoricity to function as a fluorous tag in fluorous separations. The perfluorooctylsulfonate group exhibited good solubility in organic solvents and was thermostable under the reaction conditions used. The example illustrated was a quite challenging coupling where the product was isolated in a useful yield (Scheme 57). The use of the fluorous sulfonyl group was further demonstrated in a multistep synthesis of a biaryl-substituted hydantoin [139].

The same group reported on a library synthesis of 3-aminoimidazo[1,2-a]-pyridines/pyrazines by fluorous multicomponent reactions. Here the overall yields, as well as the yields for the separate Suzuki–Miyaura reactions that were a part of the synthesis, were relatively low due to competing reactions and the poor reactivities of the substrates, but the speed of the microwave-mediated syntheses and ease of separation underlined the usefulness of fluorous reagents [140]. A recent paper further illustrated the use of Suzuki–Miyaura couplings of aryl perfluorooctylsulfonates in the decoration of products derived from 1,3-dipolar cycloadditions [141].

Recently, a method for the preparation of 2-aryl-3-methoxy-cycloalkenones was reported. Both cyclopentenones and cyclohexenones were prepared in good yields from the corresponding α-bromocycloalkenones (Scheme 58) [142].

Coats reported on the parallel synthesis of delta/mu agonists, as depicted in Scheme 59. Both solid- and solution-phase techniques were evaluated with respect to the reactivity of the vinyl bromide template, but solution-phase couplings gave more rapid reactions. In the latter case it was found that the

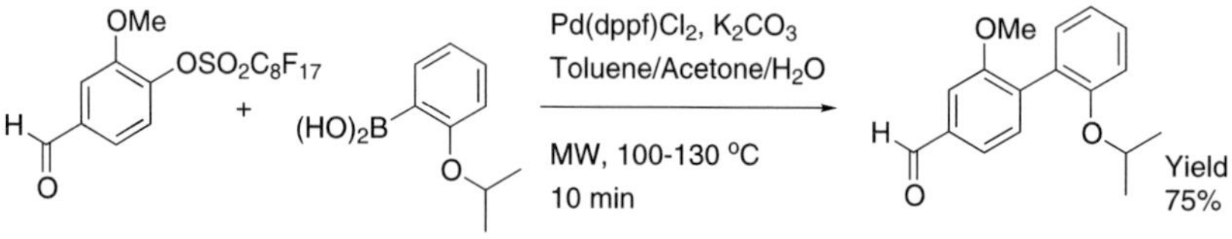

Scheme 57 Fluorous Suzuki–Miyaura reaction

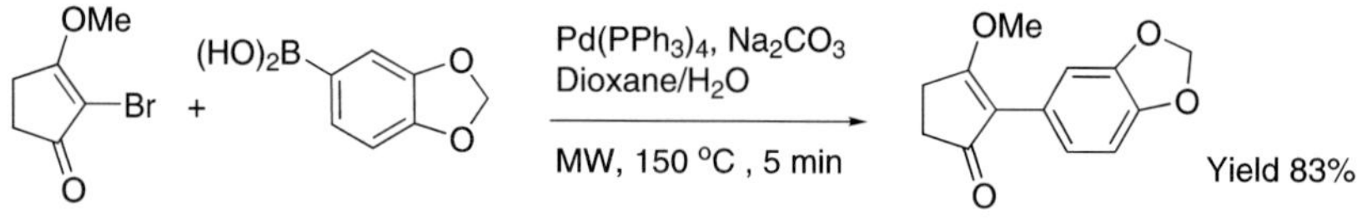

Scheme 58 Suzuki–Miyaura reaction on an α-bromocycloalkenone

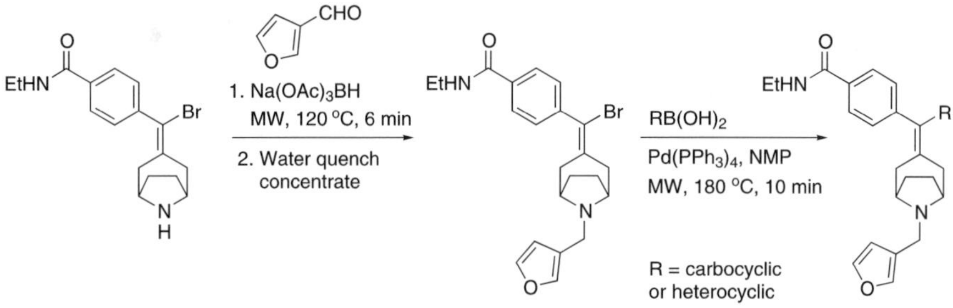

Scheme 59 Parallel synthesis of δ/μ-agonists

Suzuki–Miyaura reaction could be applied directly to the reaction mixture of the preceding reductive amination, thus ensuring a relatively fast and easy synthetic route to a library of 192 compounds [143].

Antifungal 3-aryl-5-methyl-2,5-dihydrofuran-2-ones were reported, where the 3-aryl group was introduced by palladium chemistry. The yields were generally moderate, possibly due to the instability of the core structure at high temperatures (Scheme 60) [144].

Antimicrobial oxazolidinones were successfully synthesized with the help of single-mode microwave heating on a polystyrene resin. In this case, the use of commercial multimode ovens was associated with inconsistent yields and purities, presumably due to the nonhomogeneity of the heating and a lack of sufficient temperature and pressure controls. A representative reaction is presented in Scheme 61. These solid-supported reactions proceeded smoothly in 5–10 minutes, with the boronic acid added in six equivalents and a small library with variations in both the N-acyl and the biaryl functionalities was created [145].

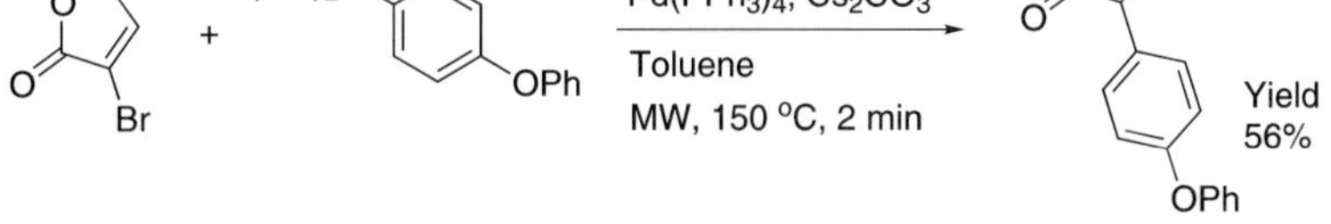

Scheme 60 Preparation of 3-aryl-5-methyl-2,5-dihydrofuran-2-ones

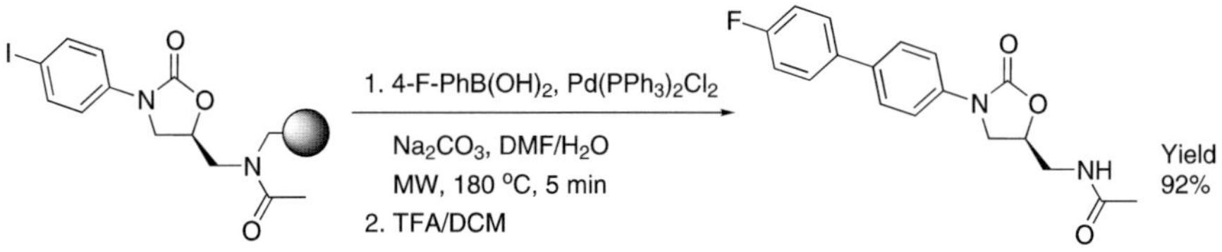

Scheme 61 Polymer-supported synthesis of antimicrobial oxazolidinones

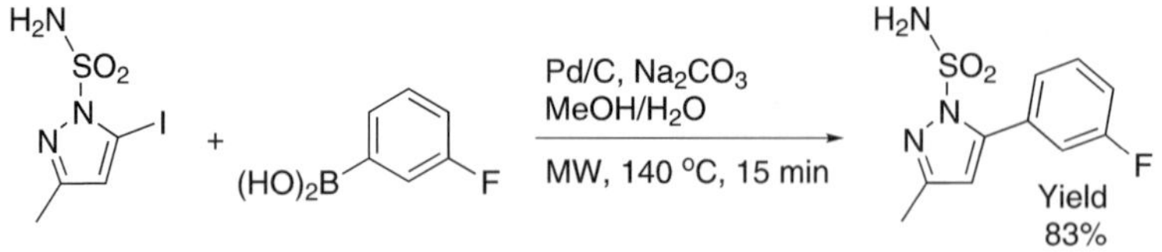

Scheme 62 Palladium on carbon-catalyzed synthesis of COX-inhibitors

Pyrazole-based COX-inhibitors were synthesized using Pd/C as a heterogeneous and ready-filterable palladium source. Electron-deficient boronic acids coupled well while *ortho*-substituted and electron-rich boronic acid were less reactive (Scheme 62) [146]. The same team also developed a two-step and one-pot procedure for the synthesis of styrene-based nicotinic acetylcholine receptor antagonists.

For a number of years Hallberg has implemented microwave-promoted reactions in the optimization of different types of aspartyl protease inhibitors [118]. Recently, the Suzuki–Miyaura coupling was used to introduce biaryl moieties into cyclic and linear sulfonamide HIV-1 protease inhibitors. A series of sixteen reactions was presented in fair to moderate yields, and the reaction times were, in all examples except two, limited to only 5 minutes. Similarly, eight relatively complex C2-symmetric plasmepsin I and II inhibitors against the malaria-causing protozoa *Plasmodium falciparum* were effectively synthesized using a microwave method by substitution of two vinyl bromide functionalities [147]. Suzuki–Miyaura reactions with aryl bromides as well as triflates have been reported in the synthesis of plasmepsin I and II inhibitors using an hydroxyethylamine transition state-mimicking scaffold [148, 149].

11.2
The Stille Reaction

The defining feature of the Stille cross-coupling reaction (also known as Migita–Kosugi–Stille coupling) is the use of an organotin moiety and an organohalide (or pseudo-halide) in combination with palladium catalysts [150]. This base-free reaction is, just like the Suzuki–reaction, very reliable, high-yielding and tolerant of many functionalities. The main drawback is the modest reactivity of the organotin reactants. The unreacting tin substituents are usually methyl or butyl, although newer dummy ligands have been developed. Typically, the transferable fourth ligand on tin is an unsaturated moiety. The group migration order is believed to be alkynyl > vinyl > aryl > alkyl.

The Stille reaction was one of the earliest transition metal-catalyzed reactions to be accelerated with microwave-heating. Single-mode irradiation with very short reaction times was easily applied to Stille–reactions in solution [130] as well as on resin support [130] (Scheme 63).

2(1*H*)-Pyrazinones had a higher reactivity on polystyrene resins using the Stille reaction compared to the Suzuki–Miyaura coupling (Scheme 64) [151].

The Stille reaction was also reported in the synthesis of melatonin derivatives. Two heating cycles were employed to reach a yield comparable to oil bath heating. The reaction time, two irradiation cycles of 20 minutes, was notably shorter than the 24-hour reaction with standard heating (Scheme 65) [58].

Different substrates for the Stille reaction were used in two one-pot microwave-assisted hydrostannylation–Stille coupling sequences. The isolated yields reported in both of these papers were high (Scheme 66, see also Scheme 69) [152, 153].

Fluorous chemistry has been applied to Stille couplings as well as Suzuki reactions, as described previously. One of the many applications reported is Stille couplings of tin reagents with fluorinated dummy ligands, where the products and the excess of the toxic tin-containing reagents can be easily separated from the reaction mixture and, in the case of the reagents, recycled. One example of the use of the $-CH_2CH_2C_6F_{13}$ (shortened F-13)-tagged organostannanes is presented in Scheme 67 [154].

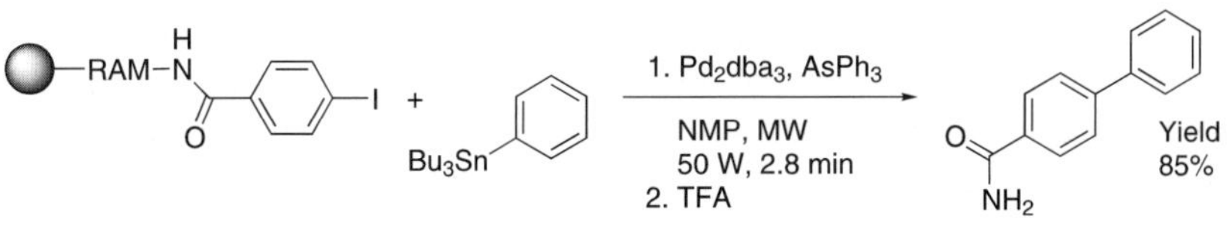

Scheme 63 Stille coupling utilizing RAM-linker on polymer support

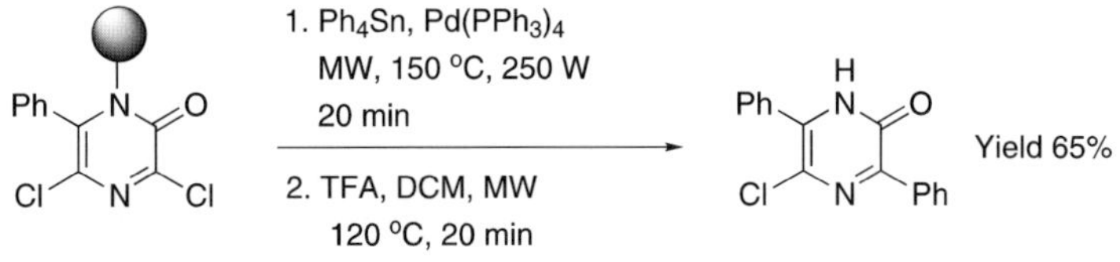

Scheme 64 Solid-phase reaction of 2(1*H*)-pyrazinones

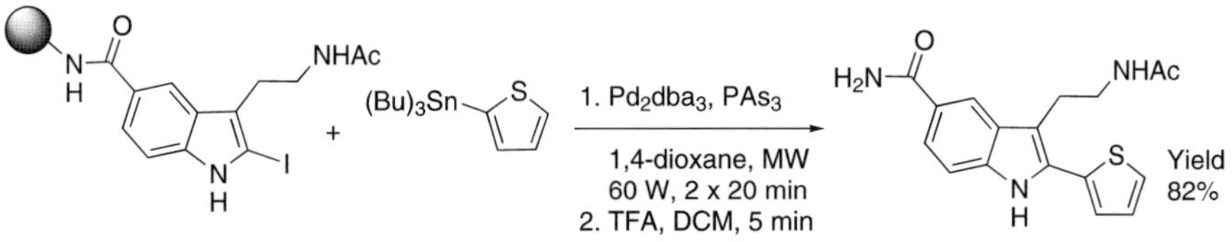

Scheme 65 Solid-phase synthesis of a melatonin analog

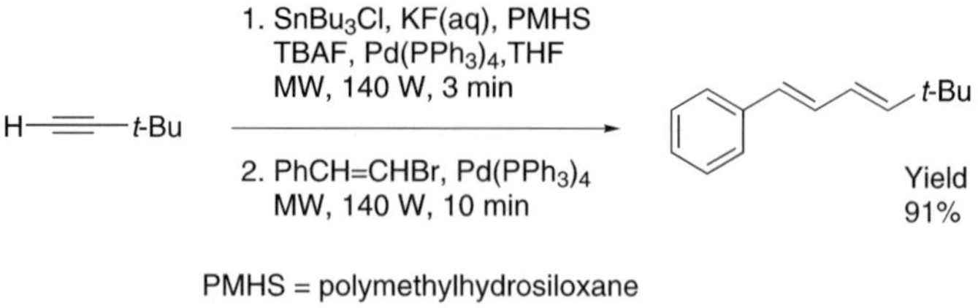

1. SnBu3Cl, KF(aq), PMHS
TBAF, Pd(PPh3)4,THF
MW, 140 W, 3 min
H—≡—t-Bu
2. PhCH=CHBr, Pd(PPh3)4
MW, 140 W, 10 min
t-Bu
Yield
91%
PMHS = polymethylhydrosiloxane

Scheme 66 One-pot hydrostannylation and Stille coupling

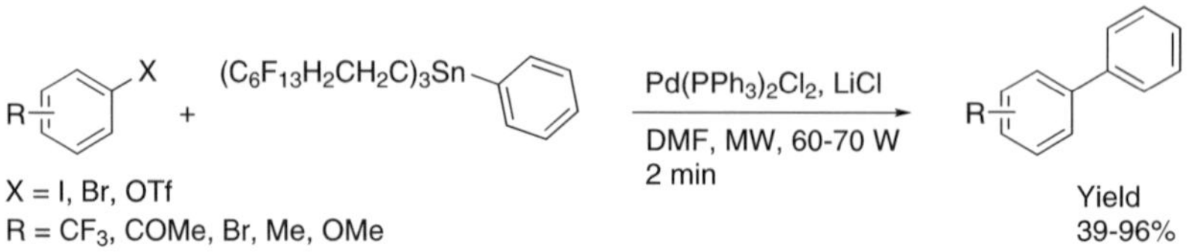

X
R
+
(C6F13H2CH2C)3Sn
Pd(PPh3)2Cl2, LiCl
DMF, MW, 60-70 W
2 min
R
X = I, Br, OTf
R = CF3, COMe, Br, Me, OMe
Yield
39-96%

Scheme 67 Stille reactions with the F-13 tagged phenyl stannane reagent

In some cases it was apparent that the fluoricity provided with the F-13 tags was not enough to provide a full partitioning of the products to the liquid fluorous phase. The concept of using more heavily fluorinated tags, such as the $-CH_2CH_2C_{10}F_{21}$ (F-21) tag, was easily envisioned, but proved to be preparatively elusive, as the solubility of these compounds was very poor. Classic heating of the reactions to 80 °C in fluorinated solvents resulted in very sluggish and unreproducible reactions. However, the application of single-mode heating to these reactions allowed fast and efficient reactions in DMF (Scheme 68) [152]. The insolubility of the F-21 tagged compounds at room temperature presented a very convenient method of removing the fluorous tin compound by filtration.

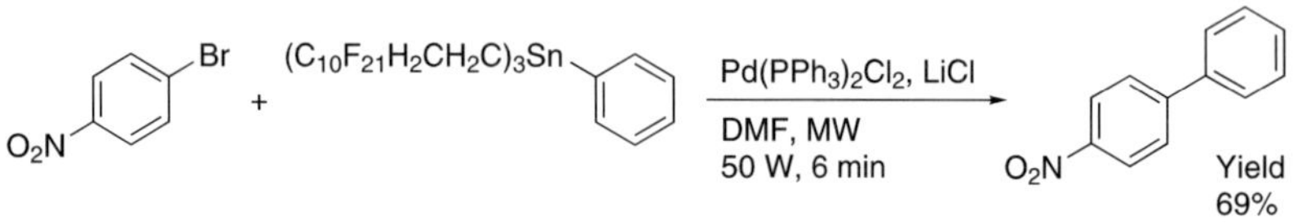

Br
O2N
+
(C10F21H2CH2C)3Sn
Pd(PPh3)2Cl2, LiCl
DMF, MW
50 W, 6 min
O2N
Yield
69%

Scheme 68 Stille reaction with the F-21 tagged phenyl stannane reagent

O
O
EtO
OEt
+
HSn(CH2CH2C10F21)3
1. AIBN, Benzotrifluoride
MW, 60 W, 10 min
2. PhI, Pd(OAc)2, CuO
DMF/Benzotrifluoride
MW, 60 W, 8 min
Ph
O
OEt
EtO
O
E/Z = 5:1
Yield 44%

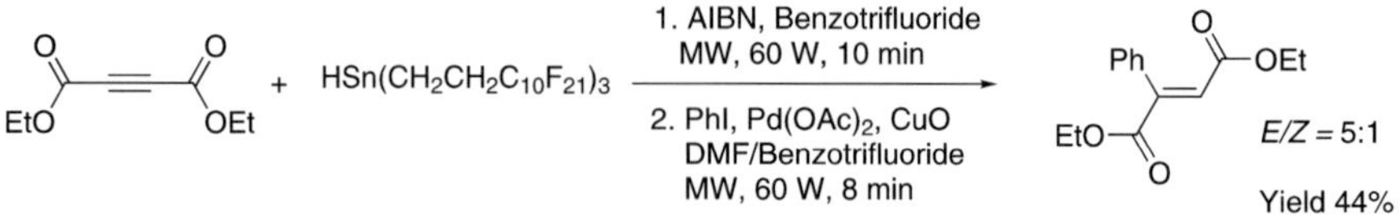

Scheme 69 One-pot hydrostannylation and Stille reaction with F-21 tagged reagents

In the same paper, the fluorous Stille procedure was applied to a one-pot hydrostannylation of an acetylene in the hybrid fluorous/organic solvent benzotrifluoride (BTF), with the subsequent cross-coupling of the product in BTF/DMF, as shown in Scheme 69 [152].

11.3
The Negishi Reaction

The first examples of microwave-assisted cross-couplings with organozinc compounds were reported in 2001, as shown in Scheme 70. Aryl- as well as alkylzinc bromides were effectively coupled with short reaction times [155].

Kappe has published a general method for the microwave-heated Negishi coupling. The organozinc reagents were here prepared from activated Rieke zinc and aryl bromides or iodides. Nickel-catalyzed reactions were reported to give high degrees of homocouplings and could not be driven to completion with electron-rich, deactivated aryl chlorides. However, palladium in combination with electron-rich phosphines was found more effective with both electron-rich and -poor aryl chlorides (Scheme 71). n-Butylzinc chloride and resin-bound aryl chlorides could also be coupled and the products could be isolated in good yield [156].

Enantiopure 1,1'-binaphthyl derivatives were prepared starting from binaphthyl iodides or -triflates without loss of enantiomeric purity. The same reaction performed under oil-bath heating was associated with slower reactions and lower yields [157].

Different pyridinyl pyrimidines were prepared with Negishi couplings (Scheme 72). The reported procedure took advantage of the lower hygroscopicity and higher solubility of zinc iodide in ethereal solvents, as compared

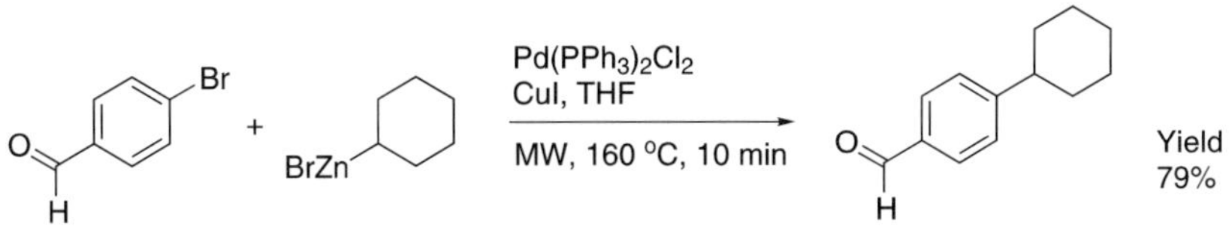

Scheme 70 Negishi coupling of an unprotected benzaldehyde bromide

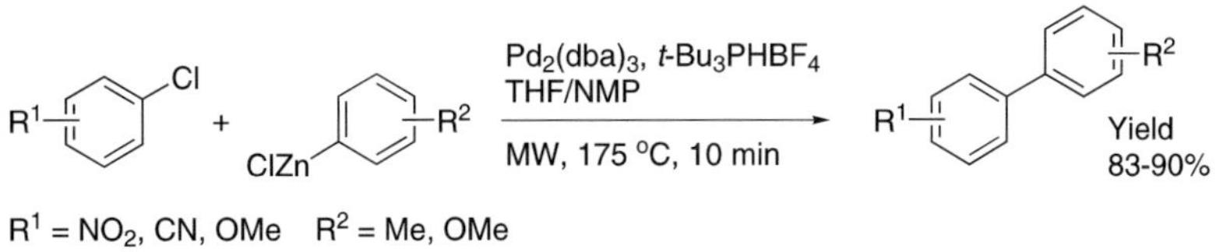

R^1 = NO$_2$, CN, OMe R^2 = Me, OMe

Scheme 71 Negishi couplings with aryl chlorides

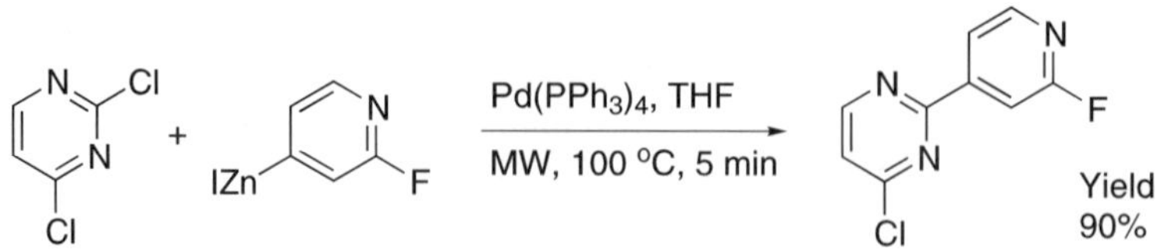

Scheme 72 Synthesis of a pyridinyl pyrimidine through a Negishi reaction

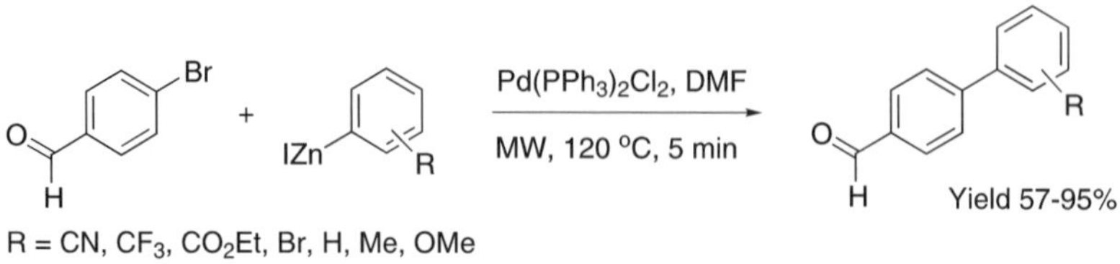

R = CN, CF$_3$, CO$_2$Et, Br, H, Me, OMe

Scheme 73 A Negishi coupling using reagents derived from zinc dust

to zinc chloride, when the organozinc substrates were first prepared via conventional lithiation and subsequent transmetallation. The latter reactant generally gave a higher amount of undesired homocoupling products in the coupling step [158].

A different method of preparing the arylzinc reagents is the reaction between activated zinc dust and aryl iodides. The generated arylzinc reagent was then used in Negishi cross-couplings to generate thirteen biaryl formaldehydes in good-to-excellent yields (Scheme 73). Both nickel and palladium catalysts could be used, but palladium was chosen due to its superior performance in DMF [159].

The same group later developed this methodology so that arylmagnesium compounds could be used in the generation of arylzinc complexes for Negishi couplings [160].

The Negishi reaction was also found to be applicable to large-scale micro-wave-heated reactions. A previously published small-scale reaction (1 mmol) was easily transferred to a larger scale (2 × 20 mmol) and made to go to completion after only 1 minute of hold time with a very good isolated yield [90].

11.4
The Kumada Reaction

Microwaves were utilized both in the preparation of the Grignard reagent and in the Kumada couplings [161] with aryl chlorides. It was noted in this case that a higher amount of homocoupling side-products was typically formed when microwaves were used as a heating source than when the reaction was carried out by employing ultrasound at ambient temperature (Scheme 74) [156].

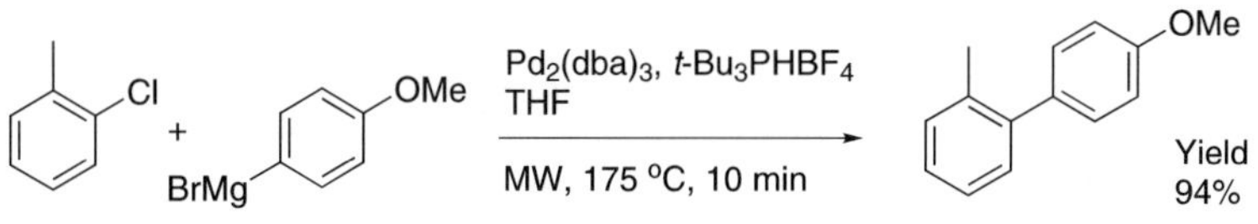

Scheme 74 Kumada reaction with aryl chlorides

Grignard reagents have been generated from sluggish aryl chlorides and bromides using controlled microwave heating. In the synthesis of a novel HIV-1 protease inhibitor, microwave irradiation was used both to generate the starting arylmagnesium halide and to promote a subsequent Kumada coupling [162].

11.5
The Hiyama Reaction

Clarke recently published the first microwave-accelerated Hiyama coupling [163, 164]. It was noted that the availability and nontoxic attributes of the organosilicon reactants make them very attractive in synthesis, but their low nucleophilicity limits their potential. Microwave heating allowed aryl bromides and activated aryl chlorides to react under palladium catalysis using an electron-rich *N*-methyl piperazine/cyclohexyl phosphine ligand (Scheme 75). A vinylation reaction with vinyltrimethoxysilane was also reported [164].

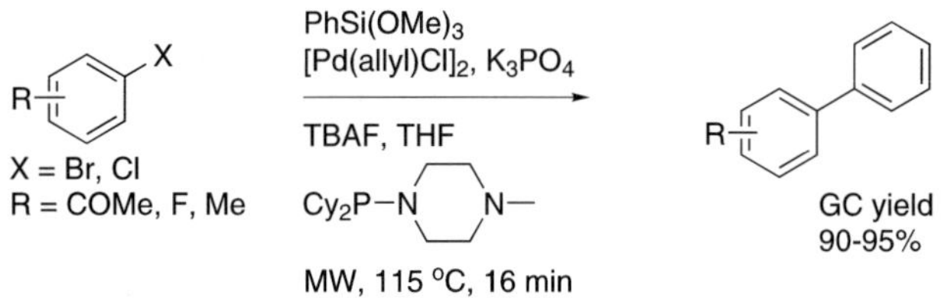

Scheme 75 Hiyama reactions with aryl halides

12
Conclusion

In this review, we have presented and discussed some of the applications of microwave heating in homogeneous transition metal-catalyzed coupling reactions proceeding via aryl or vinyl metal intermediates, highlighting the increased reaction kinetics, the high reaction control and the convenient handling of such techniques. Microwave-promoted chemistry is indeed an emerging and vibrant field for research and development. More and more chemists are becoming aware of the versatility of this energy source, and it is already clear that modern microwave synthesizers have much to offer the research chemists. It is also likely that the widespread acceptance of this technique will be important for the development of laboratory-scale environmentally conscious chemistry. A complementary and much desired future area of microwave chemistry might lie in the development of large-scale equipment, allowing the direct up-scaling of pressurized small-scale reactions into kg-scale processes.

In the opinion of the authors, we have just begun to investigate the advantages that the combination of controlled microwave-heating and metal-catalysis might offer. Nevertheless, the synthetic chemist can now take advantage of the unique carbon–carbon and carbon–heteroatom bond formations offered by transition metal activation and make the reaction happen in minutes, instead of hours, by microwave flash heating, a feat that is of high importance since many transition metal-catalyzed reactions are known to be time-consuming. Finally, we hope that despite not presenting a fully comprehensive review, we have provided some insight into the current state of the art of metal-catalyzed microwave chemistry. We eagerly await future developments in this exciting field.

Acknowledgements We would like to express our sincere gratitude to the Swedish Research Council, to Knut and Alice Wallenberg's Foundation, Medivir AB, Mr Gunnar Wikman and Biolipox AB.

References

1. Mingos DMP, Baghurst DR (1991) Chem Soc Rev 20:1
2. Lidström P, Tierney J, Wathey B, Westman J (2001) Tetrahedron 57:9225
3. Loupy A (2002) Microwaves in organic synthesis. Wiley, Weinheim
4. Kappe CO (2004) Angew Chem Int Ed 43:6250
5. Kappe CO, Stadler A (2005) Microwaves in organic and medicinal chemistry. Wiley, Weinheim
6. Kuznetsov DV, Raev VA, Kuranov GL, Arapov OV, Kostikov RR (2005) Russ J Org Chem 41:1719
7. Strauss CR, Trainor RW (1995) Aust J Chem 48:1665
8. Larhed M, Moberg C, Hallberg A (2002) Acc Chem Res 35:717
9. Leadbeater NE (2005) Chem Commun p 2881
10. Roberts BA, Strauss CR (2005) Acc Chem Res 38:653
11. Gronnow MJ, White RJ, Clark JH, Macquarrie DJ (2005) Org Process Res Dev 9:516
12. Larhed M, Hallberg A (2001) Drug Discov Today 6:406
13. Lew A, Krutzik PO, Hart ME, Chamberlin AR (2002) J Comb Chem 4:95
14. Baxendale IR, Lee AL, Ley SV (2005) In: Tierney JP, Lidström P (eds) Microwave assisted organic synthesis. Blackwell, Oxford, UK, p 133
15. Kappe CO, Dallinger D (2006) Nat Rev Drug Discovery 5:51
16. Sarko CR (2005) In: Tierney JP, Lidström P (eds) Microwave assisted organic synthesis. Blackwell, Oxford, p 222

17. Cornils B, Herrmann WA (eds) (2002) Applied homogeneous catalysis with organo-metallic compounds, vol 3: Developments, 2nd edn. Wiley, Weinheim
18. Tucker CE, de Vries JG (2002) Top Catal 19:111
19. Negishi E-i (2002) Handbook of organopalladium chemistry for organic synthesis, vol 1. Wiley, New York
20. Ley SV, Thomas AW (2003) Angew Chem Int Ed 42:5400
21. Olofsson K, Larhed M (2005) In: Tierney JP, Lidström P (eds) Microwave assisted organic synthesis. Blackwell, Oxford, p 23
22. Olofsson K, Hallberg A, Larhed M (2002) In: Loupy A (ed) Microwaves in organic synthesis. Wiley, Weinheim, p 379
23. Beller M, Cornils B, Frohning CD, Kohlpaintner CW (1995) J Mol Catal A 104:17
24. Morimoto T, Kakiuchi K (2004) Angew Chem Int Ed 43:5580
25. Skoda-Földes R, Kollár L (2002) Curr Org Chem 6:1097
26. Kaiser NFK, Hallberg A, Larhed M (2002) J Comb Chem 4:109
27. Wannberg J, Larhed M (2003) J Org Chem 68:5750
28. Georgsson J, Hallberg A, Larhed M (2003) J Comb Chem 5:350
29. Herrero MA, Wannberg J, Larhed M (2004) Synlett 2335
30. Lagerlund O, Larhed M (2006) J Comb Chem 8:4
31. Wan YQ, Alterman M, Larhed M, Hallberg A (2003) J Comb Chem 5:82
32. Marradi M (2005) Synlett 1195
33. Wu X, Wannberg J, Larhed M (2006) Tetrahedron 62:4665
34. Wannberg J, Dallinger D, Kappe CO, Larhed M (2005) J Comb Chem 7:574
35. Sinou D (1999) Top Curr Chem 206:41
36. Li C-J (2005) Chem Rev 105:3095
37. Andrade CKZ, Alves LM (2005) Curr Org Chem 9:195
38. Wu X, Larhed M (2005) Org Lett 7:3327
39. Wu X, Ekegren JK, Larhed M (2006) Organometallics ASAP
40. Wu X, Nilsson P, Larhed M (2005) J Org Chem 70:346
41. Larhed M, Hallberg A (2002) In: Negishi E-i (ed) Handbook of organopalladium chemistry for organic synthesis, vol 1. Wiley, New York, p 1133
42. De Meijere A, Meyer FE (1994) Angew Chem Int Ed Engl 33:2379
43. Beletskaya IP, Cheprakov AV (2000) Chem Rev 100:3009
44. Larhed M, Hallberg A (1996) J Org Chem 61:9582
45. Nájera C, Botella L (2005) Tetrahedron 61:9688
46. Arvela RK, Leadbeater NE (2005) J Org Chem 70:1786
47. Vallin KSA, Emilsson P, Larhed M, Hallberg A (2002) J Org Chem 67:6243
48. Littke AF, Fu GC (2002) Angew Chem Int Ed 41:4176
49. Datta GK, Vallin KSA, Larhed M (2003) Mol Div 7:107
50. Stadler A, von Schenck H, Vallin KSA, Larhed M, Hallberg A (2004) Adv Synth Catal 346:1773
51. Svennebring A, Garg N, Nilsson P, Hallberg A, Larhed M (2005) J Org Chem 70:4720
52. Andappan MMS, Nilsson P, Larhed M (2003) Mol Div 7:97
53. Sonogashira K (2002) In: Negishi E-i (ed) Handbook of organopalladium chemistry for organic synthesis, vol 1. Wiley, New York, p 493
54. Kabalka GW, Wang L, Namboodiri V, Pagni RM (2000) Tetrahedron Lett 41:5151
55. Erdelyi M, Gogoll A (2001) J Org Chem 66:4165
56. Kabalka GW, Wang L, Pagni RM (2001) Tetrahedron 57:8017
57. Erdelyi M, Gogoll A (2003) J Org Chem 68:6431
58. Berthault A, Berteina-Raboin S, Finaru A, Guillaumet G (2004) QSAR Comb Sci 23:850

59. Xia M, Wang Y-G (2002) J Chem Res Synop p 173
60. Xia M, Wang YG (2002) Chin Chem Lett 13:1
61. Yan J, Wang Z, Wang L (2004) J Chem Res p 71
62. He H, Wu YJ (2004) Tetrahedron Lett 45:3237
63. Appukkuttan P, Dehaen W, Van der Eycken E (2003) Eur J Org Chem p 4713
64. Hopkins CR, Collar N (2004) Tetrahedron Lett 45:8631
65. Leadbeater NE, Marco M, Tominack BJ (2003) Org Lett 5:3919
66. Arvela RK, Leadbeater NE, Sangi MS, Williams VA, Granados P, Singer RD (2005) J Org Chem 70:161
67. Kohara Y, Kubo K, Imamiya E, Wada T, Inada Y, Naka T (1996) J Med Chem 39:5228
68. Liljebris C, Larsen SD, Ogg D, Palazuk BJ, Bleasdale JE (2002) J Med Chem 45:1785
69. Tschaen DM, Desmond R, King AO, Fortin MC, Pipik B, King S, Verhoeven TR (1994) Synth Commun 24:887
70. Alterman M, Hallberg A (2000) J Org Chem 65:7984
71. Arvela RK, Leadbeater NE (2003) J Org Chem 68:9122
72. Srivastava RR, Collibee SE (2004) Tetrahedron Lett 45:8895
73. Zhang A, Neumeyer JL (2003) Org Lett 5:201
74. Cai L, Liu X, Tao X, Shen D (2004) Synth Commun 34:1215
75. Arvela RK, Leadbeater NE, Torenius HM, Tye H (2003) Org Biomol Chem 1:1119
76. Leadbeater NE, Torenius HM, Tye H (2003) Tetrahedron 59:2253
77. Bengtson A, Hallberg A, Larhed M (2002) Org Lett 4:1231
78. Hartwig JF (1998) Angew Chem Int Ed 37:2046
79. Sharifi A, Hosseinzadeh R, Mirzaei M (2002) Monatsh Chem 133:329
80. Wan Y, Alterman M, Hallberg A (2002) Synthesis p 1597
81. Jensen TA, Liang X, Tanner D, Skjaerbaek N (2004) J Org Chem 69:4936
82. Tundel RE, Anderson KW, Buchwald SL (2006) J Org Chem 71:430
83. Maes BUW, Loones KTJ, Lemiere GLF, Dommisse RA (2003) Synlett 1822
84. Heo J-N, Song YS, Kim BT (2005) Tetrahedron Lett 46:4621
85. Brain CT, Steer JT (2003) J Org Chem 68:6814
86. Weigand K, Pelka S (2003) Mol Div 7:181
87. McCarroll AJ, Sandham DA, Titcomb LR, Lewis AKdK, Cloke FGN, Davies BP, Perez de Santana A, Hiller W, Caddick S (2003) Mol Div 7:115
88. Maes BUW, Loones KTJ, Hostyn S, Diels G, Rombouts G (2004) Tetrahedron 60:11559
89. Arvela RK, Leadbeater NE, Collins MJ (2005) Tetrahedron 61:9349
90. Stadler A, Yousefi BH, Dallinger D, Walla P, Van der Eycken E, Kaval N, Kappe CO (2003) Org Process Res Dev 7:707
91. Lehmann H, LaVecchia L (2005) Jala 10:412
92. Loones KTJ, Maes BUW, Rombouts G, Hostyn S, Diels G (2005) Tetrahedron 61:10338
93. Wu YJ, He H, L'Heureux A (2003) Tetrahedron Lett 44:4217
94. Combs AP, Saubern S, Rafalski M, Lam PYS (1999) Tetrahedron Lett 40:1623
95. Lange JHM, Hofmeyer LJF, Hout FAS, Osnabrug SJM, Verveer PC, Kruse CG, Feenstra RW (2002) Tetrahedron Lett 43:1101
96. Poondra RR, Turner NJ (2005) Org Lett 7:863
97. Pabba C, Wang HJ, Mulligan SR, Chen ZJ, Stark TM, Gregg BT (2005) Tetrahedron Lett 46:7553
98. He H, Wu YJ (2003) Tetrahedron Lett 44:3385
99. Burton G, Cao P, Li G, Rivero R (2003) Org Lett 5:4373
100. Harmata M, Hong XC, Ghosh SK (2004) Tetrahedron Lett 45:5233

101. Appukkuttan P, Dehaen W, Fokin VV, Van der Eycken E (2004) Org Lett 6:4223
102. Andersen J, Madsen U, Bjorkling F, Liang XF (2005) Synlett 2209
103. Manbeck GF, Lipman AJ, Stockland RA, Freidl AL, Hasler AF, Stone JJ, Guzei IA (2005) J Org Chem 70:244
104. He H, Wu YJ (2003) Tetrahedron Lett 44:3445
105. Villemin D, Jaffres PA, Simeon F (1997) Phos Sulf Silic Rel Elem 130:59
106. Villemin D, Elbilali A, Simeon F, Jaffres PA, Maheut G, Mosaddak M, Hakiki A (2003) J Chem Res Synop 436
107. Stadler A, Kappe CO (2002) Org Lett 4:3541
108. Wu YJ, He H (2003) Synlett 1789
109. Lengar A, Kappe CO (2004) Org Lett 6:771
110. Alexandre FR, Berecibar A, Wrigglesworth R, Besson T (2003) Tetrahedron Lett 44:4455
111. Besson T, Dozias MJ, Guillard J, Rees CW (1998) J Chem Soc Perkin Trans 1:3925
112. Marquie J, Laporterie A, Dubac J, Roques N, Desmurs JR (2001) J Org Chem 66:421
113. Grushin VV, Alper H (1994) Chem Rev 94:1047
114. Kabalka GW, Varma RS (1989) Tetrahedron 45:6601
115. Elander N, Jones JR, Lu SY, Stone-Elander S (2000) Chem Soc Rev 29:239
116. Arvela RK, Leadbeater NE (2003) Synlett 1145
117. Miyaura N, Suzuki A (1995) Chem Rev 95:2457
118. Ersmark K, Larhed M, Wannberg J (2004) Curr Opin Drug Discov 7:417
119. Comer E, Organ MG (2005) J Am Chem Soc 127:8160
120. Leadbeater NE, Marco M (2003) J Org Chem 68:888
121. Leadbeater NE, Marco M (2002) Org Lett 4:2973
122. Bai L, Wang JX, Zhang YM (2003) Green Chemistry 5:615
123. Bedford RB, Butts CP, Hurst TE, Lidstrom P (2004) Adv Synth Catal 346:1627
124. Navarro O, Kaur H, Mahjoor P, Nolan SP (2004) J Org Chem 69:3173
125. Miao GB, Ye P, Yu LB, Baldino CM (2005) J Org Chem 70:2332
126. Arvela RK, Leadbeater NE (2005) Org Lett 7:2101
127. Najera C, Gil-Molto J, Karlstrom S (2004) Adv Synth Catal 346:1798
128. Bai L, Wang JX (2004) Chin Chem Lett 15:286
129. Namboodiri VV, Varma RS (2001) Green Chem 3:146
130. Larhed M, Lindeberg G, Hallberg A (1996) Tetrahedron Lett 37:8219
131. Stadler A, Kappe CO (2001) Eur J Org Chem 2001:919
132. Wang Y, Sauer DR (2004) Org Lett 6:2793
133. Bai L, Zhang YM, Wang JX (2004) QSAR Comb Sci 23:875
134. Gil-Molto J, Karlström S, Najera C (2005) Tetrahedron 61:12168
135. Dawood KM, Kirschning A (2005) Tetrahedron 61:12121
136. Solodenko W, Schon U, Messinger J, Glinschert A, Kirschning A (2004) Synlett 1699
137. Gladysz JA, Curran DP, Horváth IT (eds) (2004) Handbook of fluorous chemistry. Wiley, Weinheim
138. Zhang W (2004) Chem Rev 104:2531
139. Zhang W, Chen CHT, Lu YM, Nagashima T (2004) Org Lett 6:1473
140. Lu YM, Zhang W (2004) QSAR Comb Sci 23:827
141. Zhang W, Chen CHT (2005) Tetrahedron Lett 46:1807
142. Song YS, Kim BT, Heo JN (2005) Tetrahedron Lett 46:5987
143. Coats SJ, Schulz MJ, Carson JR, Codd EE, Hlasta DJ, Pitis PM, Stone DJ Jr, Zhang S-P, Colburn RW, Dax SL (2004) Bioorg Med Chem Lett 14:5493
144. Mathews CJ, Taylor J, Tyte MJ, Worthington PA (2005) Synlett 538
145. Combs AP, Glass BM, Jackson SA (1999) Methods Enzymol 369:223

146. Organ MG, Mayer S, Lepifre F, N'Zemba B, Khatri J (2003) Mol Div 7:211
147. Ersmark K, Feierberg I, Bjelic S, Hamelink E, Hackett F, Blackman MJ, Hulten J, Samuelsson B, Åqvist J, Hallberg A (2004) J Med Chem 47:110
148. Nöteberg D, Hamelink E, Hultén J, Wahlgren M, Vrang L, Samuelsson B, Hallberg A (2003) J Med Chem 46:734
149. Nöteberg D, Schaal W, Hamelink E, Vrang L, Larhed M (2003) J Comb Chem 5:456
150. Espinet P, Echavarren AM (2004) Angew Chem Int Ed 43:4704
151. Kaval N, Dehaen W, Van der Eycken E (2005) J Comb Chem 7:90
152. Olofsson K, Kim SY, Larhed M, Curran DP, Hallberg A (1999) J Org Chem 64:4539
153. Maleczka RE Jr, Lavis JM, Clark DH, Gallagher WP (2000) Org Lett 2:3655
154. Larhed M, Hoshino M, Hadida S, Curran DP, Hallberg A (1997) J Org Chem 62:5583
155. Öhberg L, Westman J (2001) Synlett 1893
156. Walla P, Kappe CO (2004) Chem Commun p 564
157. Krascsenicsova K, Walla P, Kasak P, Uray G, Kappe CO, Putala M (2004) Chem Commun p 2606
158. Stanetty P, Schnurch M, Mihovilovic MD (2003) Synlett 1862
159. Mutule I, Suna E (2004) Tetrahedron Lett 45:3909
160. Mutule G, Suna E (2005) Tetrahedron 61:11168
161. Hassan J, Sevignon M, Gozzi C, Schulz E, Lemaire M (2002) Chem Rev 102:1359
162. Gold H, Larhed M, Nilsson P (2005) Synlett 1596
163. Hiyama T, Shirakawa E (2002) In: Negishi E-I (ed) Handbook of organopalladium chemistry for organic synthesis, vol 1. Wiley, New York, p 285
164. Clarke ML (2005) Adv Synth Catal 347:303

Top Curr Chem (2006) 266: 145–166
DOI 10.1007/128_045
© Springer-Verlag Berlin Heidelberg 2006
Published online: 10 June 2006

Microwave-Enhanced High-Speed Fluorous Synthesis

Wei Zhang

Fluorous Technologies, Inc., University of Pittsburgh Applied Research Center,
970 William Pitt Way, Pittsburgh, PA 15238, USA
w.zhang@fluorous.com

Abstract Increasing reaction speed and simplifying product purification are two major ways to improve the efficiency of organic synthesis. A new technology for high-speed solution-phase synthesis has been developed by combination of microwave heating and fluorous purification. This review describes different techniques for microwave-enhanced fluorous synthesis and their applications in Pd-catalyzed cross-coupling reactions, free-radical reactions, multicomponent reactions, and compound library synthesis.

Keywords Fluorous chemistry · Microwave synthesis · Solid-phase extraction · High-throughput synthesis · Multicomponent reaction

Abbreviations
μw	microwave
HTS	high-throughput synthesis
SPOS	solid-phase organic synthesis
FTI	Fluorous Technologies, Inc.
Rf$_n$h$_m$	$C_nF_{2n+1}(CH_2)_m$
F-LLE	fluorous liquid–liquid extractions
F-SPE	fluorous solid-phase extraction
HPLC	high-performance liquid chromatography

TLC   thin-layer chromatography
IR   infrared spectroscopy
NMR   nuclear magnetic resonance
LCMS   liquid chromatography and mass spectrometry
BINAP   2,2′-bis(diphenylphosphino)-1,1′-binaphthyl
BTF   benzotrifluoride
MCR   multicomponent reaction
DOS   diversity-oriented synthesis

1
Introduction

Over the last decade, a significant amount of effort in high-throughput synthesis (HTS) has been dedicated to the development of solid-phase organic synthesis (SPOS) and production of compound libraries [1, 2]. SPOS simplifies the purification process and has the capability to synthesize small quantities of very large libraries by the "split and pool" protocol. However, SPOS has several well recognized limitations [3, 4] such as slow reaction kinetics caused by heterogeneous reactions, long method development times, and limitations of reaction scope and scale. Furthermore, the large libraries generated by SPOS usually lack molecular diversity, and compound purities can be inadequate.

In many pharmaceutical and biotechnology companies, the current effort on HTS has been redirected to solution-phase and diversity-oriented syntheses [4]. Compound libraries with high molecular diversity are given more consideration for discovery programs. Small targeted (focused) libraries (roughly 50–200 compounds) are increasingly popular for lead optimization projects. Quantities for testing or deposition in compound collections are increasing to the 10–50 mg range, and requirements for purity have stiffened (> 90% is now common). New synthetic methods are currently needed which are better integrated with medicinal chemistry, less dependent on specialized instruments, easy to combine with existing reaction and separation methods, applicable for both small and large library synthesis, and capable of quickly producing pure compounds. Fluorous chemistry has been developed as a broad-based technology platform that addresses many challenges in the modern drug discovery environment [5].

2
Fluorous Technologies

2.1
Fluorous Tagging Strategy

The development of fluorous technologies provides an opportunity to combine solution-phase reactions with phase tag-based separations [6–11]. Perfluorinated (fluorous) chains such as C_6F_{13} and C_8F_{17} are employed as phase tags to facilitate the separation process. The fluorous chain is usually attached to the parent molecule through a $(CH_2)_m$ segment to insulate the reactive site from the electron withdrawing fluorines. A fluorous chain $C_nF_{2n+1}(CH_2)_m$ can be abbreviated to Rf_nh_m when it is presented in the reaction equation. In principle, any synthetic method developed in conventional solution-phase or in polymer-bound chemistry can be adapted to fluorous synthesis. Fluorous groups can be used to tag reagents (including scavengers and catalysts) (Scheme 1A) or to tag substrates (Scheme 1B) [12]. Fluorous reagents are commonly used for single or short-step parallel synthesis, while fluorous-tagged substrates are more suitable for multistep parallel synthesis [9].

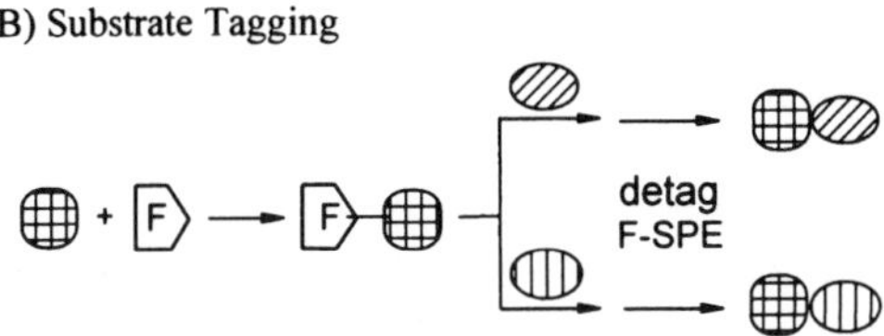

Scheme 1 Fluorous tagging strategies

2.2
Fluorous Separation

Fluorous separations rely on the strong and selective affinity interaction between fluorous molecules and fluorous separation media. The separation medium can be fluorous solvents which are immiscible with common organic solvents at room temperature and thus can be used for liquid–liquid extractions. A "heavy fluorous" tag (60% or more fluorine by molecular weight) is required to drive the fluorous molecule to a fluorous phase from a non-fluorous phase. Perfluorinated alkanes such as FC-72 (perfluorohexanes) and highly fluorinated ethers such as HFC-7100 ($C_4F_{17}OCH_3$) are good solvents for F-LLE. Fluoro*Flash* silica, a commercial product from FTI, with a $-SiMe_2CH_2CH_2C_8F_{17}$ stationary phase is another kind of fluorous separation media which can be used for F-SPE or chromatography [13, 14]. Fluorous silica gel separates "light fluorous" molecules which have relatively low fluorine content from compounds with no or different fluorine content. F-SPE is a binary technique that divides mixtures into two fractions based on the presence or absence of a fluorous tag. Fluorophobic solvents such as 80 : 20 $MeOH-H_2O$ is used to elute non-fluorous components, the fluorous component retains on the SPE cartridge until it is eluted with a more fluorophilic solvent such as MeOH. The F-SPE technique can also be used in "plate-to-plate" format [15] for parallel separations. Fluorous-HPLC can be used for analysis and separation of fluorous molecules. Common solvent systems used for reverse-phase HPLC such as $MeOH-H_2O$ and $MeCN-H_2O$ are commonly used for F-HPLC.

2.3
Features of Fluorous Synthesis

Fluorous synthesis integrates the characters of solution-phase reactions and solid-phase-type separations [12]. It has good "combinatorial" capabilities and offers the following advantages: 1) Fluorous reactions can be easily followed by common analytical methods such as TLC, HPLC, IR and NMR; 2) Fluorous components can be separated by fluorous methods as well as conventional methods such as distillation, recrystallization, and chromatography; 3) Compounds with fluorous tags (such as C_8F_{17}) have good solubility in a range of organic solvents, no fluorous solvents are needed for reactions and separations; 4) More than one fluorous reagent can be used in a single reaction; 5) Fluorous methods can easily be adaptation from non-fluorous procedures from the literature; 6) Fluorous synthesis can be combined with other methods such as microwave technology, multi-component reactions, and solid-phase synthesis; 7) Fluorous materials can be recovered after fluorous separation.

The most important characteristics that make fluorous synthesis superior to solid-supported synthesis is the favorable reaction kinetics associated with the solution-phase reactions. Comparison reactions using fluorous vs. solid-supported thiols to scavenge a bromide are shown in Fig. 1 [16]. Using 1.5 equiv of F-thiol **1**, more than 95% bromide was quenched in less than 40 min (top line). Under the same conditions and using 1.5 equiv PS-thiol **2**, only 50% of the halide was quenched after 80 min (bottom line). By doub-

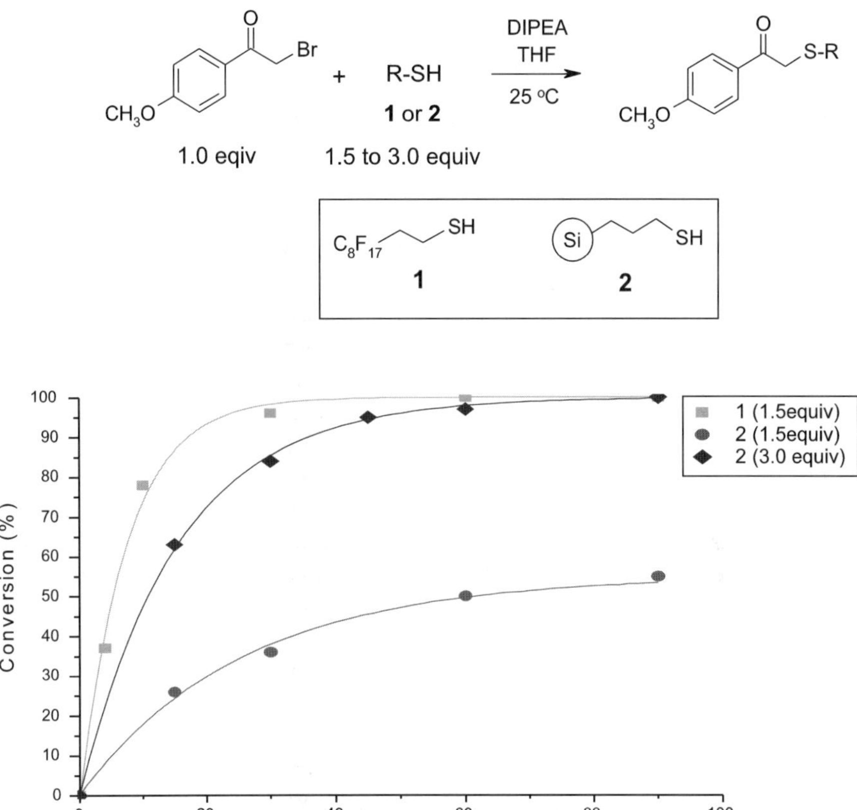

Fig. 1 Comparison of fluorous and PS-scavenging reactions

ling the amount of PS-thiol to 3.0 equiv, the conversion was improved to 95% after 60 min (middle line). The comparison experiment clearly shows that the fluorous reagent reacts much faster than the solid-supported reagent. Fluorous reagents are real molecules, they can be used in a stoichiometric manner and can be analyzed by standard methods such as LCMS and NMR. Solid-supported reagents are functionalize materials, their chemical and physical properties vary from batch to batch and are difficult to be analyzed by the user. It is not uncommon that a large excess (3.0 equiv or even more) of solid-supported reagents is used to "over kill" the reactions. This kind of practice results in high costs on reagents as well as high volume of solvent used for the reaction and the resin washes after the reaction. These issues can be avoided in fluorous synthesis.

3
Microwave-Assisted Fluorous Synthesis

3.1
Integrated Reaction and Separation System

Microwave irradiation as a powerful and controlled heating source has been widely employed in organic synthesis to reduce the reaction time, improve yield or selectivity, and reduce the amount of solvent used for reactions [17–21]. The blending of microwave heating and polymer-supported reactions to simplify purification is a logical consequence to address the separation issue. Significant progress has been made in this area [22–25]; however, there are still limitations related to the physical and chemical stability of polymer support under microwave heating and to the heterogeneous nature of solid-phase reactions. The $C-F$ bond of perfluorinated molecules has excellent chemical stability (Teflon is a fluorous polymer), which makes fluorous molecules well suited for microwave reactions at high temperature. Microwave-assisted fluorous synthesis as a new HTS technology could improve both the reaction and purification efficiency [26, 27]. An integrated system shown in Fig. 2 is constituted by an automated microwave reactor, a "plate to plate" SPE unit for parallel separations, and a centrifuge system for concentration of products in a plate format [15].

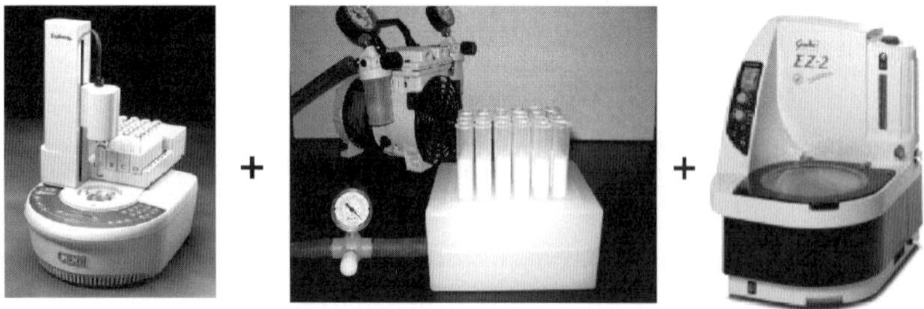

Fig. 2 A microwave-assisted fluorous synthesis system, microwave reactor (*left*), plate-to-plate SPE (*middle*), and a plate centrifuge unit (*right*)

3.2
Metal-Catalyzed Reactions

3.2.1
Fluorous Substrates

One of the most important applications of microwave reactions is for transition-metal-catalyzed transformations [28]. The Curran and Hallberg groups first introduced microwave heating to fluorous Stille reactions (Scheme 2) [29]. The reactions employed $(C_6F_{13}CH_2CH_2)_3$ attached tin reagents **3** as substrates, $PdCl_2(PPh_3)_2$ as a catalyst, and DMF as a solvent. A slight excess of tin reagent (1.2 equiv) was used to ensure the consumption of the halides or triflates (1.0 equiv). Reactions reached completion under microwave heating in less than 2 min. The reaction mixtures were subjected to a three-phase extraction with FC-84 (perfluoroheptanes), dichloromethane, and water. Fluorous tin derivatives and unreacted tin reagent were distributed to the bottom-phase of the solvent FC-84, the product and the catalyst to the middle-phase of dichloromethane, and LiCl salt to the water phase on top. The product isolated from the fluorous layer was further purified by flash chromatography. To improve the efficiency of fluorous liquid–liquid extraction, a tin reagent with heavier fluorous tags $(C_{10}F_{21}CH_2CH_2)_3$ has been developed for the Stille reactions [30].

As the longer-chain analogs of triflates and nonaflates, perfluorooctanesulfonates **4** are also highly reactive, stable for room temperature storage, and resistant towards hydrolysis. They can be easily prepared by reaction of various phenols with perfluorooctanesulfonyl fluoride using K_2CO_3 or Cs_2CO_3 as a base. Perfluorooctanesulfonates have been used in Pd-catalyzed transformations including Suzuki–Miyaura, Heck, and Buchwald–Hartwig type reactions to form aryl carbon–carbon, carbon–hydrogen, and carbon–heteroatom bonds (Scheme 3). Using perfluorooctanesulfonates for multicomponent reactions compound libraries synthesis are described in Sects. 3.3 and 3.4.

A general method for fluorous Suzuki reaction of perfluorooctanesulfonates has been developed using $[Pd(dppf)Cl_2]$ (dppf = 1,1'-bis(diphenyl-

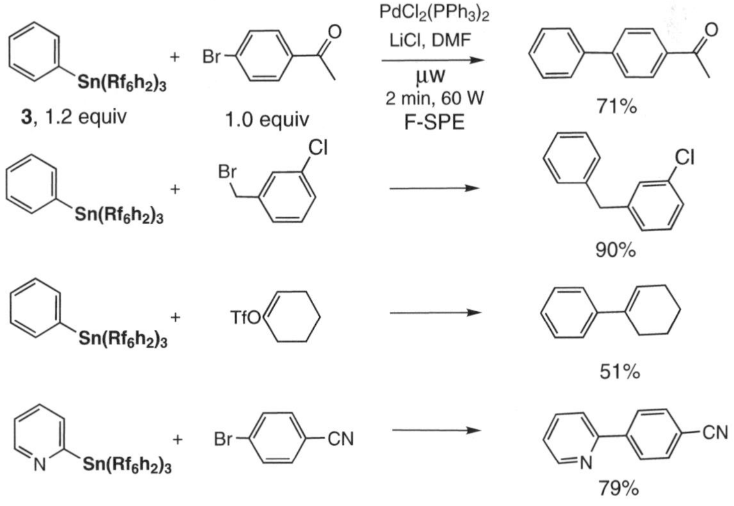

Scheme 2 Stille reactions with fluorous substrates

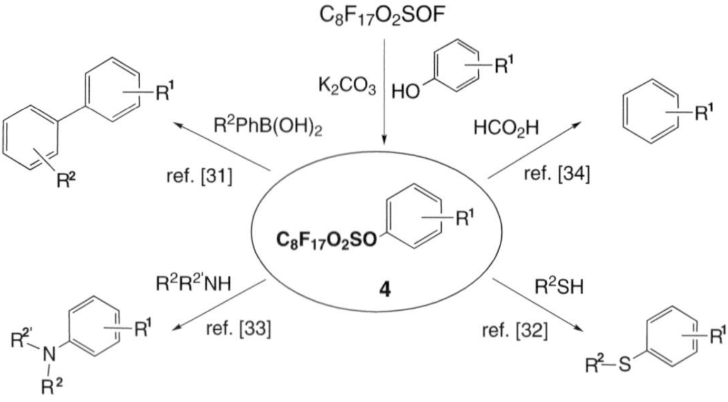

Scheme 3 Fluorous arylsulfonate-based coupling reactions

phosphano)ferrocene) as a catalyst, K_2CO_3 as a base, and $4:4:1$ toluene/acetone/H_2O as a solvent system (Scheme 4) [31]. Reactions were conducted under microwave heating at $100–130\,^\circ\mathrm{C}$ for 10 min and the reaction mixtures were directly loaded onto a Fluoro*Flash* cartridge for F-SPE. The biaryl product was collected in the fraction of $80:20$ MeOH/H_2O, while the cleaved fluorous tag remained on the cartridge until it was washed out with MeOH.

Using conditions similar to those developed for fluorous Suzuki reactions, aryl sulfides were synthesized by reaction of fluorous sulfonates with thioles (Scheme 5) [32]. The reaction mixtures were in this case also purified by F-SPE.

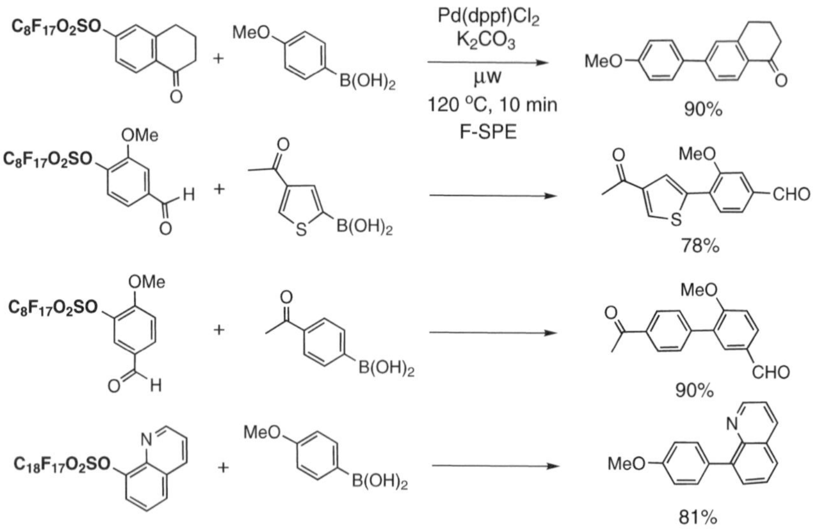

Scheme 4 Fluorous Suzuki reactions

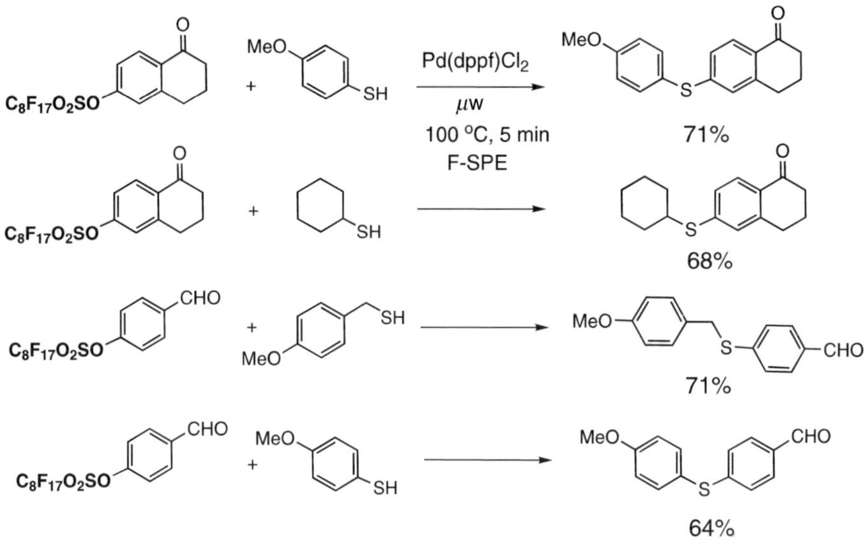

Scheme 5 Formation of aryl sulfides

Fluorous Buchwald–Hartwig type amination reactions using $Pd(OAc)_2$ and BINAP as a catalyst, Cs_2CO_3 as a base, and toluene as a solvent were examined under microwave and oil-bath heating conditions (Scheme 6) [33]. Under microwave irradiation at $120-150\,^\circ C$ for up to 30 min, reactions did not reach completion. Formation of a dark-brown precipitate suggested that the incomplete reactions could be caused by rapid decomposition of the catalyst under the microwave heating. Under oil-bath heating at $80-90\,^\circ C$ for 48 h, the coupling reactions proceeded smoothly.

Fluorous arylsulfonates have also been employed for deoxygenation reactions with formic acid with $Pd(dppf)Cl_2$ as a catalyst, K_2CO_3 as a base, and $4:4:1$ $MePhH/Me_2CO/H_2O$ as a solvent system (Scheme 7) [34]. The

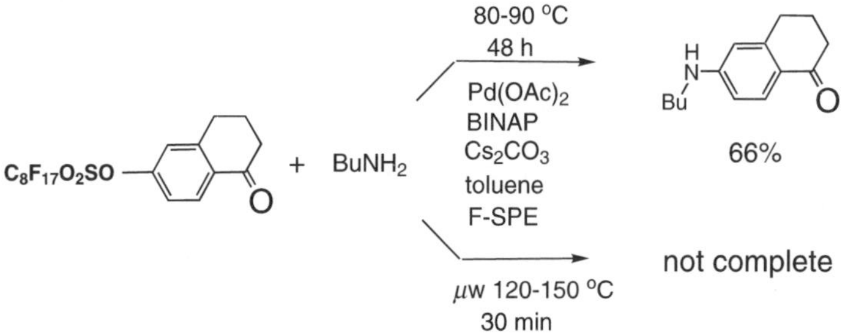

Scheme 6 Amination reactions of fluorous arylsulfonates

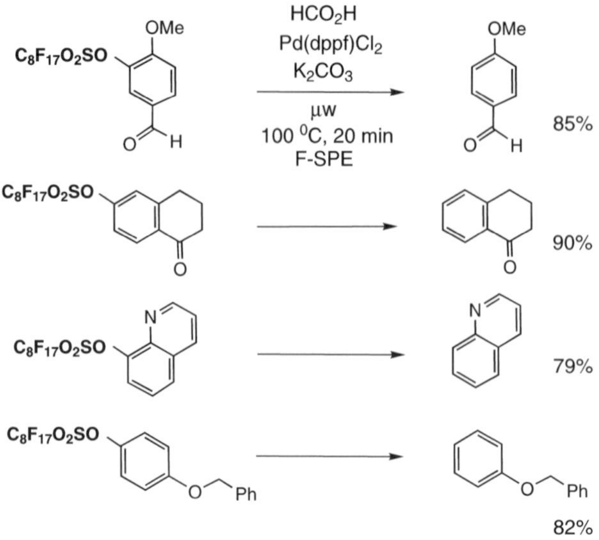

Scheme 7 Deoxygenation of fluorous arylsulfonates

reactions were performed under microwave irradiation at $100\,^{\circ}C$ for 20 min followed by F-SPE separations to provide purified final products.

3.2.2
Fluorous Catalysts

In last decade, significant progress has been made in fluorous catalysis reactions and recovery of the catalysts by fluorous biphasic separations [14]. The Larhed and Curran groups employed bidentate fluorous-tagged 1,3-bis(diphenylphosphino)propane ligand (F-dppp) for the Heck vinylation reactions. Reactions with fluorous and non-fluorous dppps were examined under microwave and conventional heating conditions (Scheme 8) [35]. Results generated from different reactions are comparable in regioselectivity and yields. The microwave-assisted reaction with F-dppp has advantages of reaction speed, purification, and catalyst recovery; the reaction finished in 15 min and the reaction mixture was directly loaded onto the SPE cartridge for purification, while the reaction with normal dppp and under conventional heating finished in 18 h and was purified by extraction with diethyl ether from the basic aqueous layer.

Larhed and coworkers employed fluorous triphenylphosphine as a ligand and $Mo(CO)_6$ as a CO source for palladium-catalyzed hydrazidocarbonylations of aryl iodides and bromides (Scheme 9) [36]. The fluorous ligand was recovered by extraction with fluorous solvent FC-84.

Fluorous distanoxane **7** and tin oxide **8** have been evaluated in catalytic transesterification reactions by the Panek and Porco groups for the pur-

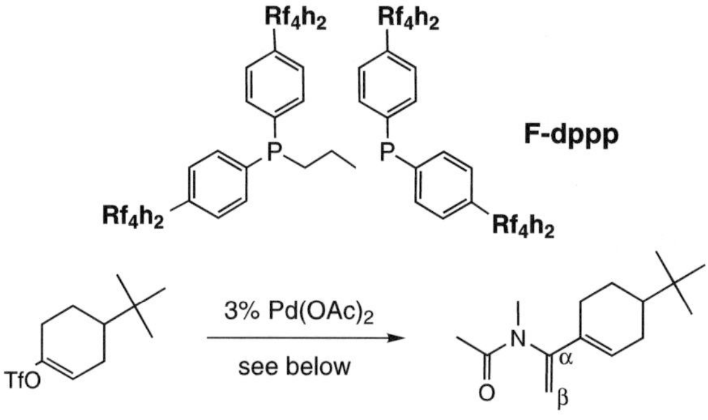

	α/β	isolated yield
dppp, $\triangle$ 60 °C, 18 h	96/4	53%
dppp, μw 90 °C, 15 min	91/9	51%
F-dppp, $\triangle$ 60 °C, 18 h	94/6	46%
F-dppp, μw 90 °C, 15 min	90/10	46%

Scheme 8 Fluorous catalyst for Heck reactions

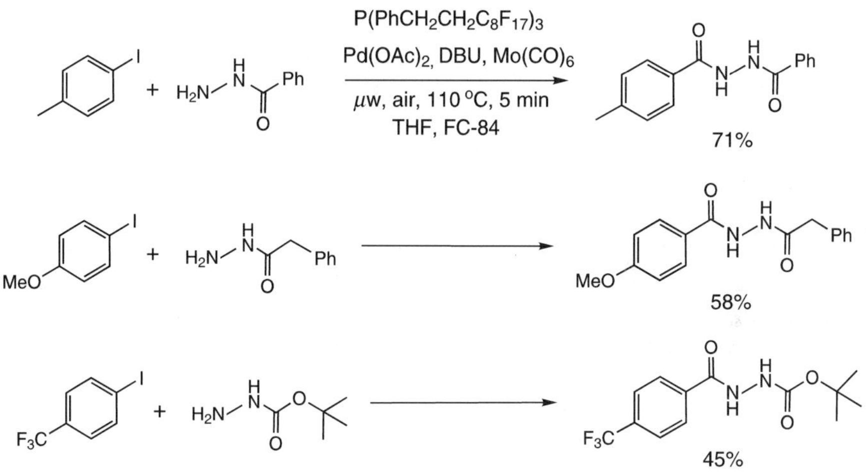

Scheme 9 Fluorous catalyst for hydrazidocarbonylation reactions

pose of easy removal of tin catalysts by F-SPE from the reaction mixture (Scheme 10) [37]. For homodimerization of a monomer under microwave irradiation, it was found that the fluorous tin oxide **8** gave the best yield (80%), while distanoxane **7** enhanced catalytic activity (78% yield) as compared to non-fluorous distanoxanes **5a–c** (42–60% yields) and tin oxide **6** (27% yield).

In addition to fluorous catalysis, the preparation of fluorous catalyst itself can benefit from microwave-assisted synthesis. The Wood group reported the preparation of rhodium perfluorobutyramide ($Rh_2(pfm)_4$), a catalyst for

catalyst (10%)
C6H5Cl
μw
200 °C, 10 min
F-SPE
5a
5b
5c
7
[[Me(CH2)3]2SnO]n
6
[[CF3(CF2)5CH2CH2]2SnO]n
8

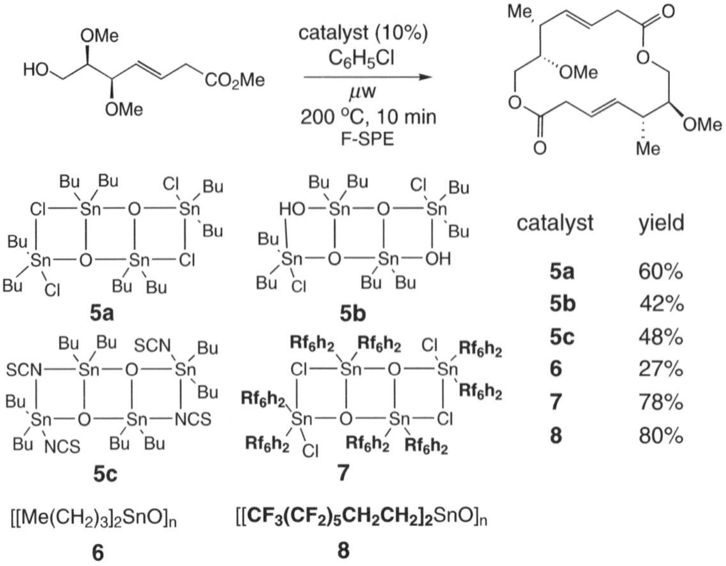

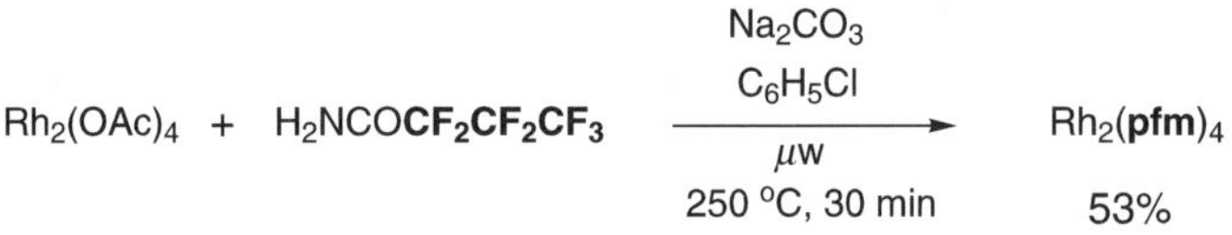

catalyst	yield
5a	60%
5b	42%
5c	48%
6	27%
7	78%
8	80%

Scheme 10 Transesterifications with tin oxide reagents

Rh2(OAc)4 + H2NCOCF2CF2CF3
Na2CO3
C6H5Cl
μw
250 °C, 30 min
Rh2(pfm)4
53%

Scheme 11 Synthesis of fluorous catalyst

olefin aziridinations, under microwave heating (Scheme 11) [38]. The product was isolated from the reaction mixture by FC-72 extraction.

3.3 Free Radical Reactions

Organotin hydrides are popular reagents for free radical reactions. However, the removal of their derivatives from the reaction mixture is not always an easy task. Potential tin residues left in the final product limit the application of tin reagents in medicinal chemistry. One way to address this issue is to use fluorous tin reagents. The Curran and Hallberg groups developed a heavy fluorous tin hydride **9** and used it under continuous microwave heating conditions for halide reduction and cyclization reactions (Scheme 12). Reactions were conducted in benzotrifluoride (BTF) to increase the solubility of the tin reagent [30]. Reaction mixtures were worked up by a three-phase extraction with FC-84, dichloromethane, and water. The products were further purified by circular chromatography.

$(C_{10}F_{21}CH_2CH_2)_3SnH$ **9**

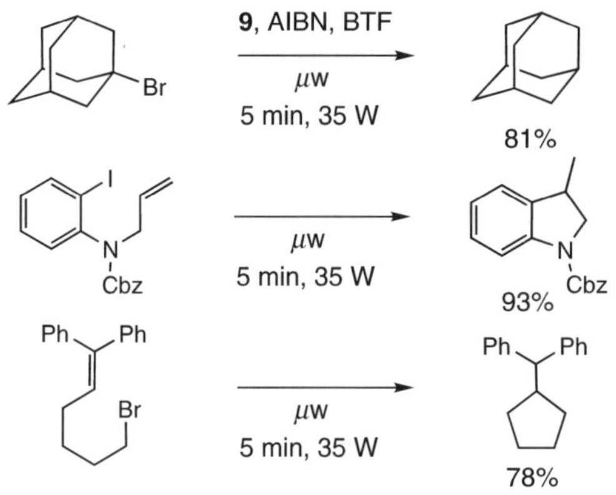

Scheme 12 Fluorous tin-promoted free radical reactions

3.4
Multicomponent Reactions

Multicomponent reactions (MCRs) have high efficiency in the construction of complex molecules [39, 40]. In general practice, more than one reagent is used in excess to push the reaction to go to completion. The unreacted components left in the reaction mixture may complicate the product purification. The employment of a fluorous component as the limiting agent for the MCRs is a good way to simplify the purification.

The Ugi reaction involves components of aldehyde, acid, amine, and isocyanide. Scheme 13 shows an Ugi reaction using fluorous benzyl-attached

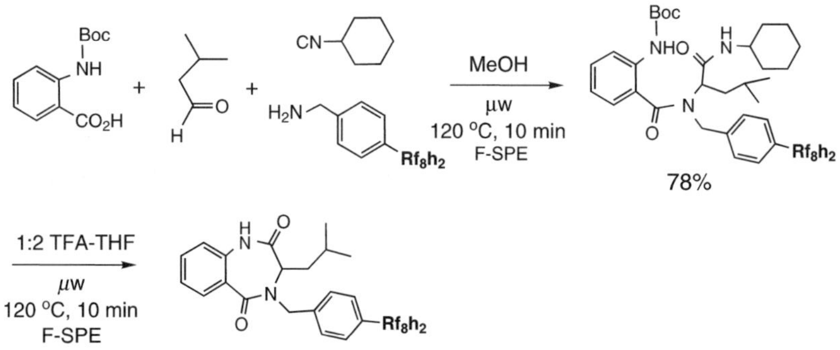

Scheme 13 Fluorous Ugi reaction

aniline as the limiting component [41]. The condensed product **10** was fished out from the reaction mixture by F-SPE and use for further derivatization to construct the benzodiazepinedione ring system.

The synthesis of proline-fused heterocyclic systems by 1,3-dipolar cycloaddition has been well-established in solution-phase synthesis (Scheme 14) [42]. It is usually performed as a one-pot, three-component reaction of a dipolarophile with an in situ prepared azomethine ylide. Perfluoroalkanesulfonyl protected hydroxybenzaldehydes [43] or fluorous alcohol protected amino esters [44] have been developed as two different fluorous components for the synthesis of proline derivatives **11** and **12**.

A one-pot, double intramolecular 1,3-dipolar cycloaddition reaction of azomethine ylides was developed by reaction of 4 equiv of an O-allyl salicyladehyde with a fluorous amino ester under microwave heating to generate a novel hexacyclic ring system **13** that contains seven stererocenters (Scheme 15) [45].

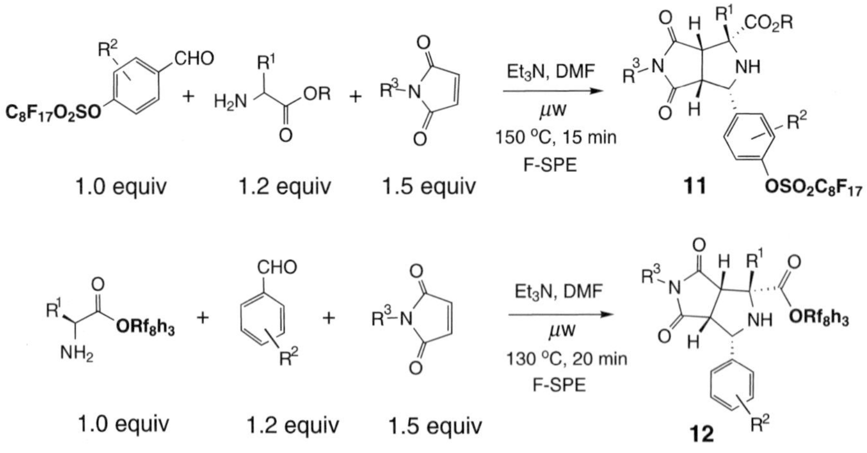

Scheme 14 Fluorous 1,3-dipolar cycloaddition reactions

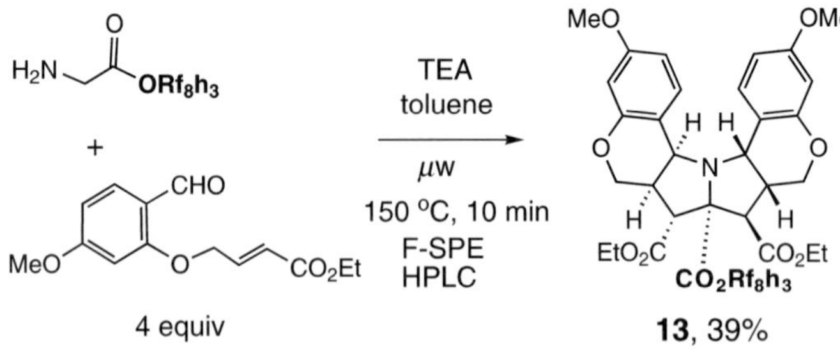

Scheme 15 Double 1,3-dipolar cycloaddition reaction

3.5
Compound Library Synthesis

Compound library synthesis relies heavily on integrated technologies to address both reaction and separation issues. The good "combinatorial" capability of the fluorous technology has demonstrated its great potential in this area [9].

Perfluorosulfonates have been demonstrated as a valuable fluorous synthon. In multistep synthesis, the perfluorooctanesulfonyl tag plays three roles: 1) as a protecting agent for the hydroxyl group; 2) as a fluorous tag to facilitate intermediate purification; and 3) as an activating group to promote the coupling reaction.

Scheme 16 shows parallel syntheses of cyclic and acyclic amide compounds. Fluorous benzaldehydes were first subjected to reductive amination reactions. The resulting amines were then reacted with isocyanates to form substituted hydantoin rings **14** or with benzoyl chlorides to form amides **15**. Purified F-sulfonates were used for palladium-catalyzed cross-coupling reactions to form corresponding biaryl **16** [31] and arylsulfide **17** [32] products, respectively.

Scheme 17 illustrates another fluorous sulfonate-based synthesis of library scaffolds. The tagged substrates were taken through aldol condensation and cycloaddition reactions to form the pyrimidine ring **18**. The intermediates were then reacted with boronic acids for Suzuki reactions to form biaryl compounds **19** [31], reacted with HCO_2H to give traceless detagged products **20** [34], or reacted with amine to form products **21** [33].

Perfluoroalkanesulfonyl protected hydroxybenzaldehydes have been used together with isonitriles and 2-aminopyridines or 2-aminopyrazines for

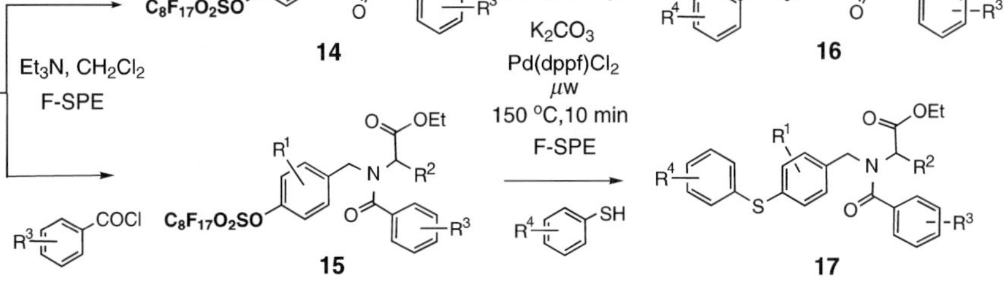

Scheme 16 Synthesis of diversified cyclic and acyclic amide compounds **16** and **17**

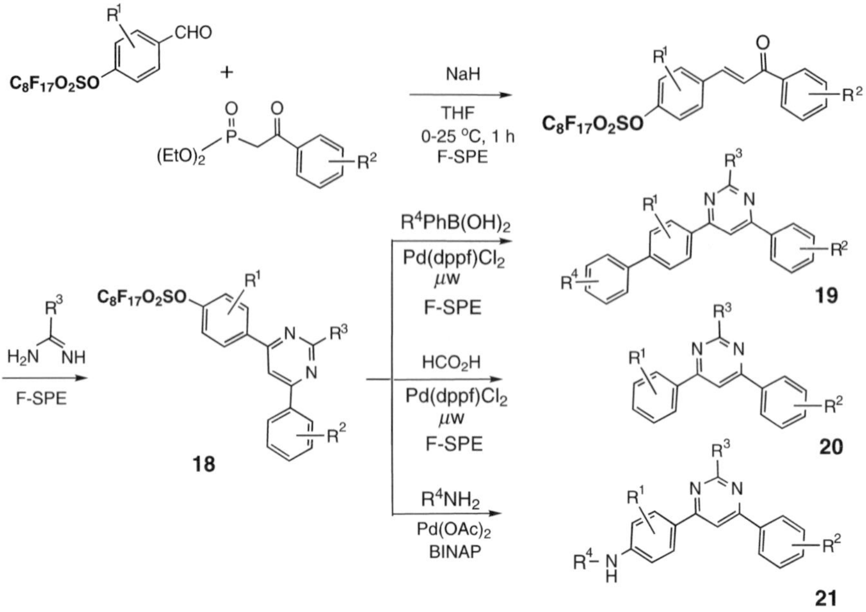

Scheme 17 Multi-step synthesis of substituted pyrimidines **19–21**

3-component condensation reactions to form imidazo[1,2-*a*]pyridine or imidazo[1,2-*a*]pyrazine ring systems **22** (Scheme 18) [46]. The condensed products were employed for Pd-catalyzed cross-coupling reactions with boronic acids or thiols to produce compounds **23** or **24**, respectively.

Zhang and Tempest optimized the Ugi/de-Boc/cyclization synthesis for quinoxalinones **27** and benzimidazoles **28** developed by Humles (Scheme 19) [47]. The original thermal reactions required 36–48 h for the Ugi reaction and the condensed products were purified by double scavenging with an immo-

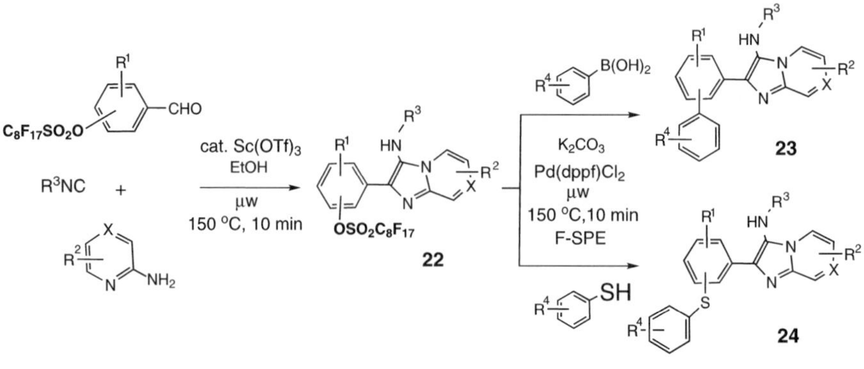

Scheme 18 MCRs for synthesis of heterocyclic libraries **23** and **24**

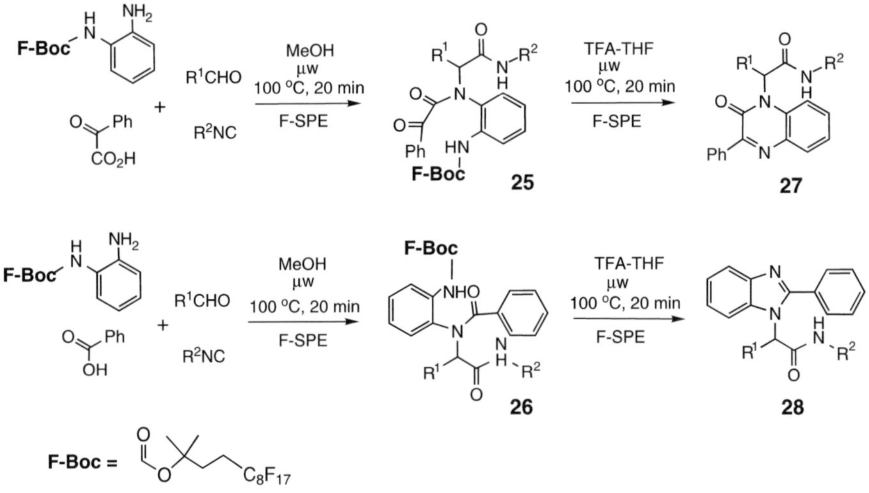

Scheme 19 Synthesis of quinoxalinones **27** and benzimidazoles **28**

bilized tosylhydrazine and diisopropylethylamine to remove excess aldehydes and unreacted acids [48]. Boc-protected diamines were used in the fluorous approach. The Ugi reactions were finished in 20 min under microwave irradiation and the double scavenging step was replaced by a simple F-SPE. The de-Boc/cyclization step was also conducted under microwave heating and products **27** and **28** were purified by F-SPE.

Parallel synthesis of a bicyclic proline library **29** has been developed by a two-step synthesis involving 1,3-dipolar cycloaddition of fluorous benzaldehydes followed by Pd-catalyzed Suzuki coupling reactions with boronic acids (Scheme 20) [43]. Both reactions were conducted under microwave irradiation and the reaction mixtures were purified by F-SPE without the need of performing a following chromatography.

Fluorous aminoesters have been employed in 3-component 1,3-dipolar cycloaddition reactions and post-condensation modifications in the synthesis of a bridged-tricyclic ring system **30** (Scheme 21) [49].

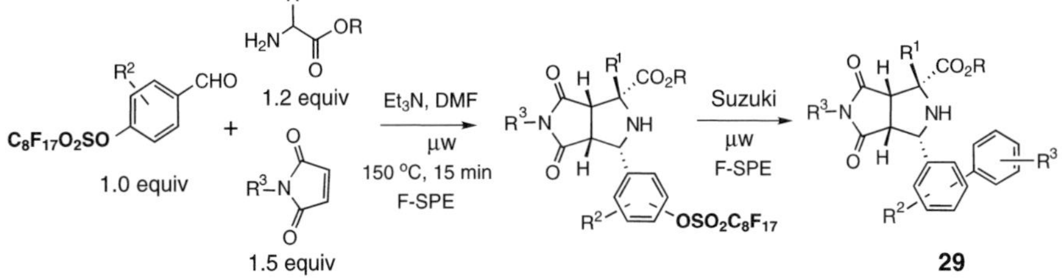

Scheme 20 Synthesis of biaryl substituted prolines **29**

Fluorous aminoesters have also been used in DOS of three unique triaza tricyclic and tetracyclic ring systems (Scheme 22) [44]. Bicyclic pyrrolidines **12** generated from one-pot, three-component 1,3-dipolar cycloaddition of azomethine ylides were further converted to hydantoin-, piperazinedione-, and benzodiazepine-fused compounds **31–33**, respectively. Each of these three heterocyclic scaffolds has four stereocenters on the central pyrrolidine ring and up to four points of diversity (R^1 to R^4). The structure of compound

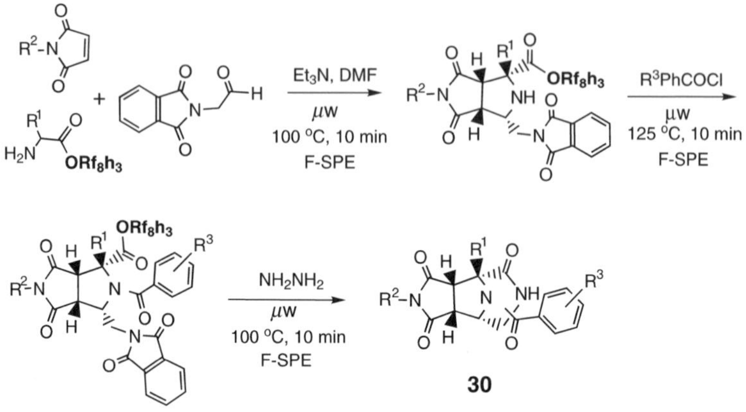

Scheme 21 Synthesis of bridged-tricyclic ring system **30**

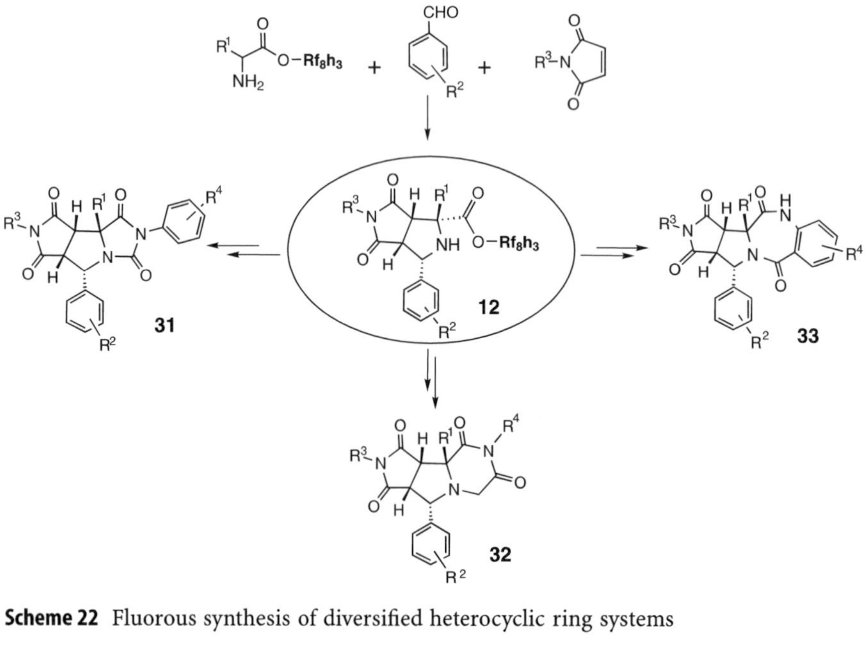

Scheme 22 Fluorous synthesis of diversified heterocyclic ring systems

31 is similar to tricyclic thrombin inhibitors [50]. The structure of compound **32** is similar to diketopiperazine-based inhibitors of human hormone-sensitive lipase [51]. Compound **33** contains a privileged benzodiazepine moiety which has a wide range of pharmaceutical utilities [52].

The synthesis of N-alkylated dihydropteridinones **34** started with a displacement reaction of 4,6-dichloro-5-nitropyrimidine with a fluorous amino-ester (Scheme 23) [53]. The compounds formed **35** were then reacted with secondary amines to yield **36**. The reduction of the nitro group of **36** was conducted by hydrogenation using Pd on charcoal as a catalyst. The cyclization reactions of **37** were promoted by microwave irradiation. The N-alkylation reaction of the cyclized products **38** with benzyl halides gave monobenzylated

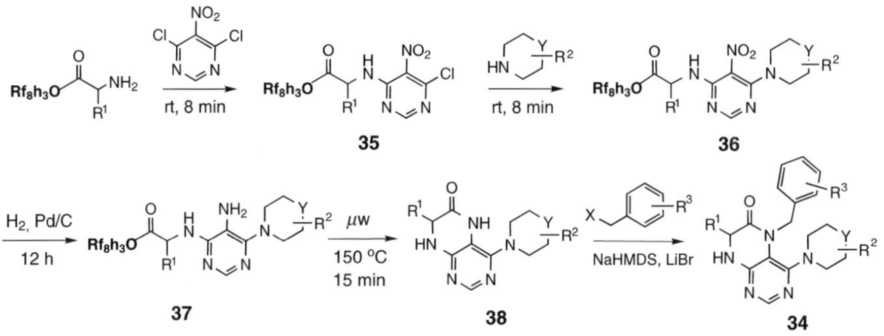

Scheme 23 Fluorous synthesis of N-alkylated dihydropteridinones **34**

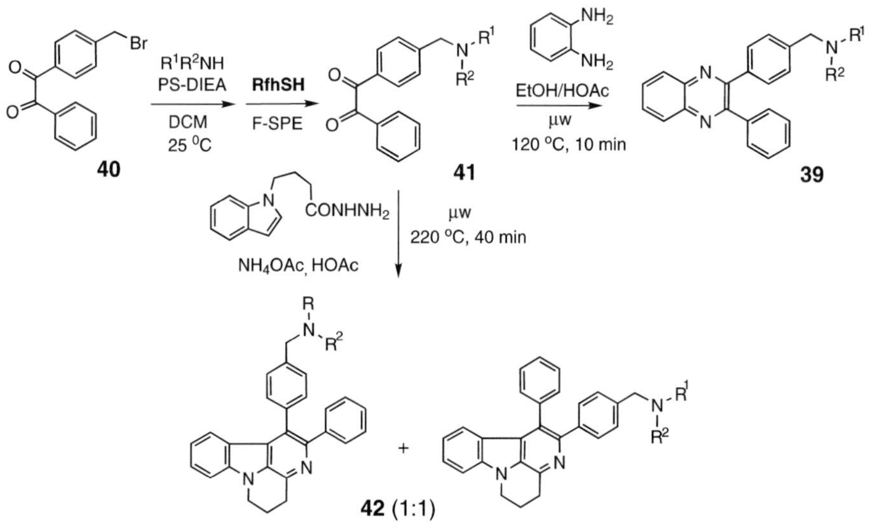

Scheme 24 Synthesis of libraries **39** and **42**

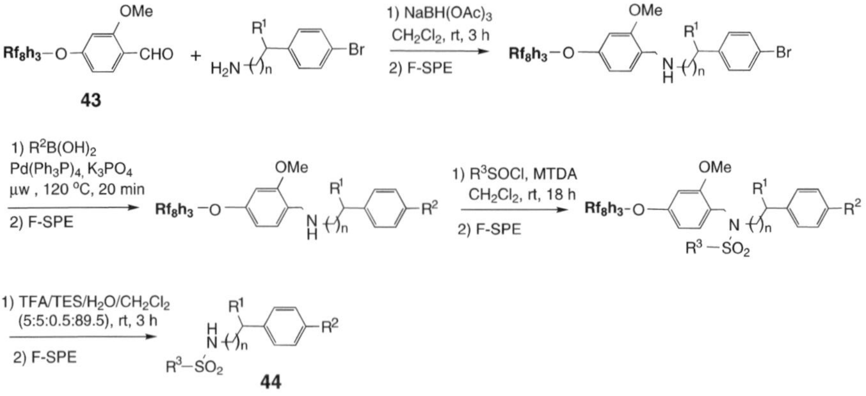

43

1) NaBH(OAc)$_3$
CH$_2$Cl$_2$, rt, 3 h
2) F-SPE

1) R^{2}B(OH)$_2$
Pd(Ph$_3$P)$_4$, K$_3$PO$_4$
μw , 120 °C, 20 min
2) F-SPE

1) R^{3}SOCl, MTDA
CH$_2$Cl$_2$, rt, 18 h
2) F-SPE

1) TFA/TES/H$_2$O/CH$_2$Cl$_2$
(5:5:0.5:89.5), rt, 3 h
2) F-SPE

44

Scheme 25 Synthesis of sulfonamides **44**

products **34** in high selectivity. All the intermediate and final products were purified by F-SPE or crystallization.

In the synthesis of a compound library of allosteric Akt kinase inhibitors **39**, Lindsley and coworkers employed different HTS techniques (Scheme 24) [54]. A polymer-supported base and a fluorous thiol scavenger were used in the alkylation reaction of **40**. F-SPE purified intermediate was then used for microwave-assisted cycloaddition of **41**. Similar intermediates have been used for generation of an unnatural canthine alkaloid library **42** by performing cycloaddition reactions with an indo-tethered acyl hydrazide [55].

Ladlow and coworkers recently developed an acid-labile fluorous benzaldehyde protecting group **43** to facilitate the parallel synthesis of sulfonamides **44** (Scheme 25) [56]. The Suzuki coupling reaction was conducted under microwave irradiation. All the intermediates and the final products were purified by F-SPE.

4
Summary and Outlooks

The integration of microwave heating and fluorous technologies has generated a powerful solution to address both the reaction and separation issues in parallel and combinatorial synthesis. With the further development of multimode microwave reactors for plate reactions and F-SPE for plate-to-plate separations, microwave-assisted fluorous synthesis will play an even more important role in compound library synthesis.

Acknowledgements Parts of author's work were supported by National Institutes of General Medical Sciences SBIR Grants (2R44GM062717-02 and 2R44GM067326-02). The Pd-catalyzed microwave reactions were conducted under license to US patent 6,136,157 and European patent No. 0901453 held by Personal Chemistry, now Biotage.

References

1. Dorwald FZ (2000) Organic Synthesis on Solid Phase. Wiley, Weinheim
2. Burgess K (2000) Solid Phase Organic Synthesis. Wiley, New York
3. Geysen HM, Schoenen F, Wagner D, Wagner R (2003) Nat Rev Drug Discov 2:222
4. Baldino CM (2000) J Comb Chem 2:89
5. Gladysz JA, Horvath IT, Curran DP (eds) (2004) Handbook of fluorous chemistry. Wiley, Weinheim
6. Horvath IT, Rabai T (1994) Science 266:72
7. Studer A, Hadida S, Ferritto SY, Kim PY, Jeger P, Wipf P, Curran DP (1997) Science 275:823
8. Curran DP (1998) Angew Chem Int Ed Eng 37:1175
9. Zhang W (2003) Tetrahedron 59:4475
10. Zhang W (2004) Chem Rev 104:2531
11. Curran DP (2006) Aldrichmica Acta 39:3
12. Zhang W (2004) Curr Opin Drug Disc Dev 7:784
13. Curran DP (2001) Synlett 1488
14. Curran DP (2004) Separation with fluorous silica gel and related materials. In: Gladysz JA, Curran DP, Horvath IT (eds) Handbook of Fluorous Chemistry. Wiley, Weinheim, p 101
15. Zhang W, Lu Y, Nagashima T (2005) J Comb Chem 7:893
16. Chen CH-T, Zhang W (2005) Mol Diversity 9:353
17. Loupy A (ed) (2002) Microwaves in Organic Synthesis. Wiley, Weinheim
18. Kappe CO (2004) Angew Chem Int Ed 43:6250
19. Kappe CO, Stadler A (2005) Microwaves in Organic and Medicinal Chemistry. Wiley, Weinheim
20. Lidstrom P, Tierney J, Wathey B, Westman J (2001) Tetrahedron 57:9225
21. Hayes BL (2002) Microwave Synthesis: Chemistry at the Speed of Light. CEM Publishing, Matthews
22. Swamy KMK, Yeh W-B, Lin M-J, Sun C-M (2003) Curr Med Chem 10:2403
23. Kappe CO (2002) Curr Opin Chem Bio 6:314
24. Al-Obeidi F, Austin RE, Okonya JF, Bond RS (2003) Mini-Rev Med Chem 3:449
25. Lidstrom P, Westman J, Lewis A (2002) Comb Chem High Throughput Screen 5:441
26. Olofsson K, Larhed M (2004) Microwave-assisted fluorous chemistry. In: Gladysz JA, Curran DP, Horvath IT (eds) Handbook of Fluorous Chemistry, chap 10.19. Wiley, Weinheim
27. Zhang W (2005) Chimica Oggi (Chemistry Today) 23:XI–XIII
28. Larhed M, Moberg C, Hallberg A (2002) Acc Chem Res 35:717
29. Larhed M, Hoshino M, Hadida S, Curran DP, Hallberg A (1997) J Org Chem 62:5583
30. Olofsson K, Kim SY, Larhed M, Curran DP, Hallberg A (1999) J Org Chem 645:4539
31. Zhang W, Chen CH-T, Lu Y, Nagashima T (2004) Org Lett 6:1473
32. Zhang W, Lu Y, Chen CH-T (2003) Mol Diversity 7:199
33. Zhang W, Nagashima T (2006) J Fluorine Chem 127:588

34. Zhang W, Nagashima T, Lu Y, Chen CH-T (2004) Tetrahedron Lett 45:4611
35. Vallin KSA, Zhang QS, Larhed M, Curran D, Hallberg A (2003) J Org Chem 68:6639
36. Herrero MA, Wannberg J, Larhed M (2004) Synlett 2335
37. Beeler AB, Acquilano DE, Su Q, Yan F, Roth BL, Panek JS, Porco JA Jr (2005) J Comb Chem 7:673
38. Keaney GF, Wood JL (2005) Tetrahedron Lett 46:4031
39. Zhu J, Bienayme H (eds) (2005) Multicomponent Reactions. Wiley, Weinheim
40. Domling A (2006) Chem Rev 106:17
41. Zhang W, unpublished results
42. Coldham I, Hufton R (2005) Chem Rev 105:276
43. Zhang W, Chen CH-T (2005) Tetrahedron Lett 46:1807
44. Zhang W, Lu Y, Chen CH-T, Curran DP, Geib S (2006) Eur J Org Chem 2055
45. Zhang W, Lu Y, Geib S (2005) Org Lett 7:2269
46. Lu Y, Zhang W (2004) QSAR Comb Sci 23:827
47. Zhang W, Tempest P (2004) Tetrahedron Lett 45:6757
48. Nixey T, Tempest P, Hulme C (2002) Tetrahedron Lett 43:1637
49. Zhang W, Nagashima T, unpublished results
50. Olsen J, Seiler P, Wagner B, Fisher H, Tschopp T, Obst-Sander U, Banner DW, Kansy M, Muller K, Diederrich F (2004) Org Biomol Chem 2:1339
51. Slee DH, Bhat AS, Nguyen TN, Kish M, Lundeen K, Newman MJ, McConnell SJ (2003) J Med Chem 46:1120
52. Horton DA, Bourne GT, Smythe ML (2003) Chem Rev 103:893
53. Nagashima T, Zhang W (2004) J Comb Chem 6:942
54. Lindsley CW, Zhao Z, Leister WH, Robinson RG, Barnett SF, Defeo-Jones D, Jones RE, Huber HE (2005) Bioorg Med Chem Lett 15:761
55. Lindsley CW, Bogusky MJ, Leister WH, McClain RT, Robinson RG, Barnett SF, Defeo-Jones D, Hartman GD (2005) Tetrahedron Lett 46:2779
56. Villard A-L, Warrington BH, Ladlow M (2004) J Comb Chem 6:611

Top Curr Chem (2006) 266: 167–198
DOI 10.1007/128_067
© Springer-Verlag Berlin Heidelberg 2006
Published online: 30 August 2006

Microwave-Accelerated Synthesis of Protease Inhibitors

Johan Wannberg[1] · Karolina Ersmark[2] · Mats Larhed[3] (✉)

[1]Department of Medicinal Chemistry, Organic Pharmaceutical Chemistry, BMC,
 Uppsala University, Box 574, 75123 Uppsala, Sweden

[2]Department of Chemistry, Université de Montréal, C.P. 6128, Succursale Centre-Ville,
 Montréal, Québec Canada

[3]Department of Medicinal Chemistry, Organic Pharmaceutical Chemistry, BMC,
 Uppsala University, Box 574, 75123 Uppsala, Sweden
 mats@orgfarm.uu.se

Abstract Controlled microwave heating has recently emerged as an enabling and productivity-enhancing tool for the medicinal chemist. With this superheating method, reaction times can often be reduced from days and hours down to minutes, and chemistry previously considered impractical or unattainable can be accessed. In this review, the search for new protease inhibitors using microwave-assisted small-scale organic transformations is presented, with a special focus on the development of inhibitors of the aspartic proteases plasmepsin I, plasmepsin II, and HIV-1 protease. A series of rapid lead-optimizations starting from transition-state mimicking core scaffolds using mainly microwave-accelerated, palladium-catalyzed coupling reactions are presented. Biological background and test results are summarized.

Keywords Cross-coupling · Heck · Microwave · Protease inhibitors · Sonogashira

Abbreviations

ADME	Absorption, distribution, metabolism, excretion
AIDS	Acquired immunodeficiency syndrome
Ala	L-Alanine
APC	Antigen-presenting cell
Asp	L-Aspartic acid
AZT	$3'$Azido-$2',3'$-dideoxythymidine

Cat   Cathepsin
CoMFA   Comparative molecular field analysis
dba   Dibenzylideneacetone
DME   1,2-Dimethoxyethane
EC_{50} Inhibitor concentration that reduces the cytopathic effect of the virus by 50%
ED_{50} Inhibitor concentration at which 50% of a given cell system is inhibited
FDA   US Food and Drug Administration
HAART   Highly active antiretroviral therapy
HIV   Human immunodeficiency virus
K_i Inhibition constant
Leu   L-Leucine
Met   L-Methionine
MMP   Matrix metalloproteinase
MW   Microwaves
Phe   L-Phenylalanine
Plm   Plasmepsin
Pro   L-Proline
SAR   Structure–activity relationship
Tyr   L-Tyrosine

1
Introduction to Microwave Chemistry

From its humble beginnings in the middle of the 1980s, microwave-heated organic synthesis has evolved rapidly in the last 10 years [1–3]. First being seen as a curiosity, the technology is now conquering new ground at a steady pace and has emerged as a productivity-enhancing tool for the medicinal chemist [4, 5]. The popularity has grown in parallel with the improvements in performance, control, reproducibility, and safety of the dedicated microwave synthesizers developed in recent years [6, 7]. Reaction times can often be reduced from hours to minutes or seconds and chemical transformations previously considered impractical or unattainable can be smoothly accessed. Competition between the manufacturers has also led to a reduction in instrument prices. If this trend continues, microwave equipment will probably become the standard tool or technique for heating small-scale organic reactions both in industry and academia. In this review, we illustrate the impact of microwave heating on drug discovery by presenting case studies and examples in the development and optimization of different classes of protease inhibitors, with a special focus on our own efforts in the areas of malarial plasmepsin inhibitors and HIV-1 protease inhibitors. After a brief background on microwave heating and aspartic protease inhibitors, selected examples including biological test data will be presented.

Electromagnetic radiation with a frequency of 0.3–300 GHz (λ = 1–0.001 m) is called microwave radiation. The microwave part of the electromagnetic spectrum lies between the more energetic infrared radiation and the less energetic radio waves (Fig. 1). Microwaves are used in radar, satellite communication, in land-based communication links spanning moderate distances (cell phones) and in other applications, including microwave ovens.

Microwaves are able to heat a chemical reaction mixture (or food) by two general mechanisms, dipolar polarization and ionic conductance [8]. All matter that contains dipoles and/or charged species can absorb microwave energy and convert it into heat. This is due to the fact that dipoles and ions are constantly trying to align themselves to the electric component of the oscillating electromagnetic field, resulting in rotation of molecules and oscillation of ions [6, 8]. The electromagnetic energy absorbed in this process is hence first converted to kinetic energy, which is then lost as heat through molecular friction. To achieve efficient heating it is important that the frequency of the applied radiation is within certain limits. If the frequency is too low, the dipoles have time to realign too quickly with the electric field and completely follow the field fluctuations, resulting in poor heating. If it is too high, the dipoles do not have time to realign themselves at all to the alternating field, which means no motion is created and therefore no heat is produced. The frequency used by domestic microwave ovens and microwave synthesizers (2.45 GHz, $\lambda = 12.2$ cm) is located between these extremes where dipoles have time to partly realign with the oscillating electric field but are not quite able to follow the field fluctuations. The result is effective generation of heat.

The microwave story in the drug discovery area began in the late 1990s, although a number of pioneers had started using microwave irradiation for organic synthesis much earlier [3, 7, 8]. In terms of personal dynamics in the integration of microwave technology with medicinal chemistry, important contributions were made by Anders Hallberg, who emphasized early the potential of combining microwave heating and transition-metal catalysis in the optimization of different classes of lead molecules [4, 9, 10]. When the benefits of combining microwave heating with solid phase chemistry and polymer-supported reagents were recognized, these concepts were investigated in detail, especially by the groups of Oliver Kappe [11] and Steven Ley [12].

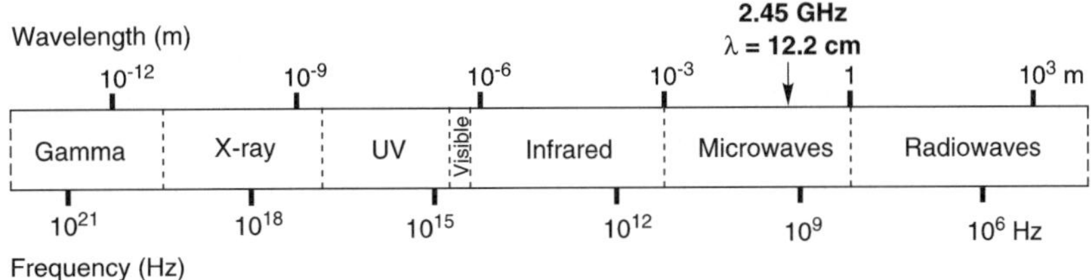

Fig. 1 The electromagnetic spectrum

The generally accepted advantages of microwave-assisted organic synthesis compared to conventional heating techniques are speed, convenience, and energy efficiency. If at least one component of the reaction mixture can interact with microwaves, a very high heating rate can be accomplished [13]. A consequence of the direct bulk heating generated by microwave irradiation is an energy-efficient and uniform heating of the whole reaction system. In conventional heating, energy must first be transferred from the heat source to the wall of the reaction vessel and then to the reaction medium. The temperature gradient created can lead to so-called wall effects, for example, catalyst deactivation on the hot vessel wall [13]. Solvents irradiated by microwaves in open vessels can generally be heated well above their boiling points at atmospheric pressure [1]. This is a result of the heating being faster than the convection to, and loss of heat from, the surface of the solvent/reaction mixture. A higher temperature is reached before bubbles form. This superheating alone (up to 20 °C above the boiling point) can increase reaction rates considerably. One should remember that according to the rate law, a 10 °C increase in reaction temperature will roughly double the reaction rate. Today, most microwave reactions are performed sequentially in septum-sealed reaction vessels. If closed vessel conditions are used the "pressure cooker" effect can lead to even more dramatic rate enhancements.

When dedicated microwave synthesizers were developed, the chemists in the pharmaceutical industry were among the first to benefit from the new technology. Time is money, so also in drug discovery and development. "Big Pharma" will therefore quickly evaluate any new time-saving tool or technique. To allow rapid fine-tuning of the molecular composition of a lead scaffold, with the ultimate highly desirable goal of obtaining improved biological activity or ADME features, expanded SAR, or promoted delineation of intellectual properties.

Combined with expedient purification techniques (e.g., scavengers, reagents on solid support, and solid phase extraction techniques) [14], microwave-assisted synthesis is leading the way towards genuine high-throughput chemistry that will hopefully ease the chemistry-related bottleneck in the drug development process.

After the first report on the utilization of microwave-promoted reactions in a medicinal chemistry context in 1999 [15], a very large number of successful examples illustrating the power of this high-speed processing technology have appeared. Rather than cover all published medicinal chemistry application examples, even though many of them describe valuable new and classical reactions conducted under microwave irradiation, we have chosen to focus primarily on transition-metal catalyzed coupling reactions. Most of the examples are palladium(0)-catalyzed processes that occur via oxidative addition complexes [16]. Although there are examples of Kumada and Negishi reactions presented, the Suzuki, Stille, and Sonogashira cross-couplings and various types of transition-metal catalyzed carbonylation, cyanation, and Heck reactions have been most frequently applied in the preparation of biologically active compounds. These transformations, we believe, will also be of great importance in future lead optimization programs, where the lack of time-efficient decoration methods of scaffolds and/or privileged structures are a potential obstruction. Furthermore, we chose to restrict the content to include only microwave-enhanced chemistry associated with the syntheses of protease inhibitors. Since the protease inhibitors are frequently structurally complex, the examples presented herein will hopefully give an accurate overview of functional group tolerability and the scope and limitations in general of microwave-accelerated transition-metal catalyzed reactions in medicinal chemistry. This review summarizes efforts reported up to the beginning of 2006 to utilize microwave energy to drive fast synthesis of novel enzyme inhibitors.

2
Microwave-Assisted Synthesis of Protease Inhibitors

2.1
Protease Inhibitors

There are four classic types of proteolytic enzymes: aspartic, serine, cysteine, and metallo proteases. They have all been shown to be validated targets for drug intervention in a wide array of diseases and syndromes, and a number of protease inhibitors are in clinical use [17]. Protease inhibitors prevent an undesired cleavage of a peptide or protein substrate by binding, reversibly or irreversibly, to the active site of the protease. The amino acid side chains of the natural protein substrate (or the inhibitor) are generally referred to as P1, P1′, P2, P2′ etc. with numbering commencing at the scissile bond. The corresponding enzyme subsites are designated correspondingly as S1, S1′, S2, S2′ etc. (Fig. 2) [18].

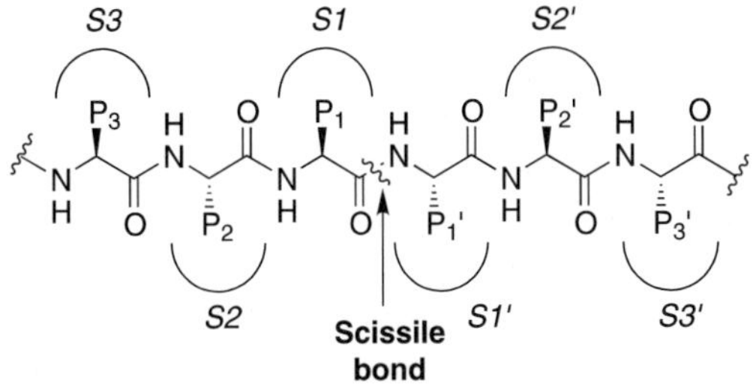

Fig. 2 Numbering of the amino acid side chains of the substrate and the corresponding enzyme binding pockets

2.2
Aspartic Protease Inhibitors

Aspartic proteases are characterized by their ability to cleave peptidic substrates with the aid of two catalytically active aspartic acid residues [19]. A water molecule is activated by one Asp and attacks the scissile amide carbonyl. The other Asp donates a proton to the amide nitrogen, creating a hydrogen-bond stabilized tetrahedral intermediate, which subsequently collapses into the carboxylic acid and amine cleavage products [20, 21].

Historically, renin was the first aspartic protease used as a target enzyme in drug discovery [22]. During the 1970s and 1980s, a lot of effort was devoted towards the development of renin inhibitors as a new class of antihypertensive drugs. During the search for renin inhibitors, effective blockers of enzyme function were discovered when the scissile peptide bonds of the natural substrate were replaced with non-hydrolyzable surrogates. This was found to be especially effective when the new structural elements were mimicking the tetrahedral intermediate of the peptide cleavage. In fact, this "transition-state" isostere strategy has been so successful that it has become the general methodology for the design of almost all aspartic protease inhibitors. The aspartic proteases that have been most frequently targeted so far are renin, the plasmepsins (malaria), the HIV-1 protease, and β-secretase (an enzyme involved in the pathogenesis of Alzheimer's disease) [23].

2.2.1
Plasmepsin Inhibitors

Malaria is a parasite-inflicted disease, which is spread throughout the tropical regions of the world by *Anopheles* mosquitoes. Four species of protozoal parasites of the *Plasmodium* genus cause malaria in humans. Of these, infection by *P. falciparum* is considered the most dangerous with 1–3 million fatal cases annually, the majority being children under the age of five [24]. Parasitic strains resistant to the currently available antimalarial drugs are evolving at an alarming rate, indicating an acute need for new therapeutic agents. The solving of the *P. falciparum* genome has revealed a number of promising new drug targets for treatment of malaria [25]. For example, the enzymes used by the parasites to degrade hemoglobin into smaller peptides and amino acids have been investigated as potential antimalarial targets. The two aspartic proteases plasmepsin I and II (Plm I and Plm II) have been the most extensively studied and a number of inhibitors comprising different transition-state mimics have been reported [26–28]. In the design of plasmepsin inhibitors, substrate selectivity is of utmost importance and particular care has to be taken to avoid inhibition of the closely related human enzyme cathepsin D (Cat D).

2.2.1.1
Hydroxyethylamines

The design of the first series of microwave-synthesized plasmepsin inhibitors was inspired by the work of Silva et al. [29] and Haque et al. [30] and incorporated hydroxyethylamine transition-state bioisosteres [31]. The basic secondary amine was intended to promote an accumulation of the inhibitors in the acidic food vacuole of the parasites. In this series, different P1′ side chains were evaluated and an important part of the work was the decoration of the *meta*- and *para*-positions of benzylic side chains. Inhibitors encompassing either a *para*-bromo or a *meta*-triflate group in the P1′-position served as starting materials in microwave-assisted Suzuki reactions with four different arylboronic acids, producing a total of eight inhibitors after 20 min of irradiation, although in moderate yields (Scheme 1). The *para*-substituted compounds turned out to be the most potent inhibitors of the series, displaying K_i values down to 63 nM for Plm I and 117 nM for Plm II, with high to moderate selectivity versus Cat D (Scheme 1). One compound in this series also exerted a relatively high activity in malaria-infected red blood cells (ED_{50} = 1.6 µM).

The results encouraged us to further optimize this class of inhibitors. Using a sequential high-throughput approach, four new series with a total of 35 novel compounds comprising different combinations of P3 and P1′ side chains were synthesized [32]. Of these, 25 compounds were synthesized applying microwave-promoted Suzuki couplings in the final step (one microwave-enhanced Negishi coupling was also conducted). The best Suzuki couplings were obtained using Na_2CO_3 as base in a DME/ethanol cock-

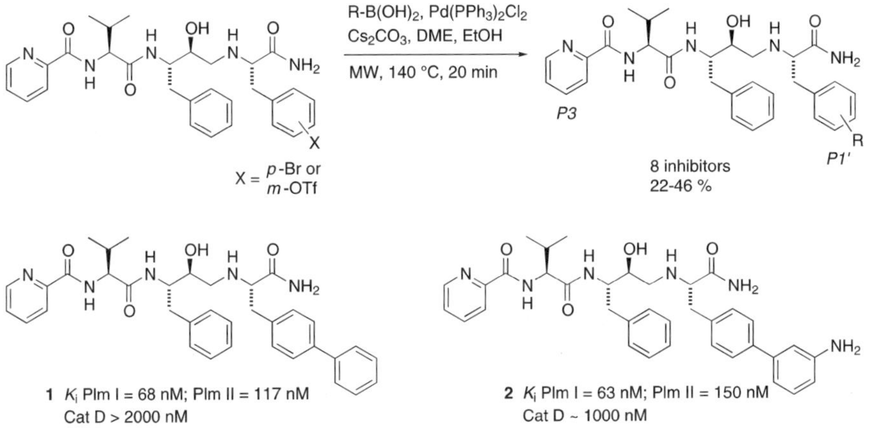

1 K_i Plm I = 68 nM; Plm II = 117 nM
Cat D > 2000 nM

2 K_i Plm I = 63 nM; Plm II = 150 nM
Cat D ~ 1000 nM

Scheme 1 Microwave-assisted Suzuki reactions and inhibition constants of hydroxyethylamine derivatives with different P1′ substituents

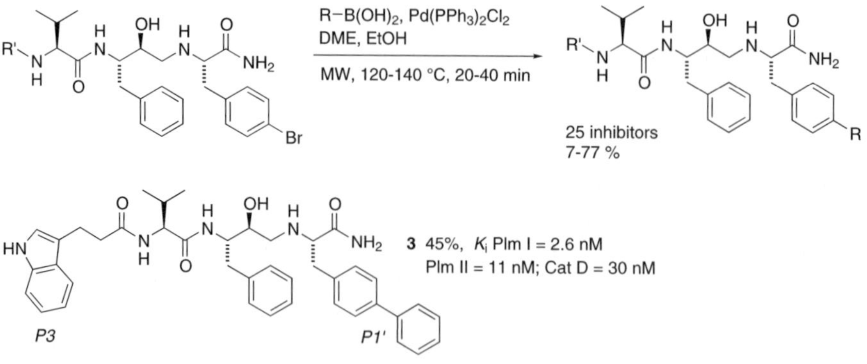

25 inhibitors
7–77 %

3 45%, K_i Plm I = 2.6 nM
Plm II = 11 nM; Cat D = 30 nM

P3

P1′

Scheme 2 Microwave-assisted synthesis of highly potent hydroxyethylamine plasmepsin inhibitors

tail, yielding up to 77% of the products after careful LC-MS purification (Scheme 2). In this study, several compounds were identified that combined Plm I and II inhibition in the low nanomolar range with a decent selectivity versus Cat D.

In an additional attempt to study the effect of P1′ modifications, compounds with the P1′ side chain attached to the nitrogen atom of the central hydroxyethylene core were investigated [33]. This alteration generated completely different structure–activity relationships, suggesting a different binding mode of the P1′ side chain to the enzyme. For example, an inhibitor with an *N*-4-phenylbenzyl P1′ side chain, synthesized from the corresponding bromide using a microwave-heated Suzuki protocol, was completely inactive whereas compounds bearing a smaller *N*-substituted P1′ group were found to be quite active as plasmepsin inhibitors. Computational docking and 3D-QSAR studies suggested that the *N*-substituted inhibitors bind to a different region of the S1′ pocket than the inhibitors where the side chain is placed on the carbon. This was confirmed by the synthesis of inhibitors with two apparent P1′ substituents (e.g., **4**, Scheme 3). These showed moderate activity, demonstrating the enzyme's ability to accommodate both P1′ side chains.

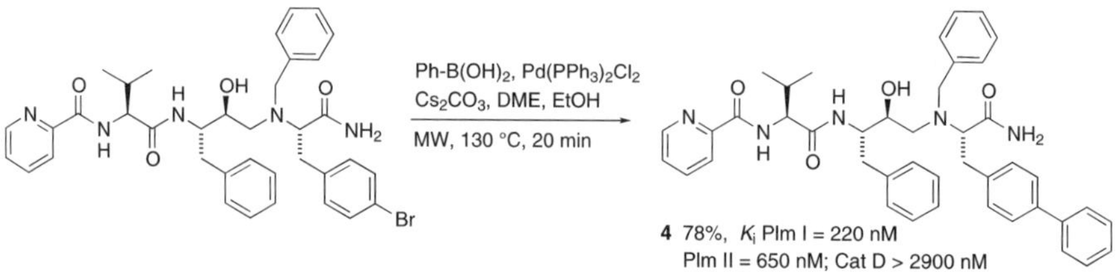

4 78%, K_i Plm I = 220 nM
Plm II = 650 nM; Cat D > 2900 nM

Scheme 3 Microwave-assisted synthesis of *N*-substituted hydroxyethylamine-based plasmepsin inhibitor

2.2.1.2
1,2-Dihydroxyethylenes

Plasmepsin I and II inhibitors containing the 1,2-dihydroxyethylene transition-state isostere were found to be active against the plasmepsins and inactive against Cat D. Furthermore, the (1S,2R)-1-amino-2-indanol moiety, present in the HIV-1 protease inhibitor indinavir, was found to be well tolerated in the S2 and S2$'$ subsites [34]. In an effort to optimize these types of C_2-symmetric inhibitors, elongations of the P1 and P1$'$ side chains were explored [35]. Previous investigations had demonstrated a flexibility of the Plm II S1$'$ subsite and a large unobstructed S1–S3 cavity, both with potential to accommodate extended P1 and/or P1$'$ side chains [29, 30]. (E)-Bromoallyl ethers were used as handles for microwave-promoted palladium(0)-catalyzed extensions using Suzuki, Heck, and Sonogashira chemistry. The Suzuki couplings were performed at 90 °C for 30 min, providing yields of 33–63% (Scheme 4). The inhibitors were generally more potent than the parent compound, with inhibitors **5** and **6** reaching subnanomolar activities against Plm I and low nanomolar activities against Plm II. Importantly, all compounds in this series, characterized by a four-atom linker between the scaffold and the P1/P1$'$ aromatic rings, demonstrated excellent selectivity versus Cat D.

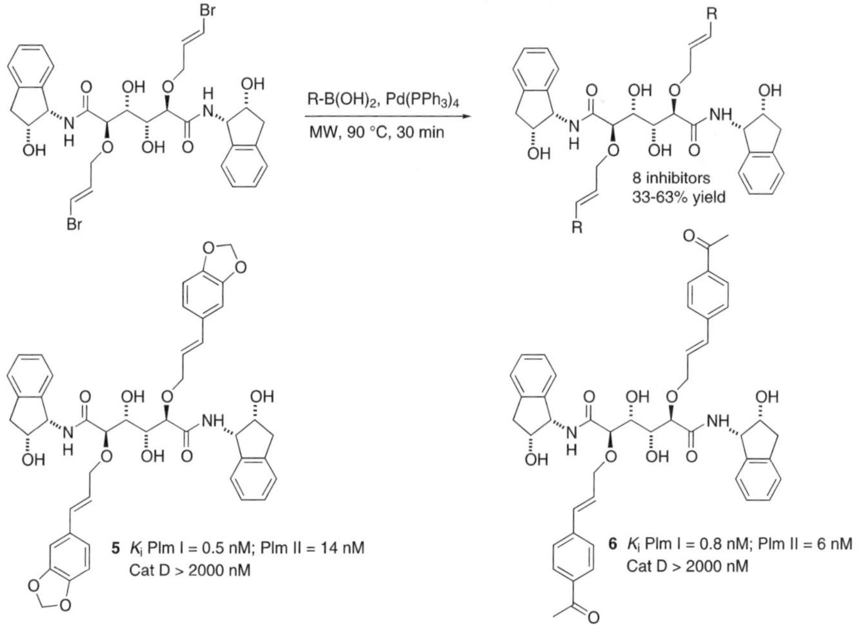

Scheme 4 Microwave-assisted synthesis of C_2-symmetric dihydroxyethylene-based plasmepsin inhibitors

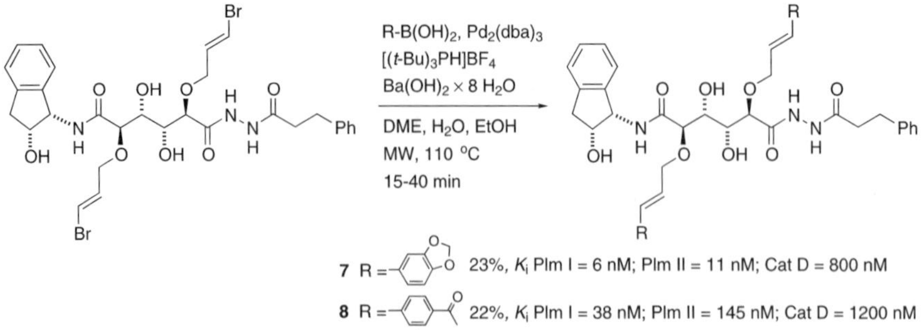

Scheme 5 Microwave-assisted synthesis of unsymmetric dihydroxyethylene plasmepsin inhibitors

In a subsequent study, the impact of replacing one or both of the amide bonds with either a diacylhydrazine or a 1,3,4-oxadiazole was explored [36]. (*E*)-Bromoallyl ethers were employed in the P1 and P1′ positions and similarly subjected to Suzuki conditions using the two aryl boronic acids that gave the most potent inhibitors in the previous study. Potent plasmepsin inhibitors were produced; however, no real improvements in activity compared to the parent compound were observed (Scheme 5). The use of the preligand, [(*t*-Bu₃)PH]BF₄ [37], in combination with barium hydroxide was found beneficial although the yields were very low in these reactions.

2.2.2
HIV-1 Protease Inhibitors

In 1981, several cases of unusual opportunistic infections (e.g., *Pneumocystis carinii* pneumonia) and a rare cancer (Kaposi's sarcoma) in previously healthy homosexual males were reported [38, 39]. Evidently, a decline in the number of circulating $CD4^+$ T cells (T-helper lymphocytes) was causing a degeneration of cell-mediated immunity leading to severe immunodeficiency [40, 41]. As the number of cases grew, the condition was given the name AIDS [42]. Clinical and epidemiological studies soon gathered convincing evidence that the syndrome was caused by an infectious agent, most likely a virus [43]. Extensive research efforts were devoted to the identification and isolation of the etiological agent behind AIDS and in 1983 a French group isolated a new retrovirus, later termed the Human Immunodeficiency Virus, HIV. The HIV virus only infects cells expressing the CD4 glycoprotein on the cell surface [44, 45]. CD4 is primarily found on helper T lymphocytes (T_H or $CD4^+$ T cells) and functions as a receptor that recognizes fragments of antigens displayed by antigen-presenting cells (APCs). Other cell types that express CD4 to a lesser extent are macrophages and dendritic cells.

Once HIV was identified as the causative agent of AIDS, an intense search for effective anti-AIDS drugs began. The first logical move was to target the conversion of the viral RNA genome to proviral DNA by the viral enzyme reverse transcriptase. A screening of compounds intended for other purposes identified the first effective antiretroviral drug, zidovudine (AZT). This nucleoside analog was originally intended as an anticancer drug but the screening also identified it as an effective inhibitor of reverse transcription. Zidovudine was the first anti-AIDS drug to be approved by the US Food and Drug Administration (FDA) in 1987 [46, 47].

As the research into the mechanisms of retroviral replication has progressed, compounds aimed at virtually all parts of the viral replicatory cycle have been evaluated and a number of valid drug targets have been identified [48].

One very highly successful strategy in fighting HIV/AIDS has been to inhibit the HIV protease enzyme [21, 49–53]. This enzyme is responsible for the cleavage of large polyproteins into functional enzymes and structural proteins in the last stage of the viral replication and is essential for the production of new mature virions. Once the three dimensional structure of the HIV-1 protease was known, the development of inhibitors progressed rapidly. The extensive knowledge gained from previous efforts towards renin inhibitors (e.g., knowledge about transition-state mimics) and the identification of the HIV protease cleavage sites (Tyr | Pro, Phe | Pro, Leu | Ala, Met | Met, Phe | Tyr, Phe | Leu, and Leu | Phe) contributed to the first generation of highly potent inhibitors designed towards the HIV-1 protease [21]. Rational, iterative drug design and development processes supported by structural studies (X-ray crystallography) and molecular modeling resulted in the first three HIV-1 protease inhibitors, approved by the FDA in short succession between December 1995 and March 1996. So far (March 2006), eight HIV protease inhibitors are in use (nine counting the prodrug fosamprenavir), all with a central hydroxyl-containing transition-state isostere unit (Fig. 3).

The introduction of HAART (highly active antiretroviral therapy) in 1996 (initially a triple combination of one protease inhibitor and two nucleoside reverse transcriptase inhibitors) resulted in a sharp decline in HIV/AIDS-related morbidity and mortality across North America [54] and Europe [55]. Plasma viral loads could now routinely and continuously be suppressed below the detectable limit in individual patients. This has led many to reevaluate HIV infections from a terminal to a chronic-but-manageable condition.

Unfortunately, there are a number of sometimes interrelated factors that can contribute to a suboptimal response to the treatment regimen (treatment failure), for example, poor adherence to the prescribed drug regimen, side-effects and toxicity [56], suboptimal potency of the regimen, pharmacokinetic problems (e.g., poor absorption and/or fast metabolism), and the emergence of drug-resistant viral strains [57]. Treatment failure generally initially involves an increase in viral load (virologic failure) followed by a de-

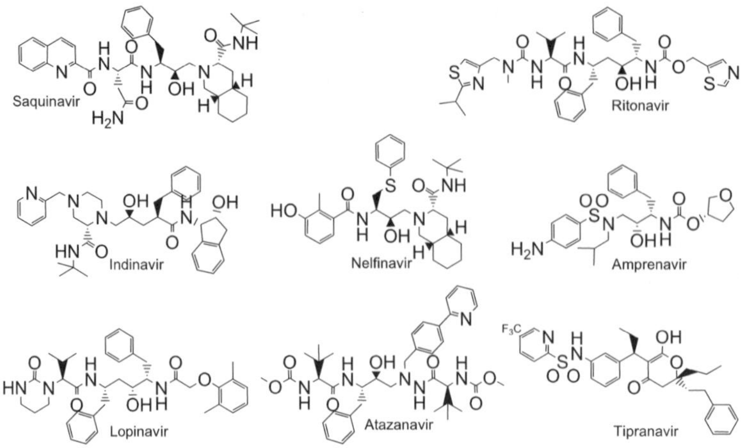

Fig. 3 Approved HIV-1 protease inhibitors

crease in circulating CD4$^+$ T cells (immunologic failure) and finally a clinical progression to AIDS and ultimately death. Despite the success of HAART, the limitations listed above indicate a continuous need for new, inexpensive antiretroviral drugs with fewer and less severe side effects and with unique resistance profiles.

2.2.2.1
Linear 1,2-Dihydroxyethylene-Based Inhibitors

About a decade ago, an HIV-1 protease inhibitor program was initiated in our department with the general aim of allowing fast, convenient, and inexpensive routes to new and potent inhibitors. Many of the inhibitors used in clinics contain a hydroxyethylamine transition-state mimicking group. Our group instead decided to base the inhibitors on suitable carbohydrates, supplying a diol-based transition-state isostere. It was found that a derivative of L-mannitol could easily be converted in a short synthetic route to highly potent C$_2$-symmetric, C-terminally duplicated inhibitors with a 1,2-dihydroxyethylene core functioning as a transition-state mimic (Scheme 6) [58].

Inspection of 3D-structures of enzyme–inhibitor complexes and molecular modeling indicated that substitution of the *para*-position of the P1/P1$'$ benzyloxy side chains, with the substituents projecting from the S1/S1$'$ through the S3–S3$'$ sites and towards the end of the active site, could provide productive interactions with the enzyme and probably increase the binding affinity. The *para*-bromo substituted inhibitor **11** was recognized as a promising starting material for simultaneous, double palladium(0)-catalyzed coupling reactions (Scheme 7). Thus, ten new compounds of varying size and polarity were produced by microwave-heated Heck, Stille, and Suzuki coupling reactions [15, 59, 60]. The reactions were performed using an early single-mode microwave synthesizer without temperature control. The microwave

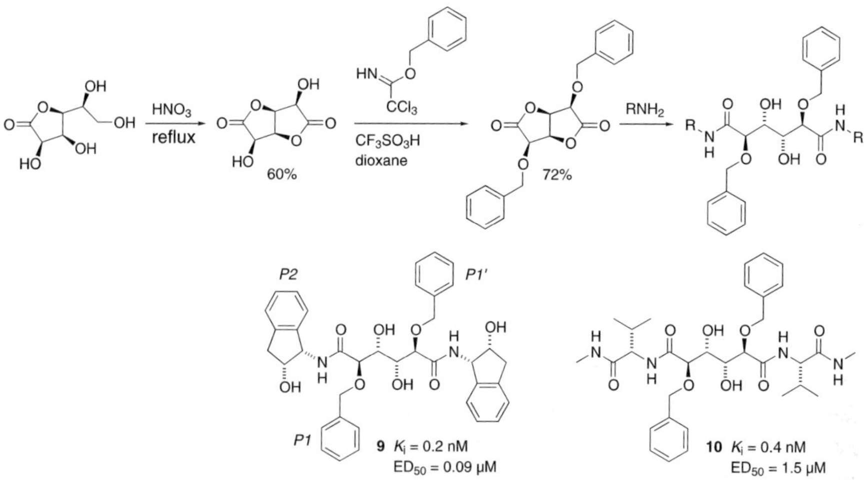

Scheme 6 Short synthetic route towards C_2-symmetric HIV-1 protease inhibitors

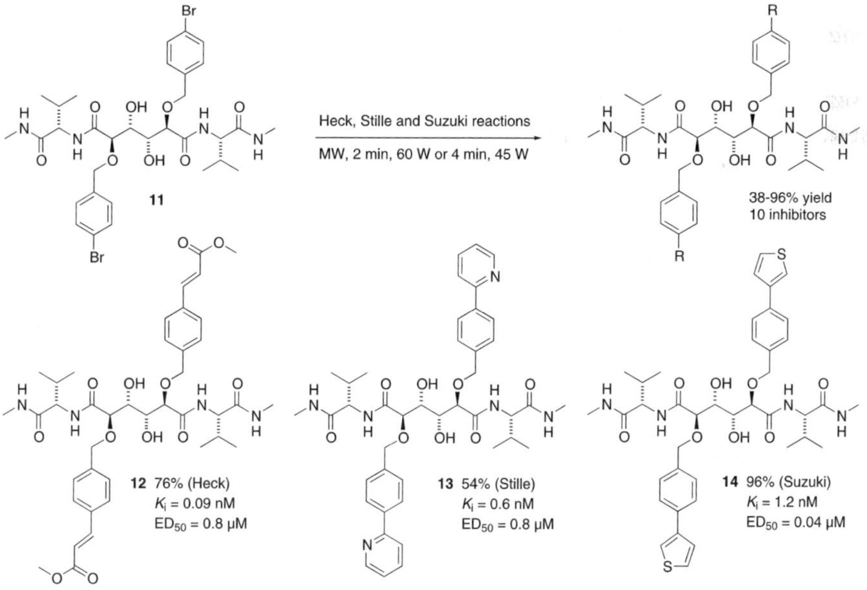

Scheme 7 Microwave methods for preparation of C_2-symmetric HIV-1 protease inhibitors modified in P1/P1′, enzyme inhibition, and antiviral activity in cell culture

settings used were a power of 45 or 60 W and the reaction times were 2 or 4 min. The synthesized compounds were in general highly potent inhibitors of the HIV-1 protease ($K_i = 0.09$–3.8 nM) and this was also in many cases reflected in impressive activities in a cell-based assay with ED_{50} values down to 0.04 µM [15].

The Uppsala team then turned their attention to the development of a fast and efficient procedure that allowed conversion of aryl halides to benzonitriles and aryl tetrazoles by microwave heating. The aryl bromide **11** was subjected to the optimized reaction conditions, which delivered the bistetrazole **15** in a two-step yield of 82% (Scheme 8, $K_i = 0.56$ nM), demonstrating the potential of the method for medicinal chemistry [61]. The tetrazole functionality is one of the most utilized carboxylic acid bioisosteres.

X-Ray structures of enzyme–inhibitor complexes revealed that the *para*-substituted benzyloxy side chains occupied both the S1/S1′ and S3/S3′ subsites of the native C_2-symmetric protease and reached water at the boundary of the active site channel. Computer modeling suggested that it might be possible to extend the *meta*-position of the P1/P1′ side chain to afford potent compounds. However, an extension of the *ortho*-position with large groups seemed less likely to furnish good inhibitors. On the other hand, there was still an interest in examining the effect of both *meta*- and *ortho*-substitution in order to get a complete picture of the binding possibilities of the inhibitor scaffold in the S1/S1′ and S3/S3′ subsites. As far as we knew, inhibitors containing large substituents in the *ortho*-positions of aromatic P1 and/or P1′ side chains had not previously been reported.

At the beginning of this decade we developed an aminocarbonylation [62, 63] method using $Mo(CO)_6$ as the carbon monoxide source [64, 65] with the intention of applying this reagent in laboratory-scale medicinal chemistry. A more precise goal was to enable fast generation of new amido-decorated HIV-1 protease inhibitors. The idea was that the microwave-promoted bis-

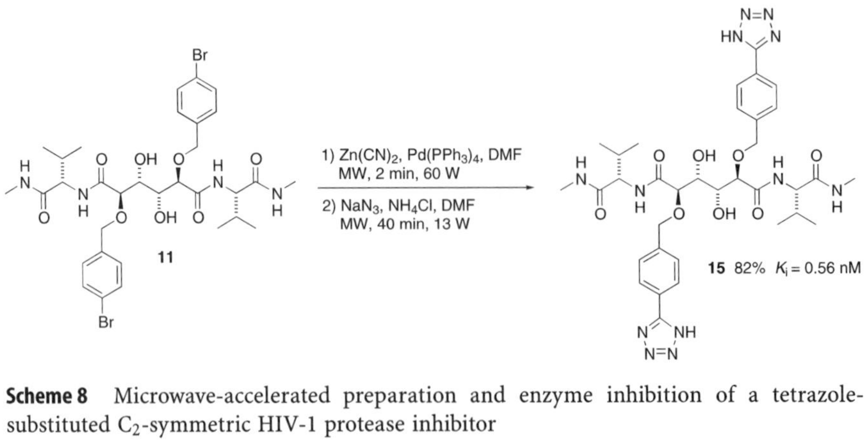

Scheme 8 Microwave-accelerated preparation and enzyme inhibition of a tetrazole-substituted C_2-symmetric HIV-1 protease inhibitor

functionalizations of our C_2-symmetric dihydroxyethylene-based inhibitor scaffold would quickly generate compounds that could help us map the impact of *ortho*- and *meta*-amide substituents on the HIV-1 protease inhibiting capacity. The introduction of two new amide bonds is generally not advisable from a drug development perspective but the ambition was that these compounds would provide information useful for the future design of potent HIV-1 protease inhibitors with unique characteristics.

The aryl halide-containing key intermediates **16** and **17** were prepared in close analogy with the previously published procedure (Scheme 9) [15, 58]. The substitutions of the halogen groups were performed in sealed vessels under non-inert palladium-catalyzed aminocarbonylative microwave conditions. The iodo-substituted **16** was used as precursor in the synthesis of the *ortho*-benzamide series of inhibitors. A total of 14 compounds were isolated in sufficient amounts after aminocarbonylations under ligandless conditions (Pd(OAc)$_2$) and with a large excess of the respective amine and 15 or 30 min of microwave heating at 110 °C [66]. The rather harsh conditions used, compared to those originally presented, were necessary in order to obtain full conversion of **16** under these diluted circumstances. The compounds tested displayed a wide range of activity but showed no evident structure–activity trend.

Aryl bromide **17** was used as precursor in the synthesis of *meta*-benzamides, analogous to the *ortho*-series but using a higher reaction temperature and employing a more advanced catalytic system (Herrmann's palladacycle) [64]. Applying slightly modified conditions for aryl bromides compared to the original procedure allowed for full conversion of **17** within 15 min at 130 °C and seven new inhibitors were prepared [66]. All compounds in the *meta*-series were highly active ($K_i = 2$–300 nM) as predicted. Secondary amides proved to be particularly potent ($K_i = 2$–20 nM).

A couple of *ortho*- and *meta*-substituted structures that were regarded as particularly interesting were chosen for re-synthesis on a slightly larger scale with thorough chemical characterization of the isolated inhibitors. In order to decrease the formation of side products detected in the initial library syntheses, it was decided to protect the central vicinal diols of the aryl halide

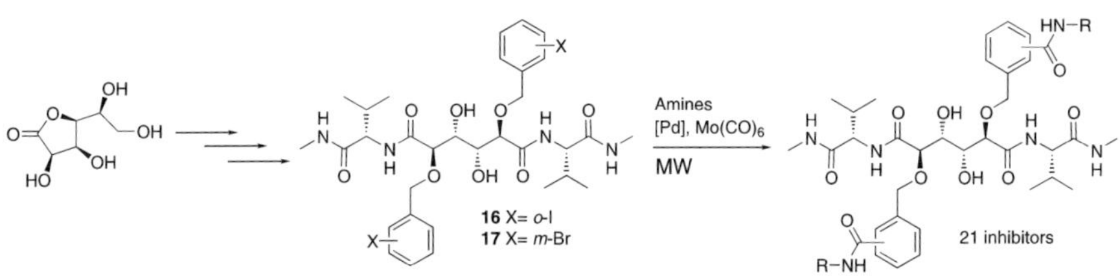

Scheme 9 Rapid synthesis of C_2-symmetric HIV-1 protease inhibitors by in situ aminocarbonylations

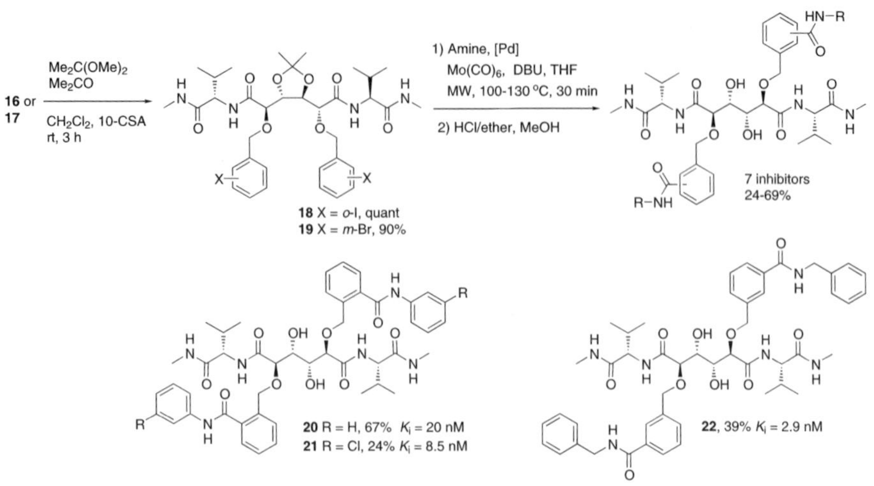

Scheme 10 Carbonylative microwave syntheses of C_2-symmetric HIV-1 protease inhibitors from protected starting aryl halides

starting materials **16** and **17** as acetonides (Scheme 10). This simplified the purification and increased the yields. Also, the larger scale involved higher substrate concentrations and CO pressure, which had a beneficial impact on the reaction.

Seven inhibitors were re-synthesized on a larger scale starting from the 1,3-dioxolanes **18** and **19**. The protected inhibitors generated by the aminocarbonylations were purified by flash chromatography, and subsequent treatment with HCl/ether in MeOH afforded the desired structures in satisfying yields. Reevaluation of the compounds showed a good correlation with the initial results, confirming the high potencies of *ortho*-anilide substituted compounds **20** and **21** [66]. The information obtained in this study suggested that *ortho*-substitution of P1 and/or P1′ benzyl side chains might provide a new binding mode, and thus a promising new approach in the search for unique HIV-1 protease inhibitors.

We intended to further explore the *ortho*-concept by introducing alternative substituents via microwave-accelerated and chemoselective carbon–carbon bond coupling reactions [9]. By avoiding the introduction of amide groups utilized in our previous investigation we wanted to obtain more drug-like inhibitors. To further reduce the peptide character it was decided to also exploit the *N*-(1*S*,2*R*)-2-hydroxy-1-indanyl group, well-known from indinavir, as P2/P2′ substituents, in addition to the valine methylamides. These two operations would lead to target compounds with two rather than six amide bonds (Schemes 11 and 12).

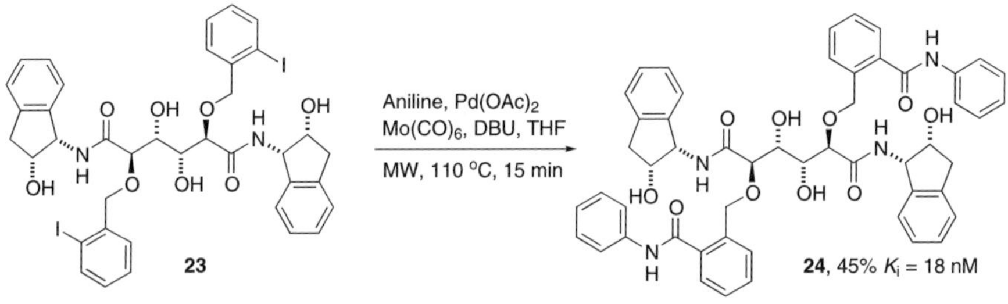

Scheme 11 Microwave-mediated preparation of an indanolamide-containing HIV-1 protease inhibitor

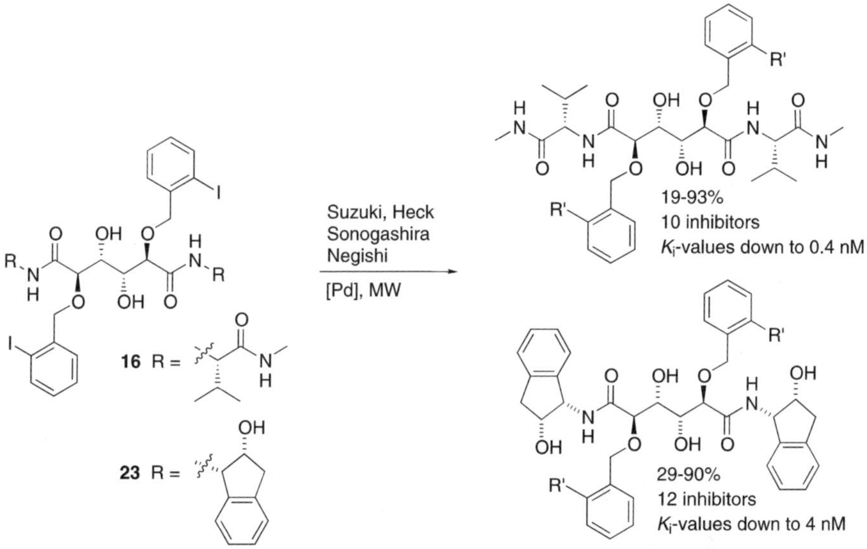

Scheme 12 Palladium-catalyzed microwave synthesis of *ortho*-substituted HIV-1 protease inhibitors

With the aryl iodide substrates **16** and **23** in hand, an abundance of possibilities to explore palladium(0)-catalyzed carbon–carbon bond-forming reactions in order to produce new HIV-1 protease inhibitors was recognized [67].

First, the indanolamino compound **24** was synthesized by aminocarbonylation of unprotected **23** with aniline as nucleophile in order to obtain a basis for comparisons between the compounds containing the valine methylamide group in the P2/P2′ and the indanolamine compounds planned in this series (Scheme 11) [68]. The *ortho*-anilide **24** proved to be a practically equipotent inhibitor to **20** (**20**, $K_i = 20$ nM; **24**, $K_i = 18$ nM), encouraging us to proceed with the synthetic efforts as planned.

In the planning of the carbon–carbon bond couplings, the specific reactants were selected both according to commercial availability and in order to furnish some degree of diversity in the size and flexibility of the isolated P1/P1'-modified inhibitors. In addition, the styrene and benzofuran group of *ortho*-substituents may be viewed as isosteres to the active *ortho*-anilides.

The *ortho*-iodobenzyloxy-containing C-terminal duplicated compounds **16** and **23** served as starting materials for microwave-enhanced palladium(0)-catalyzed carbon–carbon bond-forming reactions [68]. Eight Suzuki couplings were performed at 100–120 °C for 15–30 min, resulting in 19–65% isolated yields of the corresponding *ortho*-derivatives. The six Sonogashira reactions produced yields between 36 and 93% after 15 min at 100 °C or 120 °C. Five terminal and two internal Heck reactions were completed after 5 min at 120 °C and 30 min at 100 °C, respectively. The single Negishi coupling tried did not run to completion, with a yield of 25% after 5 min at 100 °C (Scheme 12).

Two series of inhibitors were synthesized (a total of 23 compounds, including **24**) and highly potent inhibitors equipped with large *ortho*-functionalized P1/P1' side chains as the structural theme were identified. In general, *ortho*-substitutions of the P1/P1' benzyloxy side chains consisting of small lipophilic 1- or 2-carbon linkers connected to aromatic rings provided good inhibitors as compared to the bulky substituted analogs.

Continuing on the carbonylation strategy using microwaves and with $Mo(CO)_6$ as the carbon monoxide source, we developed a convenient protocol for generating primary benzamides from aryl halides using hydroxylamine hydrochloride as an ammonia equivalent [69]. Full conversion of the starting materials was achieved after 20 min of microwave heating at 110 °C or 150 °C for aryl iodides and aryl bromides, respectively, and the corresponding benzamides were obtained in high yields. The conditions identified as appropriate for aryl iodides were exploited for the aminocarbonylation of **23** to the bis-benzamide **25** (Scheme 13). The relatively poor isolated yield of **25** (14%) was a result of purification difficulties due to problems with solubility and contamination by side products. By protection (TBDMS) of two of the

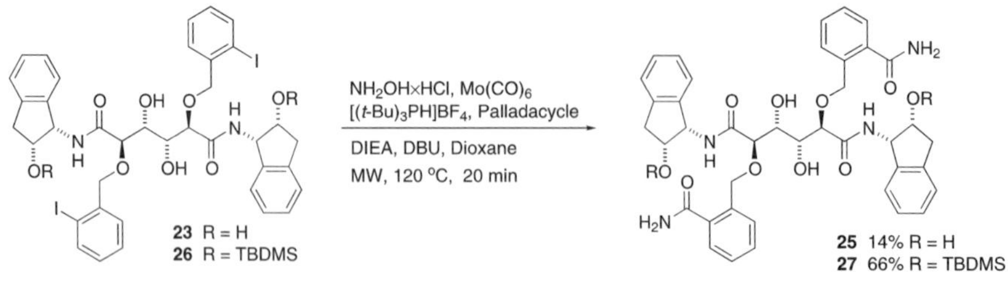

Scheme 13 Facile microwave-assisted synthesis of a primary amide HIV-1 protease inhibitor

hydroxyl groups in the starting material it was possible to decrease the formation of side products and also to improve the solubility of the product. This simplified the isolation of the bis-aminocarbonylated product **25** and after deprotection of **27** with TBAF, the indanol inhibitor **25** could be isolated in a 62% two-step yield (Scheme 13) [69].

2.2.2.2
Tertiary Hydroxyl Group-Containing Inhibitors

We recently reported a promising new class of HIV-1 protease inhibitors based on a transition-state mimic with a tertiary hydroxyl group in α-position to an amide (Fig. 4) [70]. All HIV-1 protease inhibitors in current clinical use contain a secondary hydroxy group as part of the transition-state isostere (Fig. 3). Tertiary alcohols have, up to this point, been poorly studied with few examples in the literature. The type of scaffold shown in Fig. 4 is of particular interest for two main reasons: Firstly, molecular modeling suggested that with the hydroxyl in this position, it could still interact favorably with the catalytic aspartic acids (Asp25/25′) in the enzyme. Secondly, we speculated that the α-hydroxy amide could form intramolecular hydrogen bonds; partially masking these groups and increasing the apparent lipophilicity, thereby improving the transcellular transport and positively influencing absorption properties.

The first series of inhibitors contained structural elements related to atazanavir and indinavir and provided some highly potent compounds, the best one being **28** (K_i = 2.4 nM) [70].

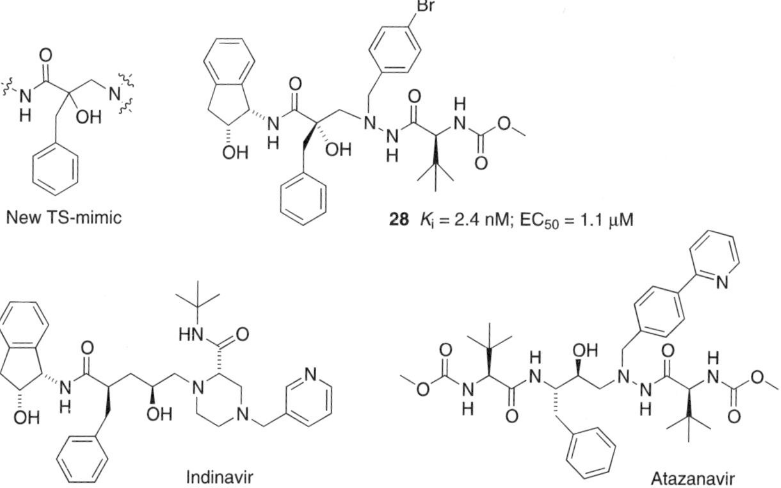

Fig. 4 Structures of the tertiary alcohol-containing HIV-1 protease inhibitor **28** and two approved inhibitors

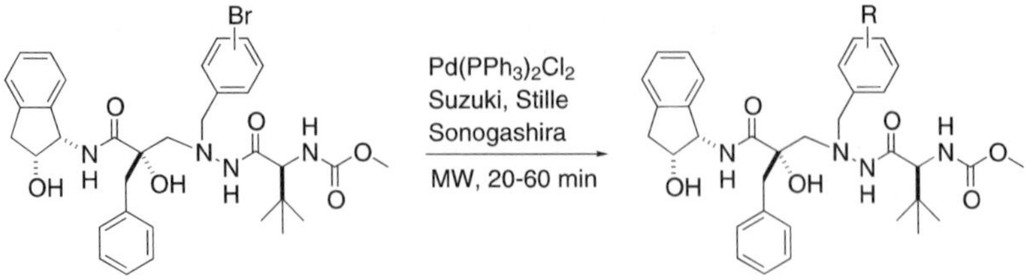

29 *p*-Br K_i = 2.4 nM; EC$_{50}$ = 1.1 µM
30 *m*-Br K_i = 18 nM; EC$_{50}$ = 3.3 µM

20 inhibitors, 17-62%

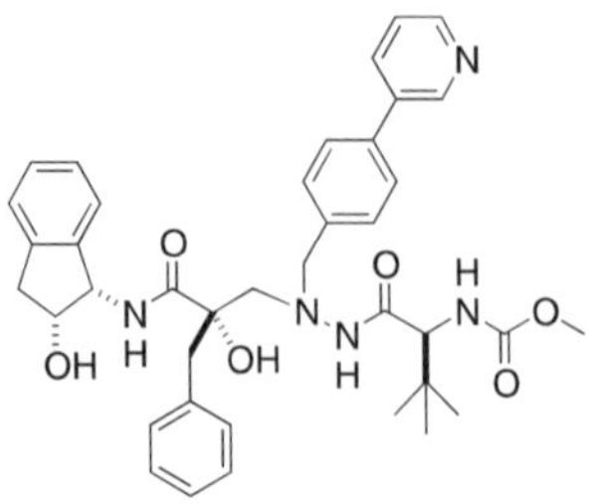

31 27%, K_i = 5.0 nM; EC$_{50}$ = 0.18 µM

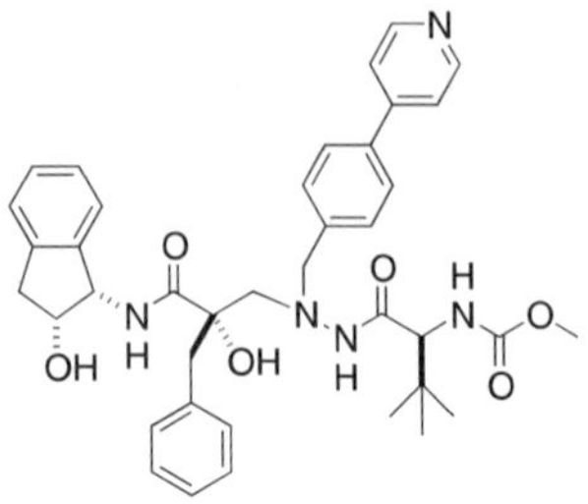

32 17%, K_i = 5.5 nM; EC$_{50}$ = 0.22 µM

Scheme 14 Microwave-promoted synthesis, enzyme inhibition, and antiviral activity of P1′-elongated tertiary alcohol-containing HIV-1 protease inhibitors

However, these compounds were only moderately active in the cell-based assay and **28** was further optimized with regard to the cell results. The *para*-bromophenyl-containing inhibitor **29** and the *meta*-bromophenyl analog **30** were exploited as starting materials for microwave-promoted Stille, Suzuki, and Sonogashira reactions (Scheme 14) [71]. High cellular antiviral potencies were obtained when the P1′ benzyl group was elongated with a 3- or 4-pyridyl substituent (**31** and **32**). The 3-pyridyl compound **31** also exhibited excellent cell permeability and was rapidly distributed across the Caco-2 cell membrane (P_{app} = 33 × 10^{-6} cm/s) [71]. A related inhibitor equipped with the tertiary hydroxyl group was also generated by a palladium-catalyzed aminocarbonylation in water [72].

2.2.2.3
Cyclic 1,2-Dihydroxyethylene-Based Inhibitors

A class of HIV-1 protease inhibitors extensively examined in our department are the cyclic sulfamides [73, 74]. These compounds are related to our linear 1,2-dihydroxyethylene inhibitors described above in that they are also derived from L-mannitol and employ a 1,2-dihydroxyethylene transition-state isostere. These cyclic inhibitors are comprised of seven-membered rings where the sulfonyl oxygens are designed to displace a structural water in the enzyme when the inhibitors bind to the active site [73]. The synthesis of the parent inhibitor **33** is summarized in Scheme 15. The cyclic sulfamides, for example **33**, were shown to bind to the C_2-symmetric HIV protease in an unsymmetric fashion while the structurally related seven-membered symmetric ureas, for example **34**, display a symmetric binding [74, 75]. The design of **34** was guided by the elegant work of Lam et al. for the synthesis of the phenyl alanine-derived corresponding carbanalogs [76]. A large number of inhibitors were synthesized by the synthetic route shown in Scheme 15 and a CoMFA model [77] was developed in an attempt to rationalize the complex structure–activity relationships of this class of compounds [78].

As part of this work, the use of high-speed Suzuki reactions in the decoration of aryl halide containing inhibitors was evaluated. Single-mode microwave irradiation set to a power of 45 W completed the reactions within 4 min and delivered yields of 48–73% [78]. The synthesized inhibitors displayed an affinity towards the HIV-1 protease in the K_i-range of 63 nM to 6.0 μM (Scheme 16).

The results inspired a search for related but simpler inhibitors, preferably with a shorter synthetic route. X-Ray structures of enzyme–inhibitor complexes and computer modeling suggested that appropriate *ortho*-substitutions of the P2/P2′ benzyl groups would make it possible to span the S1/S1′ and S2/S2′ pockets and thereby get interactions with two binding sites with one side chain [79]. This might make it possible to remove the P1/P1′ side chain completely. Starting from L-dimethyltartrate, a four-step synthesis of a new cyclic sulfamide scaffold was developed (Scheme 17). The protected aryl bromide **37** was utilized as a precursor in palladium(0)-catalyzed, microwave-heated decorations. Heck, Negishi, and Suzuki reactions were used to examine the length and type of spacers that would allow for the op-

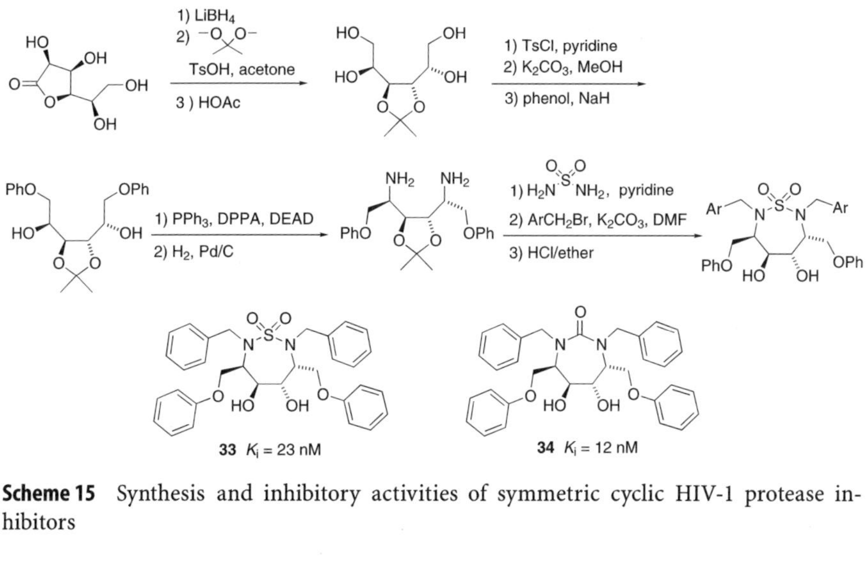

Scheme 15 Synthesis and inhibitory activities of symmetric cyclic HIV-1 protease inhibitors

timal interactions of the aromatic rings with both the S1/S1′ and S2/S2′ sites of the enzyme (examples are shown in Scheme 17) [79]. Disappointingly, the inhibitors were only moderately active and the different linker lengths produced compounds that were roughly equipotent.

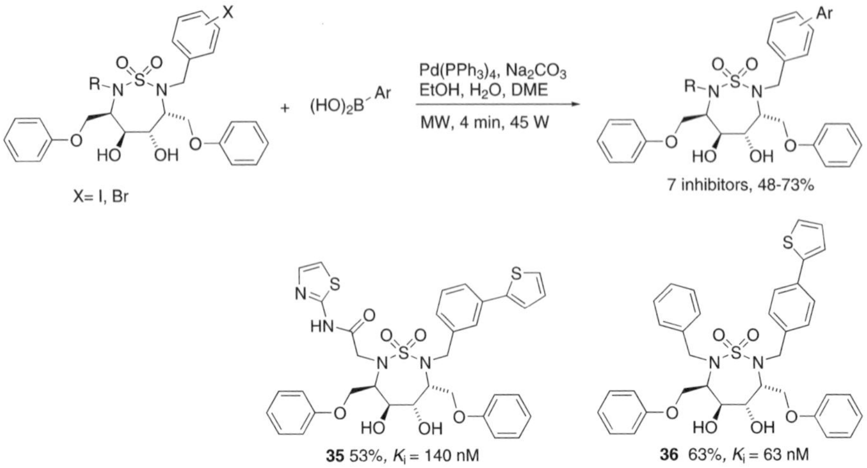

Scheme 16 Microwave-enhanced Suzuki couplings and inhibitory activities of unsymmetric HIV-1 protease inhibitors

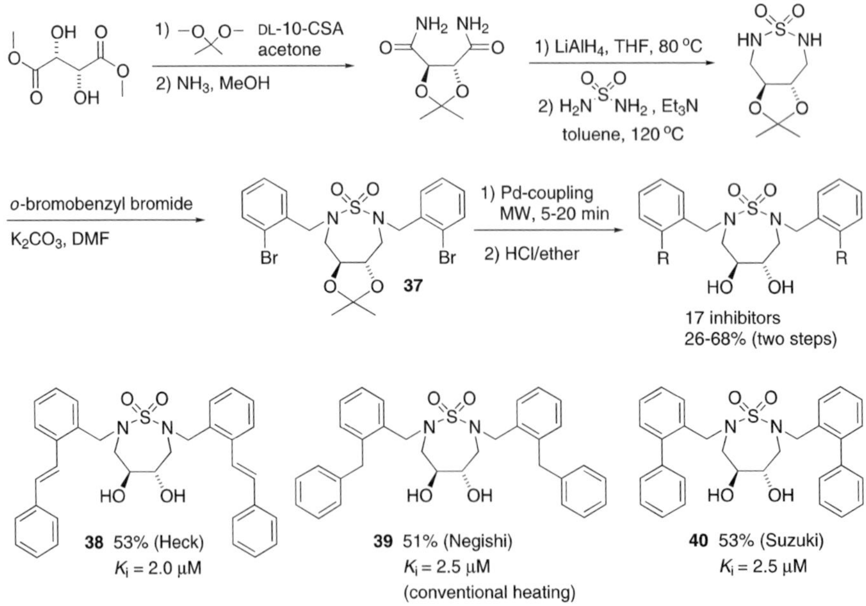

Scheme 17 Microwave-generated symmetric HIV-1 protease inhibitors

A Suzuki strategy was used for further optimization. A set of boronic acids were chosen, aiming at moderately lipophilic building blocks of varying size that would produce inhibitors with a molecular weight < 700 g/mol. A total of 16 inhibitors were synthesized by this methodology and the reaction conditions used to obtain one of the compounds, the benzofuran derivative **41**, are shown in Scheme 18. This compound was the most potent of the series with the furan oxygens suggested to interact favorably with the backbone NH of Gly48 of the protease [79].

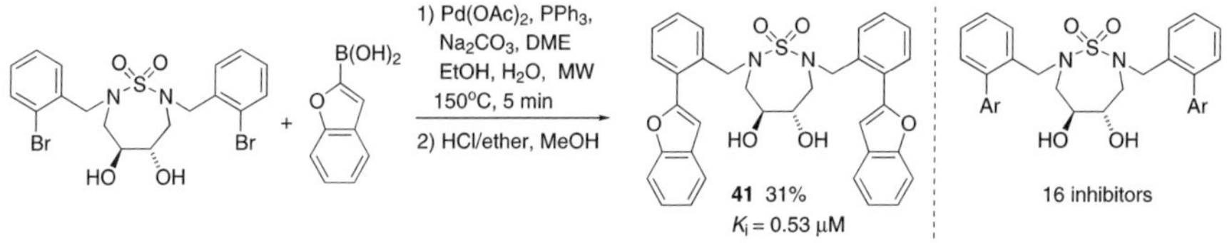

Scheme 18 Optimization of cyclic sulfamide HIV-1 protease inhibitors

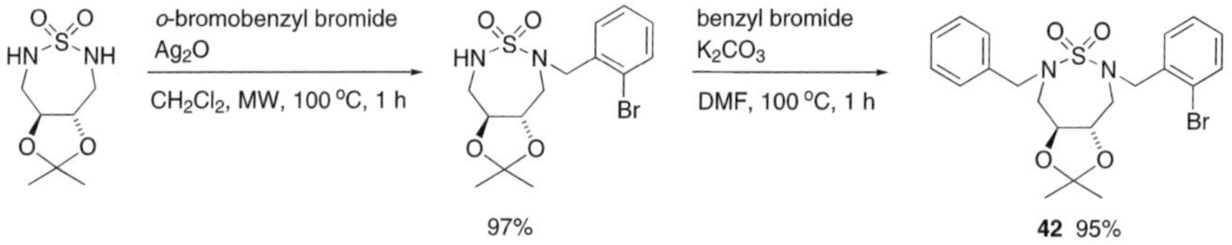

Scheme 19 Microwave-assisted, silver oxide-promoted monobenzylation

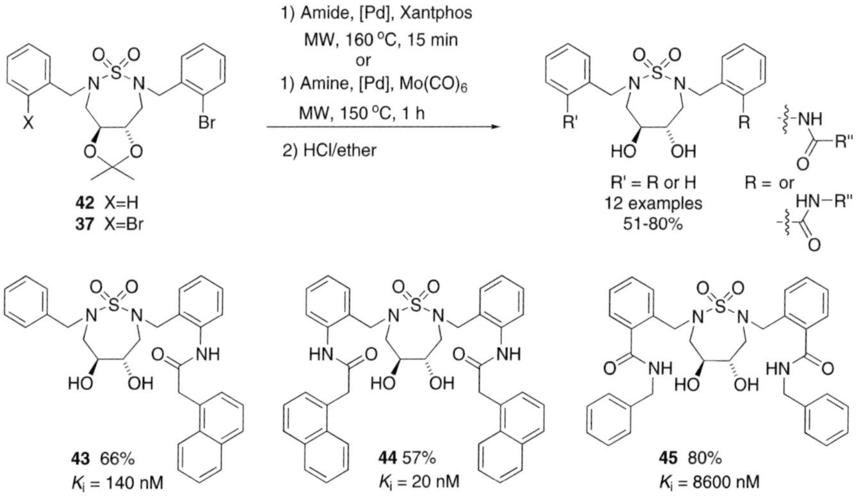

Scheme 20 Generation of cyclic HIV-1 protease inhibitors by microwave-heated amidation reactions

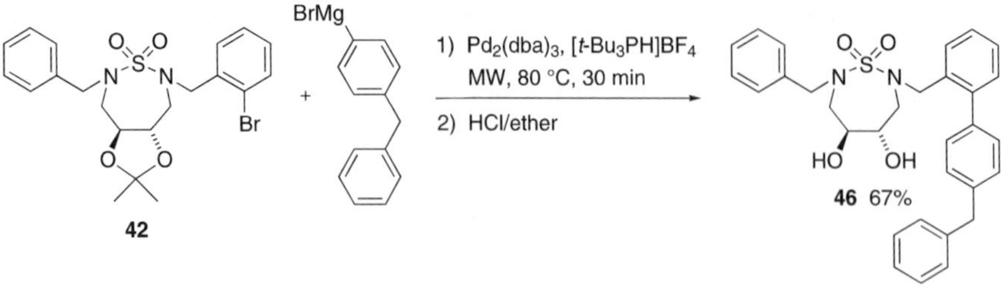

Scheme 21 Microwave-assisted Kumada cross-coupling

An efficient microwave procedure for selective monobenzylation of the same seven-membered scaffold was also developed, as exemplified with **42** in Scheme 19 [80].

This compound, as well as the symmetric **37**, was used in a series of microwave-heated aminocarbonylations and direct amide *N*-arylations, producing 12 compounds (six unsymmetric and six symmetric) with a two- or three-atom spacer between the aromatic moieties (Scheme 20) [80]. The best unsymmetric inhibitor had a K_i of 140 nM (**43**) and the best symmetric inhibitor 20 nM (**44**).

In a separate project, methods for microwave-assisted generation of Grignard reagents from aryl chlorides and aryl bromides were developed [81]. One of the aryl magnesium bromides generated was used with the unsymmetric **42**, 2% $Pd_2(dba)_3$ and 4% $[t\text{-}Bu_3PH]BF_4$ in a microwave-promoted Kumada coupling to provide **46** in 67% isolated yield, a compound unfortunately found to be a relatively poor HIV-1 protease inhibitor (Scheme 21).

2.3
Serine Protease Inhibitors

Serine proteases are involved in many important physiological processes in man, including blood coagulation, immune response, and digestion. The serine proteases are characterized by a Ser-His-Asp catalytic triad. Peptide cleavage starts with a nucleophilic attack by the serine hydroxyl on the scissile peptide bond carbonyl, forming an acyl–enzyme complex and releasing the C-terminal amine cleavage product. A water molecule then hydrolyzes the acyl–enzyme complex with release of the acid cleavage product. Enzymes of this class can be further divided into trypsin-like, elastase-like, or chymotrypsin-like serine proteases.

The chymotrypsin-like HCV NS3 protease is an essential enzyme in the life cycle of the hepatitis C virus. An estimated 3% of the world population is infected by hepatitis C and it is the leading cause of chronic liver disease and the primary indication for liver transplantation [82]. The HCV NS3 protease is therefore considered an important pharmaceutical target.

Within an HCV NS3 protease inhibitor program at Uppsala University, some highly potent compounds have been developed. As a part of this program we wanted access to intermediates of the type represented by compound **49** (Scheme 22). In the final inhibitors, the acylsulfonamides would act as carboxylic acid bioisosteres. We envisaged that carbonylation of the aryl bromide **47** using a primary sulfonamide as the nucleophile would be beneficial (Scheme 22) [83]. Again we exploited microwave heating and $Mo(CO)_6$ as the source of carbon monoxide in the development of a new amidocarbonylation method. The arylpalladium precursor **47** was then reacted with methylsulfonamide under the optimized microwave conditions, producing the target acylsulfonamide **48** in an isolated yield of 52%. The building block was subsequently coupled with a second building block to produce the complete inhibitor **49** in 76% yield. When evaluated in an in vitro assay comprising the full-length NS3 protein, **49** was demonstrated to be a highly potent inhibitor with a K_i value of 85 ± 7 nM [83].

In addition to the use in the synthesis of potential hepatitis C drugs, microwave-assisted chemistry has also been used in the synthesis of mast cell tryptase inhibitors, thrombin inhibitors, and Factor Xa inhibitors. The trypsin-like serine protease tryptase is the major secretory product of human mast cells and has been implicated as an inflammatory mediator in a number of conditions, especially asthma. Once released upon mast cell activation, the tryptase cleaves substrates that otherwise cause smooth muscle relaxation and thereby bronchi- and vasodilation. It is therefore not surprising that numerous reports on low molecular weight tryptase inhibitors have appeared.

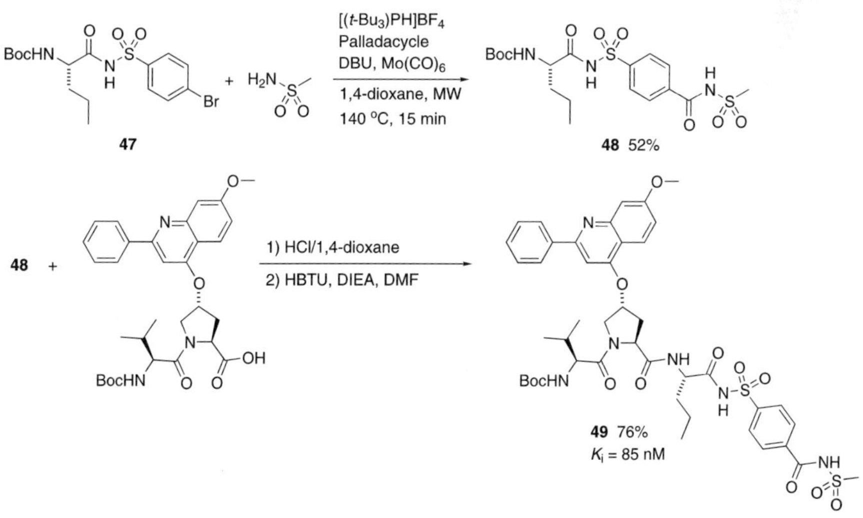

Scheme 22 Synthesis of an HCV NS3 protease inhibitor via a microwave-induced amidocarbonylation reaction

Recently, the synthesis and evaluation of a new pyrazinone class of tryptase inhibitors has been reported. One step in the preparation of these compounds involved regioselective chloride displacements from a dichloropyrazinone scaffold using amines and anilines as nucleophiles. The aniline reactions required the use of microwave-assisted heating for 12 min at 120 °C, as illustrated in Scheme 23 [84]. Although the aniline-derived compounds (e.g., **50**) were modest inhibitors, other compounds from this study, synthesized by classical heating methods, were shown to be highly potent tryptase inhibitors.

Factor Xa is a trypsin-like serine protease that plays a central role in the blood coagulation cascade. It is responsible for the cleavage of prothrombin into thrombin and is considered as an attractive target for new anticoagulants. There are, however, not so many low molecular weight Factor Xa inhibitors reported in the literature. In a lead optimization program a mesyl group was replaced by phthalimide in a large series of compounds under mi-

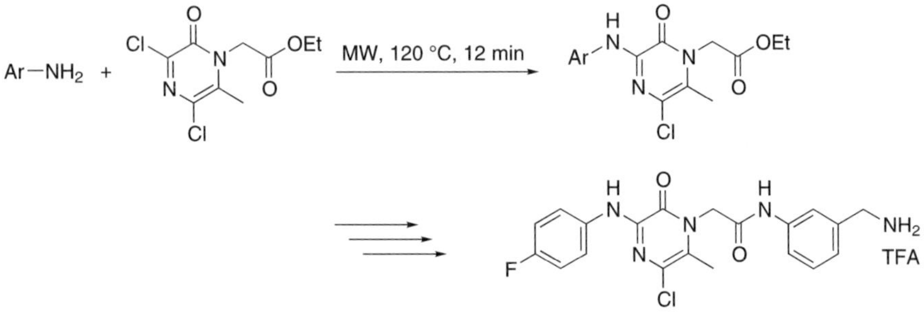

50 24% inhibition at 5 μM

Scheme 23 Microwave-accelerated nucleophilic substitution in tryptase inhibitor synthesis

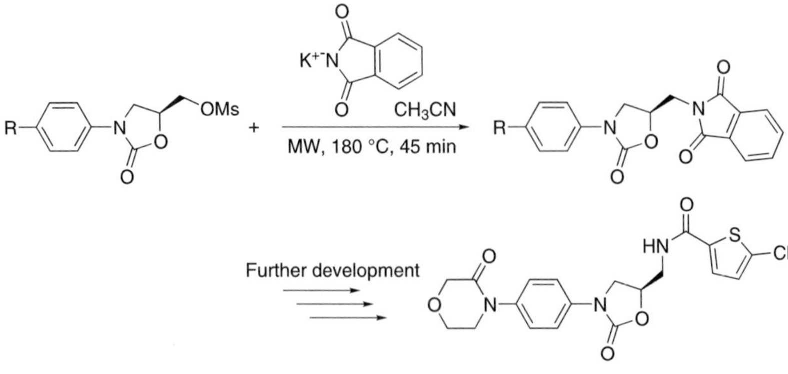

51 IC50 = 0.7 nM

Scheme 24 Microwave-assisted substitution reaction from a factor Xa inhibitor development program crowave irradiation for 45 min at 180 °C in a sealed tube (Scheme 24) [85]. The oxazolidinone **51** is a potent Factor Xa inhibitor derived from this program. This inhibitor displayed an excellent in vivo antitrombotic activity and a remarkable selectivity versus other serine proteases such as thrombin, trypsin, and plasmin; **51** is currently under clinical development.

In addition, highly active macrocyclic thrombin inhibitors have been synthesized by reaction sequences involving microwave-assisted amide bond formations [86].

2.4
Cysteine Protease Inhibitors

The cysteine proteases can be divided into three classes: the papain-like, the caspases (and related enzymes), and the picorna viral cysteine proteases. The proposed catalytic mechanism for cysteine protease peptide cleavage is related to the serine protease mechanism but with a cysteine thiol acting as the nucleophile that attacks the scissile peptide bond carbonyl.

A recent publication described the synthesis of a series of dipeptide nitriles and the evaluation of their interaction with various papain-like cysteine proteases. Different P2 side chains were explored within the series of 44 compounds and microwave-assisted Suzuki couplings were used in a limited part of the synthetic efforts. Two different biphenyl derivatives were synthesized from the corresponding aryl bromides, as shown in Scheme 25, using 30 min of microwave heating at 70 °C [87]. Especially the meta-biphenyl derivative **52** was shown to be a potent inhibitor of papain and the cathepsins L, S, and K.

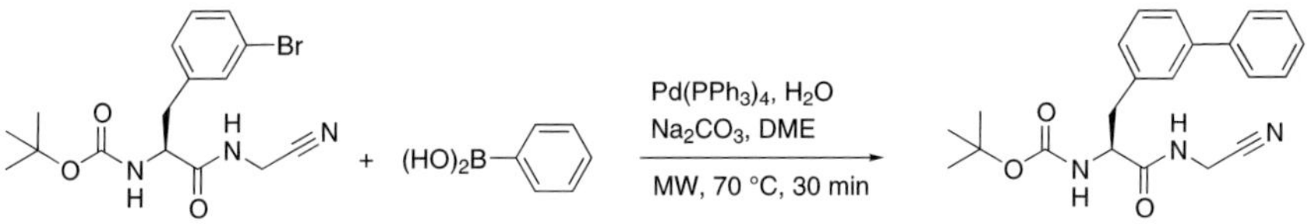

52 K_i papain = 1.2 µM; Cat L = 0.43 µM
Cat S = 0.27 µM; Cat K = 26 µM

Scheme 25 Preparation of a papain-like cysteine protease inhibitor

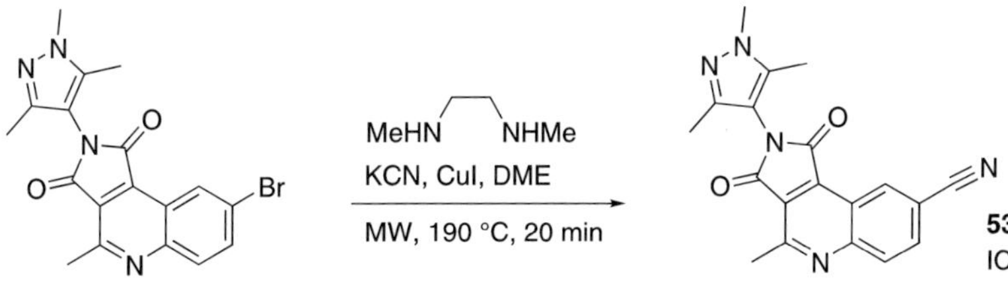

53 84%
IC_{50} = 16 nM

Scheme 26 Caspase-3 inhibitor produced by a copper-catalyzed cyanation reaction

In another recent paper, high temperature microwave heating was utilized in order to facilitate copper-catalyzed cyanation of a quinolinylbromide (Scheme 26) [88]. This was part of a larger study aimed towards the synthesis of caspase-3 inhibitors. The caspases are a family of related cystein proteases that play important roles for programmed cell death, apoptosis. Caspase-3 inhibitors could become valuable as neuro-, cardio- and hepatoprotectants. The compounds synthesized in this study included some highly potent inhibitors and among them was the microwave chemistry-derived **53**, with an in vitro caspase-3 IC_{50} of 16 nM.

2.5
Metallo Protease Inhibitors

Matrix metalloproteinase-13 (MMP-13) irreversibly cleaves type II collagen and inhibitors of this enzyme are believed to have the capacity to alter the progression of osteoarthritis. Extensive research efforts are devoted to finding agents against osteoarthritis. In a program aiming at orally active MMP-13 inhibitors, a microwave-promoted cyanation (in the presence of two equivalents of copper cyanide) was successfully executed in the synthetic route to **54**, a potent MMP-13 inhibitor (Scheme 27) [89]. This compound, with an IC_{50} value of 1.2 nM, exhibits a fair selectivity versus MMP-2.

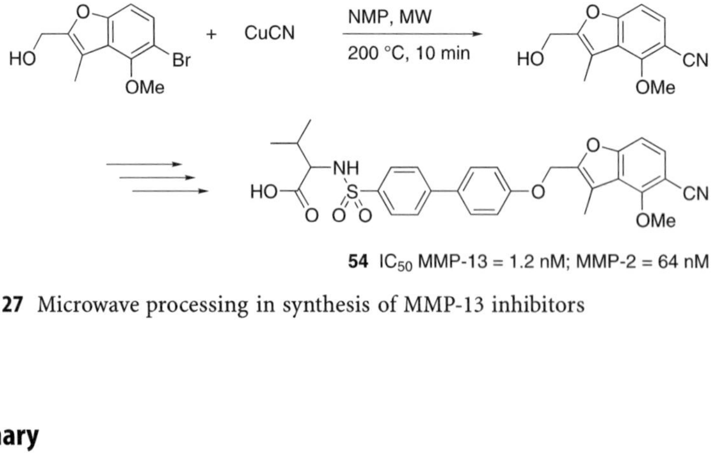

Scheme 27 Microwave processing in synthesis of MMP-13 inhibitors

3
Summary

Controlled microwave heating is a new enabling tool that helps the medicinal and combinatorial chemist to rapidly both optimize reaction conditions and perform small-scale target syntheses. In this short review, we have presented examples of microwave heating in high-speed medicinal chemistry. More specifically, we have mainly described microwave-enhanced synthesis of protease inhibitors using transition-metal catalysis. In all depicted examples, the main chemical effort was directed towards convenient and reliable procedures avoiding complicated handling of inert conditions and reactive gases. This methodology has helped in establishing structure–activity-relationships in a uniquely rapid fashion. As progress continues in microwave chemistry, new challenges will arise. It is likely that they will be met by innovative microwave approaches that can lead to increased efficiency in drug discovery research.

Acknowledgements We gratefully acknowledge the Swedish Research Council (VR), the Swedish Foundation for Strategic Research (SSF), Medivir AB, Dr Jenny Ekegren, Dr Kristofer Olofsson and Mr Gunnar Wikman for valuable help and support.

References

1. Lidström P, Tierney J, Wathey B, Westman J (2001) Tetrahedron 57:9225
2. Kappe CO (2004) Angew Chem, Int Ed 43:6250
3. Loupy A (ed) (2002) Microwaves in organic synthesis. Wiley, Weinheim
4. Larhed M, Hallberg A (2001) Drug Discov Today 6:406
5. Kappe CO, Dallinger D (2006) Nat Rev Drug Discovery 5:51
6. Kappe CO, Stadler A (2005) Microwaves in organic and medicinal chemistry. Wiley, Weinheim
7. Roberts BA, Strauss CR (2005) Acc Chem Res 38:653
8. Mingos DMP, Baghurst DR (1991) Chem Soc Rev 20:1
9. Larhed M, Moberg C, Hallberg A (2002) Acc Chem Res 35:717
10. Ersmark K, Larhed M, Wannberg J (2004) Curr Opin Drug Discov Devel 7:417
11. Kappe CO (2002) Curr Opin Chem Biol 6:314
12. Baxendale IR, Lee AL, Ley SV (2005) In: Tierney JP, Lidström P (eds) Microwave assisted organic synthesis. Blackwell, Oxford, p 133
13. Strauss CR, Trainor RW (1995) Aust J Chem 48:1665
14. Ley SV, Baxendale IR (2002) Nat Rev Drug Discovery 1:573
15. Alterman M, Andersson HO, Garg N, Ahlsen G, Lövgren S, Classon B, Danielson UH, Kvarnström I, Vrang L, Unge T, Samuelsson B, Hallberg A (1999) J Med Chem 42:3835
16. Olofsson K, Larhed M (2005) In: Tierney JP, Lidström P (eds) Microwave assisted organic synthesis. Blackwell, Oxford, p 23
17. Tong L (2002) Chem Rev 102:4609
18. Schechter I, Berger A (1967) Biochem Biophys Res Commun 27:157
19. Cooper JB (2002) Curr Drug Targ 3:155
20. Leung D, Abbenante G, Fairlie DP (2000) J Med Chem 43:305
21. Brik A, Wong C-H (2003) Org Biomol Chem 1:5
22. Greenlee WJ (1987) Pharm Res 4:364
23. Dash C, Kulkarni A, Dunn B, Rao M (2003) Crit Rev Biochem Mol Biol 38:89
24. Breman JG (2001) Am J Trop Med Hyg 64:1
25. Gardner MJ, Hall N, Fung E, White O, Berriman M, Hyman RW, Carlton JM, Pain A, Nelson KE, Bowman S, Paulsen IT, James K, Eisen JA, Rutherford K, Salzberg SL, Craig A, Kyes S, Chan M-S, Nene V, Shallom SJ, Suh B, Peterson J, Angiuoli S, Pertea M, Allen J, Selengut J, Haft D, Mather MW, Vaidya AB, Martin DMA, Fairlamb AH, Fraunholz MJ, Roos DS, Ralph SA, McFadden GI, Cummings LM, Subramanian GM, Mungall C, Venter JC, Carucci DJ, Hoffman SL, Newbold C, Davis RW, Fraser CM, Barrell B (2002) Nature 419:498

26. Berry C (2000) Curr Opin Drug Discov Dev 3:624
27. Boss C, Richard-Bildstein S, Weller T, Fischli W, Meyer S, Binkert C (2003) Curr Med Chem 10:883
28. Ersmark K, Samuelsson B, Hallberg A (2006) Med Res Rev 26(5):626
29. Silva AM, Lee AY, Gulnik SV, Maier P, Collins J, Bhat TN, Collins PJ, Cachau RE, Luker KE, Gluzman IY, Francis SE, Oksman A, Goldberg DE, Erickson JW (1996) Proc Natl Acad Sci USA 93:10034
30. Haque TS, Skillman AG, Lee CE, Habashita H, Gluzman IY, Ewing TJ, Goldberg DE, Kuntz ID, Ellman JA (1999) J Med Chem 42:1428
31. Nöteberg D, Hamelink E, Hulten J, Wahlgren M, Vrang L, Samuelsson B, Hallberg A (2003) J Med Chem 46:734
32. Nöteberg D, Schaal W, Hamelink E, Vrang L, Larhed M (2003) J Comb Chem 5:456
33. Muthas D, Nöteberg D, Sabnis YA, Hamelink E, Vrang L, Samuelsson B, Karlen A, Hallberg A (2005) Bioorg Med Chem 13:5371
34. Ersmark K, Feierberg I, Bjelic S, Hultén J, Samuelsson B, Åqvist J, Hallberg A (2003) Bioorg Med Chem 11:3723
35. Ersmark K, Feierberg I, Bjelic S, Hamelink E, Hackett F, Blackman MJ, Hulten J, Samuelsson B, Åqvist J, Hallberg A (2004) J Med Chem 47:110
36. Ersmark K, Nervall M, Hamelink E, Janka LK, Clemente JC, Dunn BM, Blackman MJ, Samuelsson B, Åqvist J, Hallberg A (2005) J Med Chem 48:6090
37. Netherton MR, Fu GC (2001) Org Lett 3:4295
38. CDC (1981) MMWR Morb Mortal Wkly Rep 30:250
39. CDC (1981) MMWR Morb Mortal Wkly Rep 30:305
40. Gottlieb MS, Schroff R, Schanker HM, Weisman JD, Fan PT, Wolf RA, Saxon A (1981) N Engl J Med 305:1425
41. Masur H, Michelis MA, Greene JB, Onorato I, Stouwe RA, Holzman RS, Wormser G, Brettman L, Lange M, Murray HW, Cunningham-Rundles S (1981) N Engl J Med 305:1431
42. Gottlieb MS, Schroff R, Schanker HM, Weisman JD, Fan PT, Wolf RA, Saxon A, Mannucci PM, Vigano S (1982) MMWR Morb Mortal Wkly Rep 31:507
43. CDC (1982) N Engl J Med 306:248
44. Dalgleish AG, Beverley PC, Clapham PR, Crawford DH, Greaves MF, Weiss RA (1984) Nature 312:763
45. Klatzmann D, Champagne E, Chamaret S, Gruest J, Guetard D, Hercend T, Gluckman JC, Montagnier L (1984) Nature 312:767
46. Fischl MA, Richman DD, Grieco MH, Gottlieb MS, Volberding PA, Laskin OL, Leedom JM, Groopman JE, Mildvan D, Schooley RT, Jackson GG, Durack DT, King D (1987) N Engl J Med 317:185
47. Richman DD, Fischl MA, Grieco MH, Gottlieb MS, Volberding PA, Laskin OL, Leedom JM, Groopman JE, Mildvan D, Hirsch MS, Jackson GG, Durack DT, Nusinofflehrman S (1987) N Engl J Med 317:192
48. De Clercq E (2005) J Med Chem 48:1297
49. Huff JR (1991) J Med Chem 34:2305
50. Swanstrom R, Erona J (2000) Pharmacol Ther 86:145
51. Tomasselli AG, Heinrikson RL (2000) Biochim Biophys Acta 1477:189
52. Huff JR, Kahn J (2001) Adv Protein Chem 56:213
53. Erickson JW, Gulnik SV, Markowitz M (1999) Aids 13 Suppl A:S189
54. Palella FJ Jr, Delaney KM, Moorman AC, Loveless MO, Fuhrer J, Satten GA, Aschman DJ, Holmberg SD (1998) N Engl J Med 338:853
55. Mocroft A, Vella S, Benfield TL, Chiesi A, Miller V, Gargalianos P, d'Arminio Monforte A, Yust I, Bruun JN, Phillips AN, Lundgren JD (1998) Lancet 352:1725

56. Fellay J, Boubaker K, Ledergerber B, Bernasconi E, Furrer H, Battegay M, Hirschel B, Vernazza P, Francioli P, Greub G, Flepp M, Telenti A (2001) Lancet 358:1322
57. Clavel F, Hance AJ (2004) N Engl J Med 350:1023
58. Alterman M, Björsne M, Muehlman A, Classon B, Kvarnström I, Danielson H, Markgren P-O, Nillroth U, Unge T, Hallberg A, Samuelsson B (1998) J Med Chem 41:3782
59. Larhed M, Lindeberg G, Hallberg A (1996) Tetrahedron Lett 37:8219
60. Larhed M, Hallberg A (1996) J Org Chem 61:9582
61. Alterman M, Hallberg A (2000) J Org Chem 65:7984
62. Beller M, Cornils B, Frohning CD, Kohlpaintner CW (1995) J Mol Cat A: Chemical 104:17
63. Skoda-Foldes R, Kollar L (2002) Curr Org Chem 6:1097
64. Wannberg J, Larhed M (2003) J Org Chem 68:5750
65. Wannberg J, Dallinger D, Kappe CO, Larhed M (2005) J Comb Chem 7:574
66. Wannberg J, Kaiser N-FK, Vrang L, Samuelsson B, Larhed M, Hallberg A (2005) J Comb Chem 7:611
67. Negishi E-i (2002) Handbook of organopalladium chemistry for organic synthesis, vol 1. Wiley, New York
68. Wannberg J, Sabnis YA, Vrang L, Samuelsson B, Karlen A, Hallberg A, Larhed M (2006) Bioorg Med Chem 14:5303
69. Wu X, Wannberg J, Larhed M (2006) Tetrahedron 62:4665
70. Ekegren JK, Unge T, Safa MZ, Wallberg H, Samuelsson B, Hallberg A (2005) J Med Chem 48:8098
71. Ekegren JK, Ginman N, Johansson Å, Wallberg H, Larhed M, Samuelsson B, Unge T, Hallberg A (2006) J Med Chem 49:1828
72. Wu X, Ekegren JK, Larhed M (2006) Organometallics 25:1434
73. Hulten J, Bonham NM, Nillroth U, Hansson T, Zuccarello G, Bouzide A, Åqvist J, Classon B, Danielson UH, Karlen A, Kvarnström I, Samuelsson B, Hallberg A (1997) J Med Chem 40:885
74. Hulten J, Andersson HO, Schaal W, Danielson HU, Classon B, Kvarnström I, Karlen A, Unge T, Samuelsson B, Hallberg A (1999) J Med Chem 42:4054
75. Backbro K, Löwgren S, Österlund K, Atepo J, Unge T, Hulten J, Bonham NM, Schaal W, Karlen A, Hallberg A (1997) J Med Chem 40:898
76. Lam PYS, Jadhav PK, Eyermann CJ, Hodge CN, Ru Y, Bacheler LT, Meek OMJ, Rayner MM (1994) Science 263:380
77. Cramer RD, Patterson DE, Bunce JD (1988) J Am Chem Soc 110:5959
78. Schaal W, Karlsson A, Ahlsen G, Lindberg J, Andersson HO, Danielson UH, Classon B, Unge T, Samuelsson B, Hulten J, Hallberg A, Karlen A (2001) J Med Chem 44:155
79. Ax A, Schaal W, Vrang L, Samuelsson B, Hallberg A, Karlen A (2005) Bioorg Med Chem 13:755
80. Gold H, Ax A, Vrang L, Samuelsson B, Karlen A, Hallberg A, Larhed M (2006) Tetrahedron 62:4671
81. Gold H, Larhed M, Nilsson P (2005) Synlett, p 1596
82. Shepard Colin W, Finelli L, Alter Miriam J (2005) Lancet Infect Dis 5:558
83. Wu X, Rönn R, Gossas T, Larhed M (2005) J Org Chem 70:3094
84. Hopkins C, Neuenschwander K, Scotese A, Jackson S, Nieduzak T, Pauls H, Liang G, Sides K, Cramer D, Cairns J, Maignan S, Mathieu M (2004) Bioorg Med Chem Lett 14:4819
85. Roehrig S, Straub A, Pohlmann J, Lampe T, Pernerstorfer J, Schlemmer K-H, Reinemer P, Perzborn E (2005) J Med Chem 48:5900

86. Nantermet PG, Barrow JC, Newton CL, Pellicore JM, Young M, Lewis SD, Lucas BJ, Krueger JA, McMasters DR, Yan Y, Kuo LC, Vacca JP, Selnick HG (2003) Bioorg Med Chem Lett 13:2781
87. Loeser R, Schilling K, Dimmig E, Guetschow M (2005) J Med Chem 48:7688
88. Kravchenko DV, Kysil VV, Ilyn AP, Tkachenko SE, Maliarchouk S, Okun IM, Ivachtchenko AV (2005) Bioorg Med Chem Lett 15:1841
89. Hu Y, Xiang JS, DiGrandi MJ, Du X, Ipek M, Laakso LM, Li J, Li W, Rush TS, Schmid J, Skotnicki JS, Tam S, Thomason JR, Wang Q, Levin JI (2005) Bioorg Med Chem 13:6629

Top Curr Chem (2006) 266: 199–231
DOI 10.1007/128_060
© Springer-Verlag Berlin Heidelberg 2006
Published online: 29 July 2006

Microwaves in Green and Sustainable Chemistry

Christopher R. Strauss[1] (✉) · Rajender S. Varma[2]

[1]CSIRO Molecular & Health Technologies, Bayview Ave., 3169 Clayton, Victoria, Australia
chris.strauss@csiro.au

[2]US Environmental Protection Agency, 26 West Martin Luther King Drive,
Mail Stop 443, Cincinnati, Ohio 45268, USA

Abstract The various roles of microwave-assisted organic chemistry in green and sustainable chemistry are discussed beginning with the strategies, technologies, and methods that were employed routinely at the time the first reports of microwave applications for organic synthesis appeared. Applications of open-vessel microwave chemistry and closed-vessel microwave chemistry are presented, with sections on solvent-free methods, reactions in high-temperature water, technology for transposing reaction conditions.

Keywords Solvent-free · Microwave · Reactor · High-temperature water ·
Reaction Conditions

Abbreviations
CMR   Continuous microwave reactor
CVMC   Closed-vessel microwave chemistry
GC   Gas chromatography
HPLC   High-performance liquid chromatography
IBD   Iodobenzene diacetate
MBR   Microwave batch reactor
MS   Mass spectrometry
MW   Microwave irradiation
NMR   Nuclear magnetic resonance spectroscopy
OVMC   Open-vessel microwave chemistry
PTC   Phase transfer catalysis
PTFE   Polytetrafluoroethylene
PFA   Perfluoroalkoxy
RX   Alkyl halide
X   Halide

1

Introduction

Since 1905 when viscose rayon was invented, major advances in synthetic polymeric materials have included phenol-formaldehyde resins (1909), polystyrene (1925), urea-formaldehyde resins (1929), neoprene rubber (1931), low-density polyethylene (1933), nylon fibre (1936) polytetrafluoroethylene (1938), polymethylmethacrylate (1938), silicone resins (1943), epoxy resins (1948), polyester fibre (1949), polycarbonate plastic (1957), polyaryl ethers (1962), polyimides (1963) and a vast range of engineered resins and composites (after 1970) [1].

This extensive, but by no means comprehensive, list of significant developments indicates the broad extent and scope of advancement in the science and technology of synthetic polymers, polymer manufacturing and polymer processing that took place during the 20th century. Now, of course, numerous practical, beneficial products and processes based upon polymers and composites are integral to everyday life and the chemistry and materials technologies underpinning them are frequently taken for granted.

Back in 1900, the US had an annual death rate of 2%. Life expectancy was 50 years, infant mortality was 1 in 18 and maternal death from childbirth was 1 per 167 live births. Approaching the end of the 20th century, through advances in nutrition and medicine brought about by chemical discoveries, the annual death rate had been lowered to 0.9%, infant mortality to 1 in 68 and maternal death from childbirth to 1 in 8000. Life expectancy had increased by 40% to 70 years [2]. Although many more could be cited, no further examples are needed to highlight the key role of chemistry in improving the quality of life, longevity and the standard of living since Perkin first began producing Tyrian purple (mauve) in 1857 [3].

Thus, superficially, it may appear surprising that the positive public perception of the chemical industry that had been built up in the first half of the 20th century was severely eroded in the latter half. The crisis of confidence coincided with growing concerns relating to human activity on the environment. In *Silent Spring*, published in 1962, Rachel Carson was the first to link the indiscriminate use of agricultural pesticides and herbicides to adverse ecological effects, predominantly on bird life [4]. She also associated aerial spraying of pesticides with kills of fish species and other aquatic organisms. Issues resulting from inappropriate disposal of chemical waste at Love Canal (in the 1970s) and at Times Beach (1971), a mass fish-kill in the Mississippi River in 1964 and an oil fire on the Cuyahoga River in 1969, possibly underscored and extended Carson's message [5].

Public opinion against chemistry continued to develop during the 1970s and 1980s. British polls indicated that it stemmed largely from a perception of indifference by the industry towards environmental issues and safety. This perception also translated into decreasing student enrollment in chemical courses, beginning a long-term trend that is disturbing in view of the ongoing need for new chemical products and processes as well as the strategic importance and enabling nature of chemistry. By 1981, disaffection with the industry had become so pronounced that a senior executive of a major chemical company coined the word *chemophobia*, defining it as "the almost spontaneous negative response that occurs when people hear the words *chemicals* and *chemical company*" [6].

Perhaps the lowest point came in late 1984 with one of the worst industrial accidents in history: In Bhopal, India, the overnight escape of methyl isocyanate from a chemical plant left an estimated 3000–10 000 nearby residents dead and an additional 50 000 permanently injured, without a warning having been raised [7]. The disaster accelerated a re-ordering of priorities within the chemical industry, with safety and the environment becoming the highest. This remains the case. Broad-ranging initiatives by governments and industry, usually acting consultatively and co-operatively, have been directed towards increased efficiency in chemical manufacturing, improved safety, reduced environmental impact, as well as enhanced communication and discussion of chemical issues within communities.

From a different (but related) environmental perspective, provision of adequate supplies of food, shelter, medicines and resources of energy and raw materials already constitute serious global challenges. Superimposed over the issues of safety and the environment introduced above are difficult and long-term problems involving sustainability. The current population of planet earth, approximately 6 billion, has tripled since 1938. Various projections indicate that by 2050 it will be about 11 billion, nearly twice the present level. Innovative methods for greater food production from diminishing tracts of available arable land and increased and efficient utilization of renewable resources will be essential.

The foregoing discussion indicates that the chemical industry will be expected to contribute significantly to the resolution of a diverse array of problems concerning sustainability, cleaner and more efficient production. It will need to create green replacement technologies and products. To do that will require specific and generic chemical tools. Within the industry, widening recognition of needs for cleaner processing, reduction or elimination of hazardous materials and increased efficiency has led to the emergence of green and sustainable chemistry as a field. Rather than remediation, which involves the cleaning up of waste after it has been produced, the main objective is to avoid waste generation in the first place. This approach will require new environmentally benign syntheses, catalytic methods and chemical products that are *"benign by design"* and that utilize renewable resources where possible [8–10].

2
Advent of Microwave-Assisted Organic Chemistry

Herein, various roles of microwave-assisted organic chemistry in green and sustainable chemistry are discussed. We begin by considering the strategies, technologies and methods that were employed routinely at the time that reports of microwave applications for organic synthesis first appeared. Before 1986, which is usually regarded as the year in which microwave-assisted organic chemistry began [11, 12], great advances were made in analytical methods for separation and structural elucidation of organic compounds. Powerful, multi-nuclear NMR instruments, capillary GC, HPLC and tandem techniques such as GC-MS and HPLC-MS were introduced. Most were computer-assisted or driven and equipped with Fourier transform. Hence, analysis of reaction mixtures and structural elucidation became much less challenging and far more convenient in the decade leading up to 1986 than previously.

In contrast, chemical synthesis and manufacture were still being conducted with traditional equipment and often by wasteful methods that changed little over the preceding half century [13]. Typical of the advances in hardware for synthetic chemistry were rotary evaporation and the introduction of ground-glass joints, adapters and stoppers instead of corks or rubber bungs for glassware. Although useful, they did not represent a quantum leap forward. Before the Bhopal incident in December 1984, environmental aspects were accorded little, if any, consideration. Yield (and for chemicals manufacturing, profit) were higher priorities than were safety, environmental or ecological effects of chemical products and processes. Reaction temperatures well below 0 °C became fashionable for synthetic protocols, which often were indirect. The use of protecting groups was lauded by peers, even though it introduced additional synthetic steps, generated waste and lowered atom economies. Significantly, Trost's concept of an atom economy was not introduced until the 1990s [14, 15]. Now, of course, it reflects contemporary priorities for high efficiency and has become an essential plank in all synthetic strategies.

Thus, Gedye and his co-workers [11, 16] introduced microwave-assisted organic chemistry at a time when development of technologies for practical organic synthesis had virtually stagnated and when there was community disquiet over possible detrimental effects of chemicals and chemical practices on the environment. The Canadian workers had found that with microwave heating, considerably shorter reaction times than normal were required for the common organic transformations that they had attempted. Their findings were supported by Giguere et al. [12], who had conducted similar studies independently. A severe drawback to broad implementation of the approach of these groups, however, was the inability to control the reaction conditions. Both groups encountered vessel deformation and explosions. Interested potential users of the technology deduced that although microwave heating could be advantageous for synthesis, first and foremost, it needed to be conducted safely.

Two sub-fields of microwave chemistry developed independently [13, 17, 18]. A common objective was the safe and clean application of microwaves in organic synthesis by capturing the benefits and minimizing the risks. The approaches differed through their use of either closed or open vessels, usually, but not exclusively with or without solvent, respectively. Classifiable as open-vessel microwave chemistry (OVMC) and closed-vessel microwave chemistry (CVMC), they tend to have specific rather than common advantages and disadvantages. Superficially, the differences may appear relatively minor at first glance. However, OVMC and CVMC enabled the development of substantially diverse chemistries that in many respects could be specific to the particular technique employed. In the main, both approaches were complementary rather than competitive although this was not often appreciated by the exponents.

General advantages of microwave heating for chemistry with either open or closed vessels, with or without solvents, now are well recognized [17, 18]. In the main:

1. Reaction times can be dramatically shorter than with conventional heating.
2. Energy can be introduced remotely, without contact between the source and the sample.
3. Energy input to the sample starts and stops immediately that the power is turned on or off.
4. Thermal inertia is lower than with conventional conductive heating.
5. Energy is delivered throughout the mass of the product not at surfaces.
6. Heating rates are higher than can be achieved conventionally if at least one of the components can couple strongly with microwaves.
7. The technique can be readily employed for sequential or parallel synthesis.

These advantages are all significant to green and sustainable chemistry as they have the potential to increase efficiency, to enhance safety in several different ways and to eliminate or avoid waste generation through cleaner processing.

3
Open-Vessel Microwave Chemistry (OVMC)

At the outset, OVMC employed domestic microwave ovens, commercially available vessels (typically glass beakers, watch glasses and conical flasks) and it was easy to implement. Indeed it was pioneered by several groups with ready adoption in mind and accordingly, it soon became widely used. Significant early contributions came from France through the groups of Bram and Loupy [19–21], Villemin [22–24], Hamelin and Texier-Boullet [25, 26], from the US through those of Varma [27, 28] and Bose [29, 30], and from Spain through that of Delgado and Alvarez [31, 32]. Many others subsequently have extended the range of transformations performed.

Environmentally attractive aspects of the technique included the employment of solvent-free conditions and "dry" media, which afforded improved safety and reduced the possibilities for uncontrolled exotherms [33–35]. Most commonly, reactants were adsorbed onto a solid support and solvents were removed before exposure of the supported reaction mixture to microwave irradiation. The samples did not need to be mixed and accurate temperature measurement was often not required. Domestic microwave ovens were typically employed as standard apparatus for reactions on scales up to several grams. For improved control, monomodal reactors with accurate measurement and monitoring of temperature were later introduced [33].

For microwave-heated reactions under solvent-free conditions, the following major benefits have been claimed, particularly with regard to green chemistry [36]:

1. Avoidance of large volumes of solvent reduces emissions and the need for redistillation
2. Simple work-up, by extraction, distillation or sublimation
3. Recyclable solid supports can be used instead of polluting mineral acids and oxidants
4. The absence of solvent facilitates scale-up
5. Safety is enhanced by reducing risks of overpressure and explosions
6. Reactions are quite often cleaner, faster and higher yielding than conventional synthesis
7. Synthesis on a scale of multi-hundreds of grams has been achieved.

Notwithstanding some of these points, although not constituents of reaction mixtures, solvents usually are employed in "dry" media processes, to load the reactants onto the support and afterwards to elute the products. For polar supports such as alumina or silica gel, which are commonly used in liquid chromatography, substantial quantities of solvent may be required. From a perspective of green chemistry, recycling of the solvent and the support are desirable.

Also, the susceptibility of a material to microwave energy can depend on variables including sample size, shape and dielectric properties as well as the location of the sample within the cavity [13]. Until recently, a lack of facilities for mixing reactions and for measuring temperature had affected the reproducibility of "dry" media reactions between microwave systems. That circumstance should be largely overcome through commercial reactors that enable sample mixing and temperature measurement.

Appropriate safety precautions should be employed when using "dry" media, which typically involve a low ratio of organic reactants to solid support. Potential hazards include toxic effects of volatiles liberated from the supports during the reactive step and also of the supports themselves. Minerals doped with inorganic or organic oxidants such as MnO_2, CrO_3, iodobenzene diacetate, and sodium periodate, or reductants like $NaBH_4$ and catalysts including KF and CsF, that have been employed as "dry" media, have large surface areas and could cause severe biological effects if inhaled [37].

Despite minor disadvantages, which can be addressed by safe chemical practices, the OVMC technique has produced a stunning array of clean chemical processes since its inception. Selected examples appear below:

4
Solvent-Free Methods

Solvent-free conditions can be employed according to three main methods: (a) using only neat reactants; (b) reactants adsorbed onto solid supports; or (c) reactants in the presence of phase transfer catalysts (in the case of anionic reactions). Besides the apparent potential benefits in solvent usage, reactions can be conducted conveniently and rapidly, often without temperature measurement in domestic microwave ovens. However, they now are often carried out under more precisely controlled conditions using monomode reactors initially introduced by the former French manufacturer Prolabo. Nowadays, several systems are available that provide facilities for the accurate measurement and monitoring of temperature throughout the reaction by modulation of emitted power with an infrared detector or an optical fibre.

Since the first reports of organic reaction accelerated in "dry media" conditions [20, 24], a large number of microwave-promoted solvent-free protocols have been demonstrated for important organic transformations such as condensation, cleavage (protection/deprotection), rearrangement, cyclization, oxidation, reduction and synthesis of heterocyclic compounds in a relatively environmentally friendly manner. A vast majority of these solvent-free reactions have been performed using unmodified household microwave ovens or commercial systems operated at 2450 MHz, in open glass containers with neat reactants. The general procedure involves simple mixing of neat reactants with the catalyst, their adsorption on mineral or "doped" supports, and subjecting the reaction mixture to microwave irradiation followed by extraction or filtration to isolate the products. Some representative microwave-promoted chemical syntheses conducted under solvent-free conditions are summarized below:

4.1
Use of Nearly Neutral Amphoteric Supports (Aluminas)

Alumina behaves as an amphoteric support due to acidity connected to OH surface groups (resulting from hydration of Al_2O_3) and basicity related to the available lone pairs on oxygen atoms.

Among a wide range of other applications, the combination of alumina-supported reactions and microwave irradiation was successfully applied to the cleavage of esters, a commonly used strategy to deprotect alcoholic groups in multi-step organic synthesis. Deacylation of alcoholic and phenolic ac-

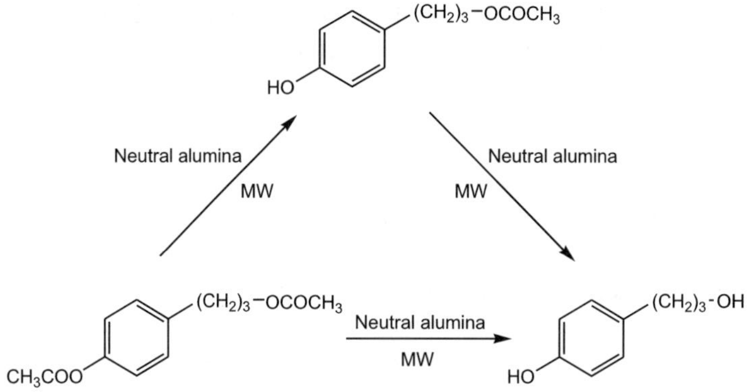

Scheme 1 Microwave-assisted deacylation of acetates on neutral alumina [27]

etates under solvent-free conditions on neutral alumina surfaces using microwave irradiation was demonstrated in a high-school science project in the early 1990s (Scheme 1) [27]. Interestingly, chemoselectivity between alcoholic and phenolic groups in the same molecule was achieved merely by adjusting the irradiation time.

The approach has found further applications in cleavage of benzyl esters, dethioacetalization, deoximation, and desilylation reactions [34].

4.2
Use of Basic Supports (KF/Alumina)

When impregnated on alumina, KF is a very strong base able to abstract a proton from rather weakly acidic carbon acids (up to pKa around 35) [38]. Reactions with basic alumina in conjunction with microwave irradiation leads to efficient procedures for base-catalyzed reactions as exemplified in the synthesis of 2-aroyl-benzofurans from α-tosyloxyketones and salicylaldehydes (Scheme 2) [39].

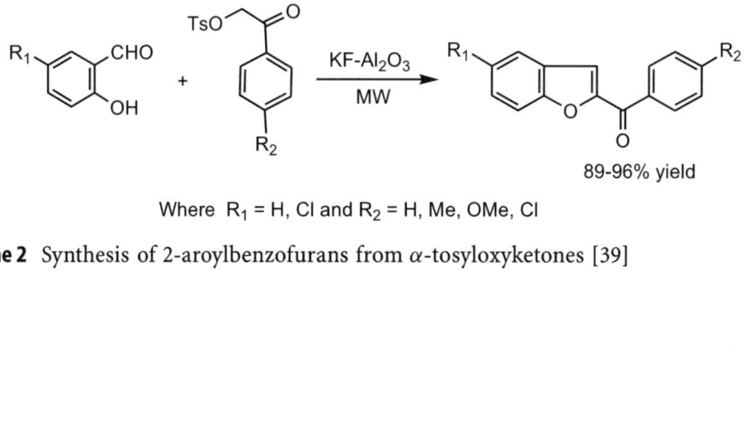

Where R_1 = H, Cl and R_2 = H, Me, OMe, Cl

Scheme 2 Synthesis of 2-aroylbenzofurans from α-tosyloxyketones [39]

4.3
Use of Acidic Supports (K10 and KSF Montmorillonites)

Clays are made up of layered silicates, the interlayer cations possibly being exchanged to contribute to their acidities. Commercially available and inexpensive K10 and KSF montmorillonites, resulting from washing the clay with mineral acids, behave as strong acids (comparable with concentrated sulfuric acid). They can be used as substitutes for HCl or H_2SO_4 [40] giving rise to dramatically improved outcomes when compared with classical reaction conditions. For example, in the acetalization of galactono-lactone with long-chain aldehydes, conventional conditions using solvent (DMF) and sulfuric acid were replaced by a solvent-free microwave procedure using K10 or KSF clays as acidic catalysts (Scheme 3). Yields up to 89% were obtained within 10 min instead of 25% after 24 h under classical conditions [41].

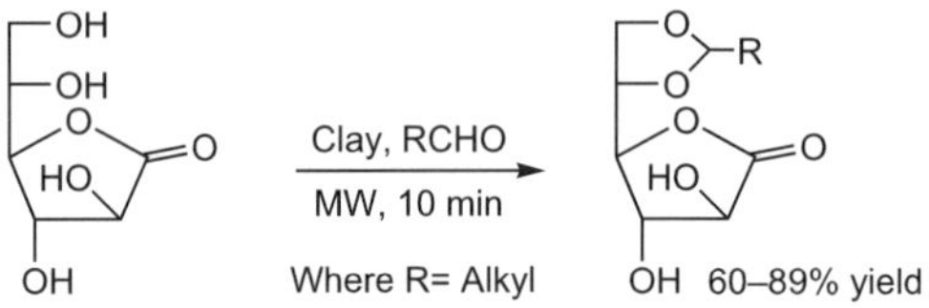

Clay, RCHO

MW, 10 min

Where R= Alkyl 60–89% yield

Scheme 3 Acetalization of *L*-galactono-1,4-lactone [41]

4.4
Oxidations on Solid Supports

The utility of oxidizing agents can be compromised by potential dangers in handling of metal complexes, inherent toxicity, cumbersome product isolation and waste disposal problems. Immobilization of metallic reagents on solid supports has overcome some of these limitations (Scheme 4) as the containment of metals on the support restricts their being leached into the environment. Examples include silica-supported manganese dioxide (MnO_2) [42] and chromium trioxide (CrO_3) immobilized on pre-moistened alumina affords efficient oxidation of benzyl alcohols to carbonyl compounds simply by mixing (Scheme 4). Remarkably, neither over-oxidation to carboxylic acids nor the usual formation of tar (a typical occurrence in many CrO_3 oxidations) was observed [43].

A protocol for the rapid oxidation of alcohols to carbonyl compounds has been reported with montmorillonite K10 clay-supported iron(III) nitrate (clayfen). The simple solvent-free experimental procedure involved mixing of neat substrates with clayfen, followed by microwave irradiation for 15–60 s [44]. The use of clayfen mixed with iron(III) nitrate on clay as an oxidant afforded higher yields (Scheme 4) and was more efficient, since the

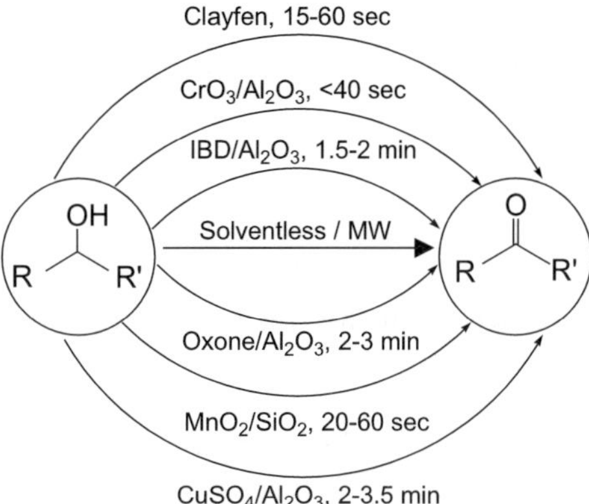

Scheme 4 Microwave-assisted solvent-free oxidation of alcohols to carbonyls under various conditions [42–44, 47]

amounts used in these protocols were half of those used in solution phase reactions by Laszlo et al. [45, 46].

Varma et al. reported the use of supported iodobenzene diacetate (IBD) as an oxidant. With alumina as a support, the yields improved markedly compared to those obtained with neat IBD (Scheme 4) [47]. IBD-alumina has also been used for the rapid, high-yielding and selective oxidation of alkyl, aryl and cyclic sulfides to the corresponding sulfoxides upon microwave irradiation [48].

Selective oxidation of sulfides to sulfoxides and sulfones was achieved over sodium periodate (NaIO$_4$) on silica (20%), under solvent-free conditions to afford either sulfoxides or sulfones as desired (Scheme 5) [49].

A low amount of the active oxidizing agent, 20% NaIO$_4$ on silica, was employed, enhancing safety, convenience and reducing waste. Among several refractory thiophenes oxidized under these conditions, benzothiophenes afforded the corresponding sulfoxides and sulfones using ultrasonic and microwave irradiation, respectively, in the presence of NaIO$_4$-silica. A notable feature of the protocol was its applicability to long-chain fatty sulfides that were insoluble in most solvents and otherwise difficult to oxidize (Scheme 6).

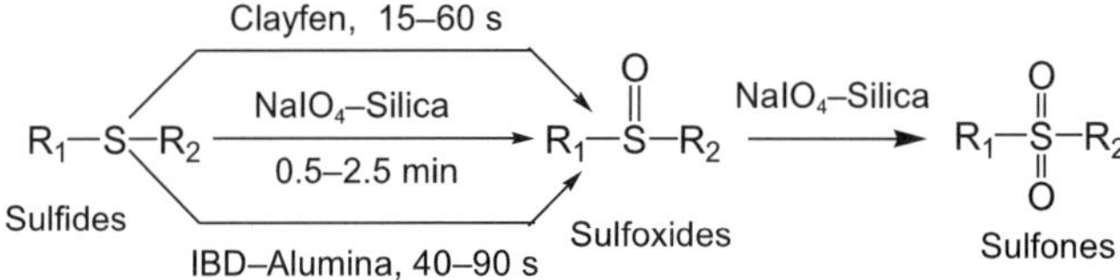

Scheme 5 Microwave-assisted oxidation of sulfides to sulfoxides and/or sulfones [49]

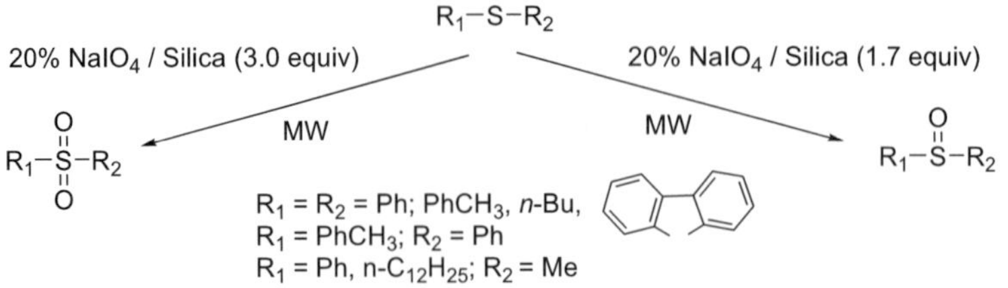

$$R_1-S-R_2$$

20% NaIO$_4$ / Silica (3.0 equiv)   20% NaIO$_4$ / Silica (1.7 equiv)

MW   MW

$$R_1-\overset{\overset{O}{\|}}{\underset{\underset{O}{\|}}{S}}-R_2 \qquad\qquad\qquad\qquad\qquad\qquad\qquad\qquad\qquad R_1-\overset{\overset{O}{\|}}{S}-R_2$$

R$_1$ = R$_2$ = Ph; PhCH$_3$, n-Bu,
R$_1$ = PhCH$_3$; R$_2$ = Ph
R$_1$ = Ph, n-C$_{12}$H$_{25}$; R$_2$ = Me

Scheme 6 Solvent-free oxidation of sulfides using sodium periodate-silica [49]

The reactions were improvements upon conventional reaction conditions for the oxidation of sulfides to the corresponding sulfoxides and sulfones. The latter conditions typically require more aggressive oxidants such as nitric acid, hydrogen peroxide, chromic acid, peracids, and periodate.

4.5
Reductions

The relatively inexpensive and safe sodium borohydride (NaBH$_4$) has been extensively used as a reducing agent because of its compatibility with protic solvents. Varma and coworkers reported a method for the expeditious reduction of aldehydes and ketones that used alumina-supported NaBH$_4$ and proceeded in the solid state accelerated by microwave irradiation (Scheme 7) [50]. The chemoselectivity was apparent from the reduction of *trans*-cinnamaldehyde to afford cinnamyl alcohol.

The reaction was improved by the presence of moisture, but did not proceed in the absence of alumina. The alumina support could be recycled and reused repeatedly for subsequent reduction by mixing with fresh borohydride without any apparent loss in activity. In terms of safety, the air used for cooling the magnetron vented the microwave cavity, thus preventing any ensuing hydrogen from reaching explosive concentrations. The process has been utilized successfully for microwave-enhanced solid-state deuteriation reactions using sodium borodeuteride-impregnated alumina [51]. Extension of these studies to specific labeling has been ex-

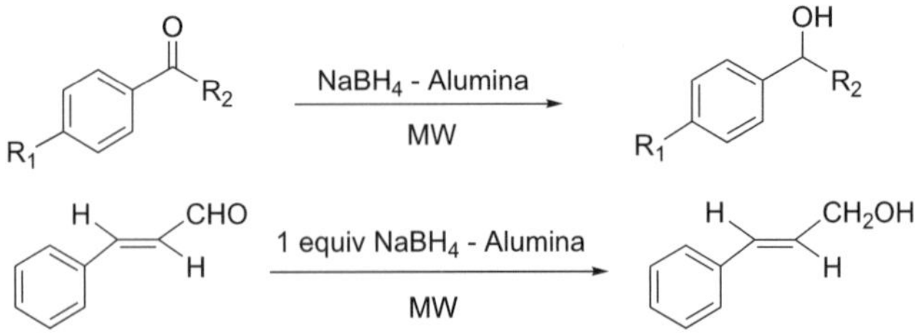

Scheme 7 Solventless reduction of carbonyls on alumina [50]

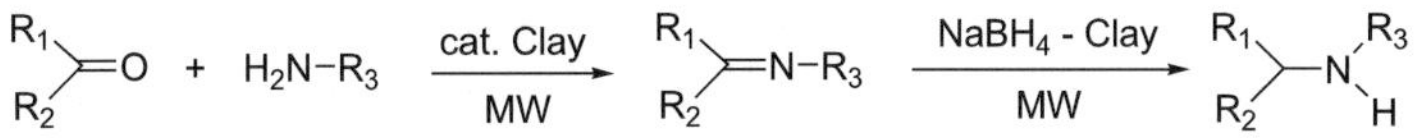

$$R_1R_2C{=}O + H_2N{-}R_3 \xrightarrow[\text{MW}]{\text{cat. Clay}} R_1R_2C{=}N{-}R_3 \xrightarrow[\text{MW}]{\text{NaBH}_4\text{ - Clay}} R_1R_2CH{-}N(R_3)H$$

Scheme 8 Reductive amination of carbonyls [54]

plored [52] including deuterium-exchange reactions for the preparation of reactive intermediates [53].

A solvent-free reductive amination of carbonyl compounds using sodium borohydride supported on moist montmorillonite K10 clay also was facilitated by microwave irradiation (Scheme 8) [54]. Clay served the dual purpose of a Lewis acid and provided water from its interlayers to enhance the reducing ability of $NaBH_4$.

4.6
Neat Reactants without Added Catalyst

The simplest solvent-free method involves irradiation of neat reactants as interfacial reactions in an open container, either in a domestic oven or in a monomode reactor [34, 55]. In the absence of reagents or supports, the scope for such processes appears to be limited to relatively straightforward condensations that can be conducted without added catalysts, to nucleophilic additions, S_N2 alkylations using neutral nucleophiles as amines or phosphines or to intramolecular thermolytic processes such as rearrangement or elimination.

4.6.1
Reactions with at Least One Liquid Reagent

An important and extensively studied case involves reactions carried out in heterogeneous media between a solid and a liquid as neat reagents. The reaction can occur as an *interfacial procedure* allowed by either a partial solubilization of the solid in the liquid or by adsorption of the liquid on the surface of the solid, essentially owing to its high specific area [34, 55].

The case of two liquid reagents is more understandable, occurring in the homogeneous phase if the compounds are mutually miscible or as an emulsion. In both cases, enhancements in reactivity are provided by concentration effects on kinetics due the absence of any dilution induced by the solvent.

It was concluded that solvent-free microwave reactions between phthalic anhydride and amino compounds needed at least one liquid phase and that reactions between two solids might not occur in this case and a high boiling point solvent might be necessary; excellent yields were obtained within short reaction times (5–10 min) (Scheme 9) [56].

In the absence of reagents or supports, the scope for such processes appears to be limited to relatively straightforward cases that can be conducted

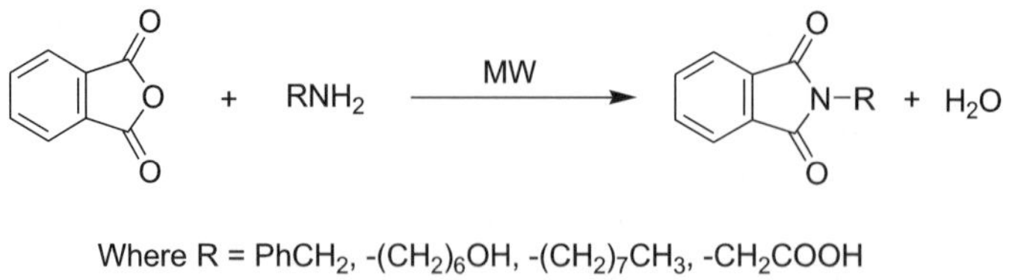
MW
RNH2
N–R
H2O
Where R = PhCH2, -(CH2)6OH, -(CH2)7CH3, -CH2COOH

Scheme 9 Synthesis of phthalimides under microwave irradiation [56]

without added catalysts, including as typical examples, nucleophilic additions such as amidations by direct reaction of amines with carboxylic acids without necessity for catalyst or any activated acyl compounds [57, 58].

Solvent-free uncatalyzed amidations of acids occurred readily under microwave conditions when conducted with a slight excess of either amine or acid (1.5 equiv). The pathways involved thermolysis of the previously formed ammonium salts (acid-base equilibrium) and proceeded by nucleophilic attack of the amine on the carbonyl moiety of the acid, with removal of water. This procedure was extended to the preparation of functionalized tartramides directly from amines and tartaric acid under solvent-free conditions and microwave activation (Scheme 10). By conventional heating, yields were lower even after 16 h.

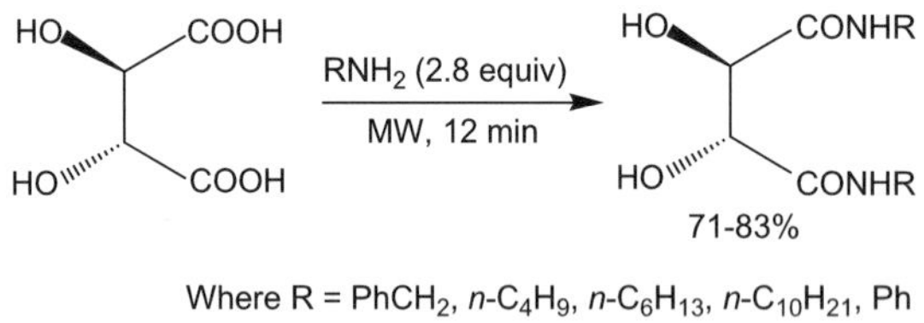
HO
COOH
HO
COOH
RNH2 (2.8 equiv)
MW, 12 min
HO
CONHR
HO
CONHR
71-83%
Where R = PhCH2, n-C4H9, n-C6H13, n-C10H21, Ph

Scheme 10 Solvent-free synthesis of tartramides by microwave irradiation [57, 58]

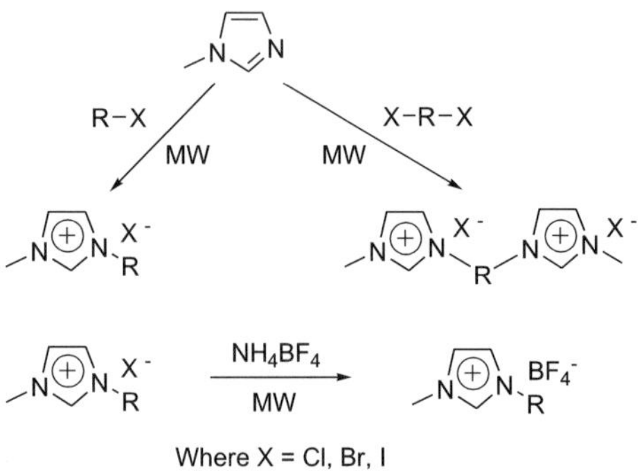
R–X
MW
MW
X–R–X
NH4BF4
MW
BF4-
Where X = Cl, Br, I

Scheme 11 Synthesis of ionic liquids via solvent-free alkylation of *N*-methyl imidazole [63]

Traditional preparations of ionic liquids usually involved heating the reactants for extended periods in refluxing solvents, with excess amounts of alkyl halides to drive the reaction to completion and using large quantities of organic solvents for product purification. These aspects affected the potential of room temperature ionic liquids (RTILs) as "green" solvents. S_N2 alkylations with amines as neutral nucleophiles, have been developed for the synthesis of RTILs [59–62]. In the preparation of 1,3-dialkylimidazolium halides as described by Varma and Namboodiri [59], the reaction time was reduced from hours to minutes, without use of a large excess of alkyl halides/organic solvents as the reaction medium. The approach was extended to the preparation of other ionic salts from N,N'-dialkylimidazolium chloride and ammonium tetrafluoroborate (Scheme 11) [63] and even indium-containing ionic liquids [64].

4.7
Microwave-Assisted Synthesis of Heterocyclic Compounds

Heterocyclic chemistry was facilitated by microwave-expedited solvent-free chemistry utilizing mineral supported reagents [65]. The scope now includes parallel synthesis [66]. A representative multi-component condensation to create a library of imidazo[1,2-a]pyridines, imidazo[1,2-a]pyrazines and imidazo[1,2-a]pyrimidines is depicted in Scheme 12 [67].

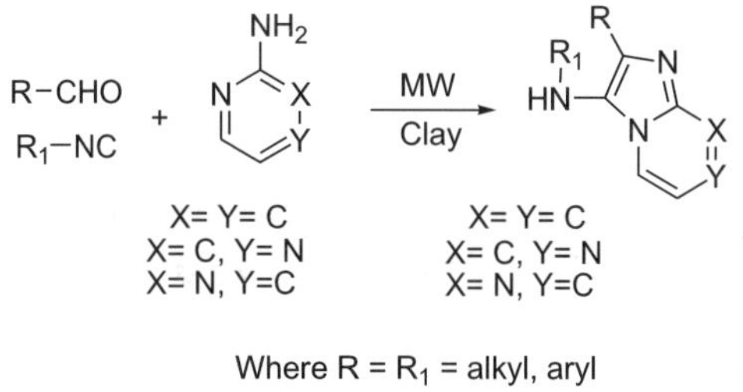

Scheme 12 Microwave-assisted three-component Ugi reaction [67]

4.8
Microwave-Assisted Cu(I)-Catalyzed
Solvent-Free Three-Component Coupling of Aldehyde, Alkyne and Amine

Propargyl amines, widely used as intermediates for the synthesis of various nitrogen-containing biologically active compounds and natural products, were prepared via a three-component coupling of aldehyde, alkyne and amine (A^3 coupling) under microwave irradiation conditions using only CuBr as catalyst (Scheme 13) [68]. In earlier work, CuOTf or $RhCl_3$ were required as

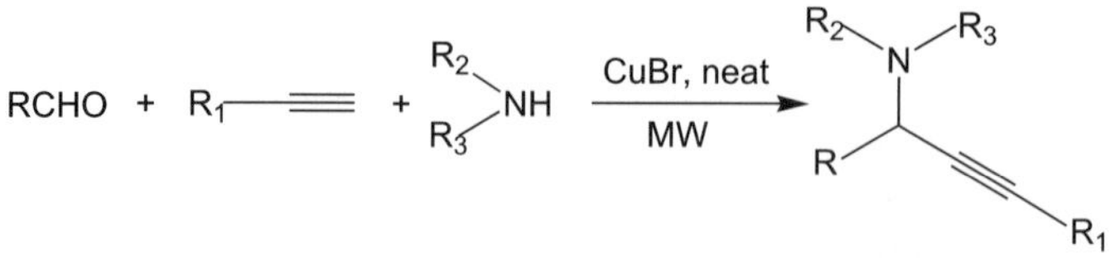

RCHO + R₁—≡ + R₂R₃NH →(CuBr, neat / MW) [propargylamine product]

Where R = aryl, alkyl; R_1= aryl, alkyl and Et_3Si; R_2 = alkyl, allyl, H;
R_3 = alkyl, allyl

Scheme 13 Microwave-assisted synthesis of propargylamines [68]

co-catalysts along with a variety of organic solvents and ionic liquids as reaction media.

4.9
Isoflavenes, Thiazoles and Benzofuran Derivatives

The estrogenic properties displayed by isoflav-3-ene derivatives have attracted the attention of medicinal chemists. Varma et al. have discovered a facile and general method for the microwave-assisted synthesis of isoflav-3-enes substituted with basic moieties at the 2-position (Scheme 14) [69]. This convergent approach exploits in situ generated enamine derivatives that are subsequently reacted with o-hydroxyaldehydes in the same pot.

Conventional preparations of thiazoles and 2-aroylbenzo[b]-furans require the use of lachrymatory α-haloketones and thioureas (or thioamides). To avoid this problem, Varma et al. have synthesized various heterocycles via solvent-free reactions of thioamides, ethylenethioureas and salicylaldehydes with α-tosyloxyketones that are generated in situ from arylmethyl ketones and [hydroxy(tosyloxy)iodo]benzene (HTIB) under microwave conditions (Scheme 15) [70].

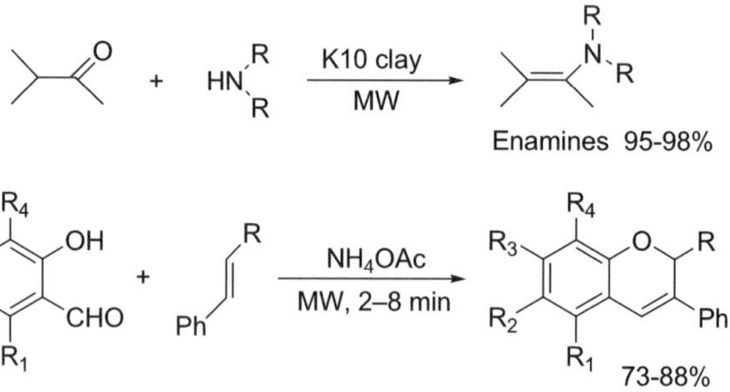

Enamines 95-98%

73-88%

Where R = morpholinyl, piperidinyl or pyrrolidinyl
and R_1 = R_3 = R_4 = H ; R_2 = H, Cl, NO_2

Scheme 14 Synthesis of 2-amino substituted isoflav-3-enes from enamines [69]

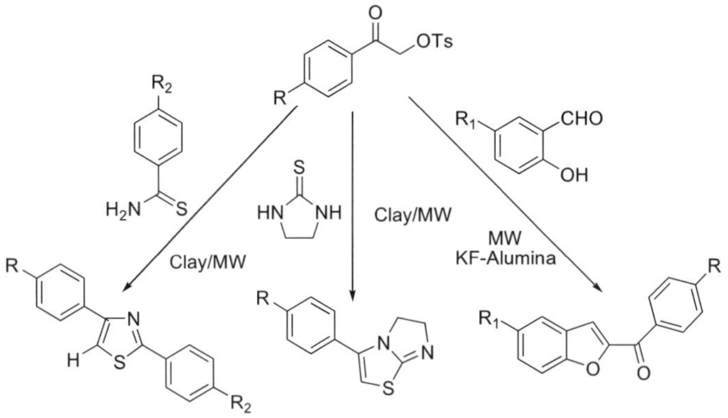

Scheme 15 Synthesis of heterocycles from in situ generated α-tosyloxyketones [70]

4.10
Reactions Involving Two Solid Reagents

As far as we are aware, the only examples of reactions between two solids in a solvent-free and catalyst-free environment were performed with neat 5 or 8-oxobenzopyran-2(1H)-ones, with a variety of aromatic and heteroaromatic hydrazines, providing rapid access to several synthetically useful heterocyclic hydrazones (Scheme 16) [71].

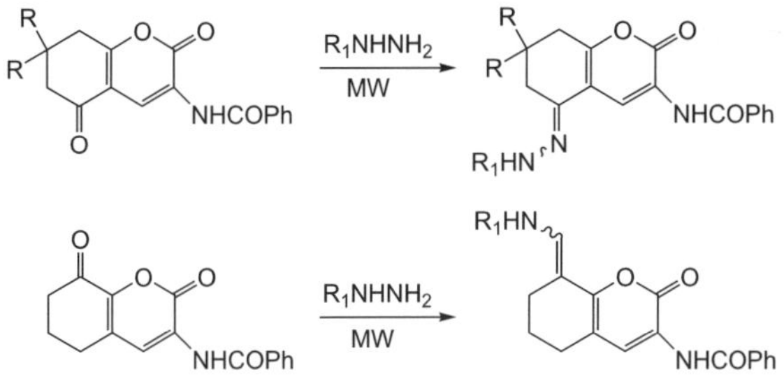

Scheme 16 Solvent-free preparation of hydrazones using microwaves [71]

4.11
Phase Transfer Catalyzed Reactions

Solid–liquid solvent-free phase transfer catalysis (PTC) is specific for anionic reactions including numerous alkylations, eliminations and anionic additions. The coupling of PTC conditions and microwave irradiation was applied to numerous alkylations, eliminations, anionic condensations, and base-catalyzed isomerisations [72]. Usually, a catalyst (typically a tetraalkyl-ammonium salt or a cationic complexing agent) is added to an equimolar mixture of an electrophile and a nucleophile, one of which serves as both a reactant and the organic phase.

5
Other Open-Vessel Methods

Other open-vessel methods, not involving "dry" media or PTC, include the use of phase transfer reagents under solvent-free conditions, the use of low-boiling solvents at reflux and microwave-induced organic reaction enhancement (MORE) chemistry [28, 29, 73]. The last mentioned method utilizes high-boiling solvents at temperatures below their boiling points. It has lost favor in recent years, presumably owing to potential safety hazards and inconvenience associated with recovery of reaction products from high-boiling solvents.

On the other hand, reflux methods now are employed conveniently in appropriate commercial microwave equipment. As shown by Baghurst and Mingos, boiling points up to 26 °C above the corresponding equilibrium values at atmospheric pressure can occur with microwave heating [74]. Solvents with relaxation times greater than 65 ps (corresponding to 2450 MHz) have loss tangents that increase with temperature and are prone to superheating, enabling some reactions to be completed more quickly than otherwise may be expected [75]. For reactions at reflux, commercial microwave systems are usually fitted with a shielded opening to prevent microwave leakage and through which a reaction vessel (usually a round-bottomed flask) can be connected to an external condenser. Instead of an external reflux condenser, early workers also employed microwave-transparent coolants including solid CO_2, within the microwave cavity [76].

6
Closed-Vessel Microwave Chemistry (CVMC)

The first reports by Gedye [11, 16] and Giguere [12] described reactions conducted by CVMC exclusively, suggesting that neither group had contemplated "dry" media reactions. Giguere used small sealed tubes or screw-cap vials that were placed in a microwave-transparent Corian box packed with vermiculite to absorb the contents in case of an explosion [12]. Heating times of up to 15 min were employed with a domestic microwave oven operating on full power. Reaction temperatures were estimated retrospectively through the use of sealed tubes containing reference materials of known melting points. Depending upon its water content, vermiculite can readily absorb microwave energy. Thus, to varying extents depending upon the specific transformation under study, heat also would have been transferred to the reaction mixture by conduction from the vermiculite. This possibly accounts for the high temperatures apparently obtained with reactions in hexane, for example. As outlined in the invited review of Strauss and Trainor [77], by 1995 Giguere's methodology had been adopted by others for applications including racemisation, intramolecular Diels–Alder reactions, decarboxylation, etherification, Ferrier rearrangement and Baylis–Hillman reactions. It had lost favour by 2000, however, when more advanced commercial systems had become available.

Gedye and his collaborators [11, 16] employed a domestic oven and commercially available screw-cap pressure vessels made from either polytetrafluoroethylene (PTFE) or perfluoroalkoxy (PFA) Teflon. Reaction pressures were measured and the final temperatures were estimated retrospectively through an infrared sensor that was directed at the vessel immediately after shutting off the microwave power. The methodology enabled the group to perform reactions on a higher scale than that of Giguere. Although it was potentially a more practical synthetic approach, the increased scale enhanced the risk for severe explosions in the event of uncontrolled exotherms including thermal runaway.

7
Dedicated Reactors

CSIRO workers in Australia considered that for microwave technology to become a generally applicable tool for synthetic organic chemists, domestic microwave (kitchen) ovens would not suffice. Dedicated reactors needed to run safely and to fail safely if necessary. They required capabilities for sample mixing and measurement and control of power, temperature and pressure while reactions were in progress. Reactions needed to be reproducible and to be able to be conducted in the presence of organic solvents at moderate pressures above atmospheric. Although such capabilities now are commonplace in modern commercial systems, in the 1990s their development presented major technical difficulties [13, 77]. For about a decade, the CSIRO group was alone in the pursuit of such microwave systems for organic synthesis. They introduced a continuous microwave reactor (CMR) in 1988 [78] and microwave batch reactors (MBRs) in 1992 [79] and 1995 [80].

The CMR, which could operate up to the pilot-scale, functioned by passage of reaction mixtures through a microwave zone under pressure, followed by rapid cooling of the pressurized effluent [78]. MBRs were introduced for small-scale (25–200 mL) reactions and were suitable for synthesis and kinetics studies. The most advanced unit, reported in 1995, operated at up to 260 °C and 10 MPa (100 atmospheres) in a laboratory environment. Its facilities included a stirrer, a cold-finger for post-reaction cooling or for concurrent heating and cooling, a valve for sample withdrawal and devices for monitoring pressure, temperature and microwave power. Heating could be programmed at high or low rates and designated temperatures could be retained for hours if desired [80].

Improvements to literature methods typically involved time-savings, higher yields, greater selectivity, the need for less catalyst, or the employment of a more environmentally benign solvent or reaction medium. Ever since the first reports, speed and convenience have been the major advantages of microwave technology for synthesis. Typically, the MBR and CMR reduced reaction times with conventional glassware by two to three orders of magnitude. Significant savings in energy, labor, management and utilization of plant can be achieved through accelerating normally sluggish reactions in particular. The MBR and CMR are portable, multi-purpose and self-contained advantages that have become significant for chemical production. Their capabilities for rapid throughput and the materials of construction enable easy cleaning for reuse and promote short turnaround times. Safety advantages include control and method of energy input, low volumes undergoing reaction at one time and opportunities for remote, programmable operation.

The CMR and MBRs provided the basis for modern commercial microwave reactors, including robotically operated automated systems that are now widely employed in synthetic research and pilot-scale laboratories in academia and industry [13]. Since 2000, commercial microwave reactors have become available. Batch systems, produced by three major companies in Italy and Germany, Sweden and the United States, typically operate on a scale from 0.5 mL up to 2 L. Other companies based in Austria, Poland and Japan have also recently entered the market. Systems possessing either multimodal or monomodal cavities are produced with one recent addition being a single unit capable of performing in either mode as required. Microwave reactors are employed extensively in chemical discovery where successive reactions can be performed rapidly in parallel or sequentially. One manufacturer recently estimated that about 10 000 reactions per week were performed in its systems alone. This indicates the extent to which microwave chemistry in closed vessels has dramatically influenced approaches to synthesis.

From the perspective of green and sustainable chemistry, speed and convenience are not the only considerations. Economic and safety factors encourage minimization in the stockpiling of chemicals and transportation of hazardous substances. In those regards, reactor size is important and miniaturization can be advantageous. In the future, individual chemical reactors may be required for diverse tasks and they may need to be readily relocatable within a chemical manufacturing facility. Benefits of microwave heating can include quick start up and shut down, high yield, low holding capacity and almost complete elimination of waste. The potential for accidents during transportation and storage could be restricted by just-in-time, point-of-use production.

8
Green and Sustainable Chemistry with MBR and CMR

8.1
Rapid Heating and Cooling

Sustainability depends heavily upon the slowing of both climate change and global warming. In turn, that appears to require the re-establishment of a tolerable steady state between CO_2 production and uptake. Such an outcome could be facilitated by the development of technologies and products reliant upon renewable resources rather than fossil fuels [81]. A key renewable resource in that regard is cellulose, which has its origins in photosynthesis and is comprised of poly-glucose units that are glycosidically linked and resistant to enzymatic cleavage. Depolymerization of cellulose to produce bio-available glucose or its oligomers from biomass represents an important strategic goal. The products could find uses as feedstock for bioconversion to low-molecular-weight short-chain organics such as ethanol, propanol, glycerol, acetic acid and lactic acid, all of which could be employed as synthetic building blocks and some as fuels. Technically, the task is challenging. High temperatures and acidic conditions are required to cleave the glycosidic bonds of cellulose, but the product monosaccharide and oligosaccharides are appreciably more acid-labile than the starting material. Typically, unreacted cellulose or polymeric tars are produced, suggesting that processes involving rapid heating and cooling could be advantageous. With 1% sulfuric acid, a useful method involved raising the temperature from ambient to 215 °C within 2 min in the MBR, maintaining this temperature for 30 s and cooling (Scheme 17) [77]. The entire operation was completed within 4 min and afforded glucose in nearly 40% yield, along with oligomeric materials. Pilot-scale processes based on those conditions are now in operation, and the reaction can even be heated conventionally.

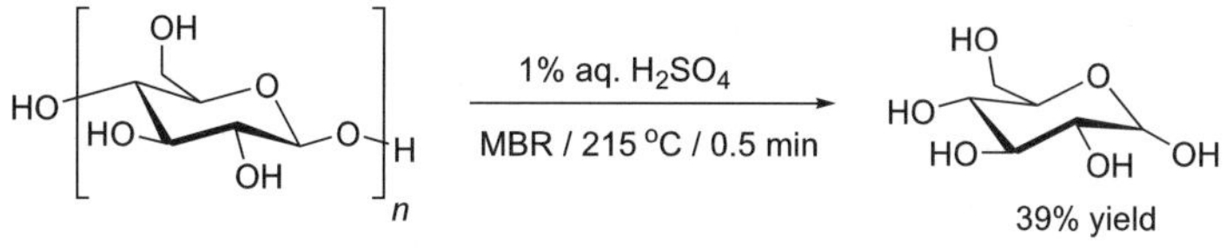

Scheme 17 Hydrolysis of cellulose in 1% sulfuric acid [77]

8.2
Green New Reactions

The MBR and CMR allowed ready access to reaction conditions that were previously only attainable with difficulty, e.g., temperatures between 120 °C and 250 °C and pressures between 1 and 5 MPa, followed by rapid cooling afterwards, if necessary. Consequently, they have assisted the discovery of new reactions that require elevated temperature. They also have been beneficial in searches toward conditions employing elevated temperatures with less catalyst, a milder catalyst or no addition of catalyst, which can be attractive alternatives to those employing aggressive reagents at lower temperatures. This approach toward greener synthetic methods was demonstrated through a catalytic, symmetrical etherification [82]. The process afforded minimal waste and was carried out near neutrality. Although not as versatile as the Williamson procedure of the 1850s, it represented a cleaner alternative toward specific ethers, since a stoichiometric amount of waste salt was not generated.

An excess of alcohol (ROH) was heated with a catalytic amount of the corresponding alkyl halide (RX). Solvolytic displacement between RX and ROH afforded R_2O and HX or its elements. The liberated HX attacked another molecule of ROH to form water and to regenerate RX. If the rates of both forward reactions were comparable, the concentration of HX was low throughout and that of RX remained relatively constant. Although HX and RX were stoichiometric reactants or products during the reaction, the net reaction comprised condensation of two molecules of ROH to give R_2O plus water (see Scheme 18 where etherification of *n*-BuOH is depicted). The process typically was suited to temperatures of 180–220 °C and reaction times of up to an hour. It was demonstrated with primary and secondary alcohols, including base and acid-labile compounds. Advantages for clean production were high atom economy, that salts were not formed, RX often was recoverable, the reaction did not require addition of strong acids or bases and that water was the major by-product.

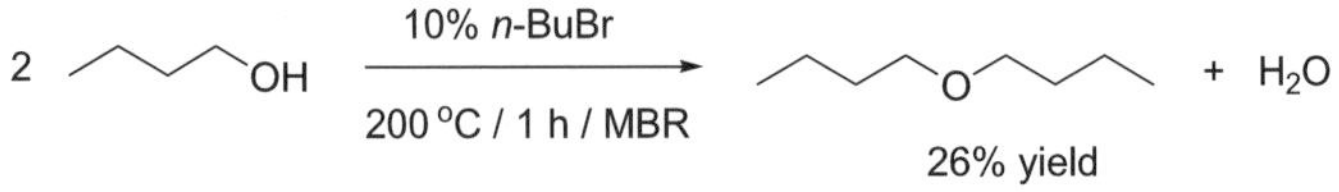

Scheme 18 Catalytic symmetrical ether synthesis [82]

8.3
High-Temperature Water as a Green Reaction Solvent or Medium

Priorities for enhanced safety, protecting the environment and improving the economics of operation (often referred to as the triple bottom line), emphasize needs for decreased usage of many organic solvents in chemical laboratories and in industrial processes. The phasing out of halogenated solvents and ongoing searches for suitable replacement media are indicative.

In contrast with its behaviour at the normal boiling point or below, water at high temperature behaves as a *pseudo-organic solvent* [77]. Its dielectric constant decreases substantially. Its ionic product increases by three orders of magnitude and above 200 °C the solvating power toward organic molecules becomes comparable with that of EtOH or Me_2CO at ambient temperature. Consequently, acid or base-catalyzed reactions in high-temperature water typically require less catalyst than normal and can proceed rapidly. Investigations into organic synthesis in high-temperature water, carried out with the MBR and CMR, showed that these properties may be exploited in several ways for green and sustainable chemistry [83–85].

Ionene, a commercial fragrance, has been prepared traditionally by treatment of α- and β-ionones with hydriodic acid containing phosphorus or by distillative heating in the presence of 0.5% iodine. In an unoptimized demonstration, cyclization occurred more cleanly and simply by MBR with β-ionone in water at 250 °C (Scheme 19) [83]. Work-up also was facilitated, as the need for removal of catalyst or the formation and separation of bi-products was avoided. At elevated temperature, the addition of water to olefins can occur readily, without adding catalyst. (S)-(+)-carvone in water for 10 min at 180 °C afforded optically pure 8-hydroxy-*p*-6-menthen-2-one as an intermediate toward carvacrol [84]. Addition of water across the 8,9-double bond of carvone occurred regioselectively, by Markovnikov addition. Carvacrol itself was obtained almost quantitatively from carvone at 250 °C after 10 min (Scheme 19) [84].

In researching the development of cleaner methods for Fischer indole synthesis, 2,3-dimethylindole was produced from phenylhydrazine and butan-2-one, by reaction in water at 220 °C for 30 min (Scheme 20) [77]. This was the first example of water as the reaction medium for Fischer indole syn-

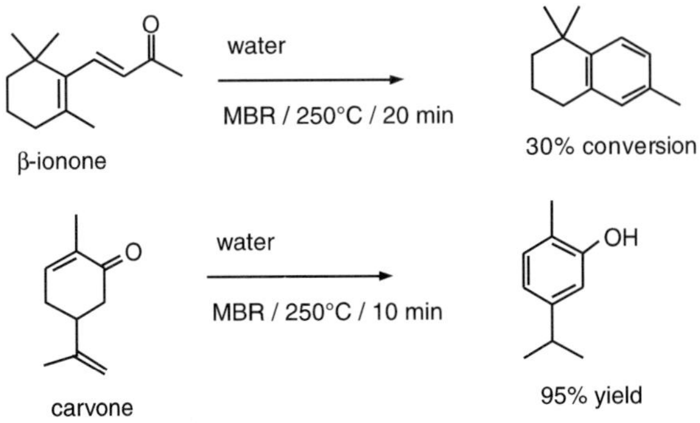

Scheme 19 Treatment of β-ionone [83] and carvone [84] in water at 250 °C

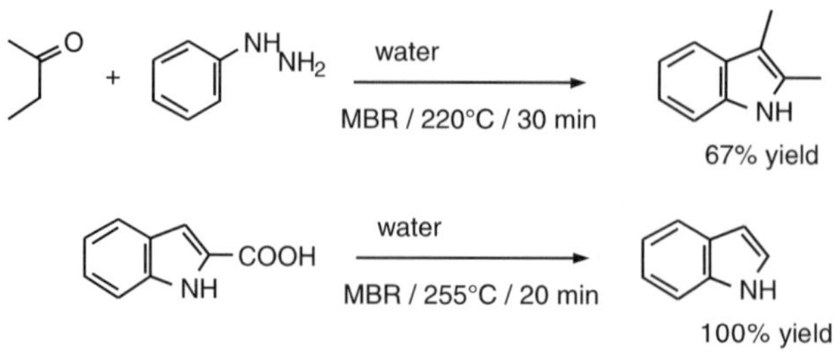

Scheme 20 Formation of 2,3-dimethylindole and decarboxylation of indole-2-carboxylic acid in water [77]

thesis and significantly, neither a preformed hydrazone nor addition of acid was required. Kremsner and Kappe have recently improved this reaction by employing an even higher temperature [86].

Published methods for decarboxylation of indole-2-carboxylic acid to form indole include pyrolysis or heating with copper-bronze powder, copper(I) chloride, "copper" chromite, "copper" acetate or copper(II) oxide, in for example, heat-transfer oils, glycerol, quinoline or 2-benzylpyridine. Decomposition of the product during lengthy thermolysis or purification affects the yields. From the perspective of green and sustainable chemistry, these methods have disadvantages associated with the choice of media and reagents as well as with the yields. Quite remarkably, however, in water at 255 °C, decarboxylation was quantitative within 20 min (Scheme 20) [77].

The processes involved in decarboxylation of indole-2-carboxylic acid and hydrolysis of the corresponding ethyl ester were optimized and found to be highly dependent on the acid/base conditions employed [87].

8.4
Aqueous *N*-Alkylation of Amines Using Alkyl Halides: Generation of Tertiary Amines and Azacycloalkanes

C – N bond formation is one of the most important transformations in organic synthesis. Amines are widely used as intermediates to prepare solvents, fine chemicals, agrochemicals, pharmaceuticals and catalysts for polymerization. The nucleophilic attack of alkyl halides by primary and secondary amines is useful for the preparation of tertiary amines but the reaction requires a longer reaction time and gives rise to a mixture of secondary and tertiary amines.

Ju and Varma envisioned that the nucleophilic substitution reaction of alkyl halides with amines may be accelerated by microwave energy because of their polar nature [88]. An environmentally friendly synthesis of tertiary amines via direct *N*-alkylation of primary and secondary amines by alkyl halides under microwave irradiation was developed (Scheme 21), that pro-

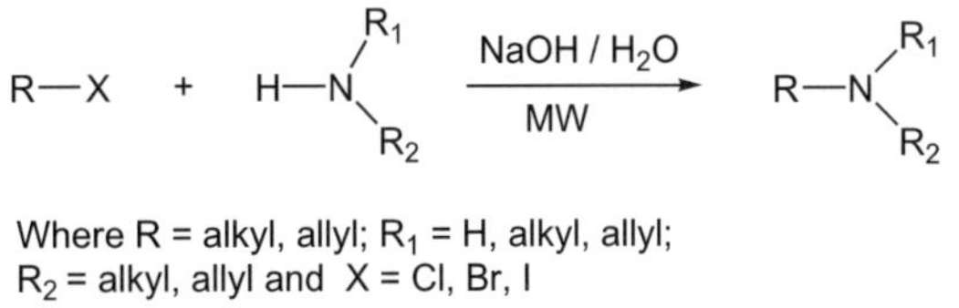

$$R\!-\!X \quad + \quad H\!-\!N\begin{array}{c}R_1\\R_2\end{array} \quad \xrightarrow[\text{MW}]{\text{NaOH / } H_2O} \quad R\!-\!N\begin{array}{c}R_1\\R_2\end{array}$$

Where R = alkyl, allyl; R_1 = H, alkyl, allyl;
R_2 = alkyl, allyl and X = Cl, Br, I

Scheme 21 Direct generation of tertiary amine in aqueous media [88]

ceeded in water in the absence of phase transfer reagent. The reaction was applicable to aliphatic halides and both aromatic and aliphatic amines.

A novel double alkylation of aniline derivatives by alkyl dihalides occurred in mildly basic aqueous media upon microwave irradiation and afforded a series of *N*-aryl azacycloalkanes (Scheme 22) [89].

This microwave-accelerated double alkylation reaction was applicable to a variety of aniline derivatives and dihalides, furnishing *N*-aryl azacycloalkanes in good to excellent yields [89]. The reaction was applicable to alkyl chlorides, bromides and iodides and was extended to include hydrazines [90]. This improved synthetic methodology provided a simple and straightforward one-pot approach to the synthesis of a variety of heterocycles such as substituted azetidines, pyrrolidines, piperidines, azepanes, *N*-substituted-2,3-dihydro-1*H*-isoindoles, 4,5-dihydro-pyrazoles, pyrazolidines, and 1,2-dihydro-phthalazines [91]. The mild reaction conditions tolerated a variety of functional groups such as hydroxyls, carbonyls, and esters.

Significantly, these reactions were not homogeneous single-phase reaction systems as neither reactant was soluble in the aqueous alkaline reaction medium. The workers postulated that selective absorption of microwaves by polar molecules and intermediates in a multi-phase system could substitute as a phase transfer catalyst without using any phase transfer reagent, thereby providing the observed acceleration similar to ultrasound irradiation [92].

In scaled-up experiments, the phase separation of the desired product in either solid or liquid form from the aqueous media facilitated product purification by simple filtration or decantation instead of by column chromatography, distillation, or extraction processes, thus reducing the use of

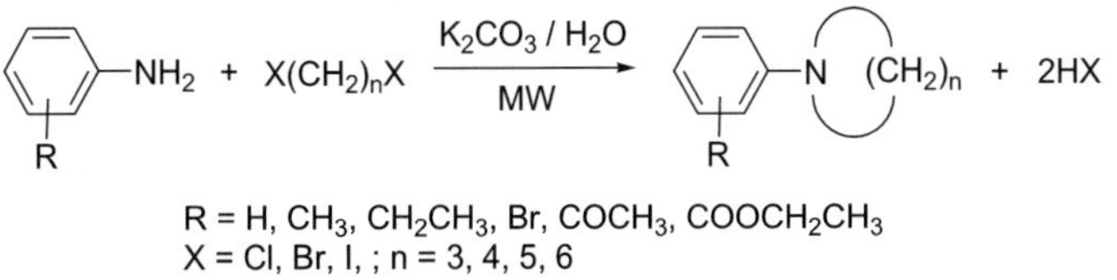

$$\text{Ar}\!-\!NH_2 \; + \; X(CH_2)_nX \quad \xrightarrow[\text{MW}]{K_2CO_3 \, / \, H_2O} \quad \text{Ar}\!-\!N\!\!\big(CH_2\big)_n \; + \; 2HX$$

R = H, CH_3, CH_2CH_3, Br, $COCH_3$, $COOCH_2CH_3$
X = Cl, Br, I, ; n = 3, 4, 5, 6

Scheme 22 Synthesis of *N*-aryl azacycloalkanes [89]

volatile organic solvent for extraction or column chromatography. In most of the reactions, completion was indicated by phase transition of lower-layered reactants to upper-layered products.

A similar strategy in aqueous media has now been applied to the nucleophilic substitution of alkyl halides or tosylates using readily available alkali azides, thiocyanates or sulfinates under microwave irradiation. The approach afforded safe and efficient preparation of azides, thiocyanates and sulfones (Scheme 23) (Ju et al., personal communications).

More than a decade ago, Sheldon observed that the manufacture of fine chemicals and pharmaceuticals generated about 25–100 times more waste than product and that it was approximately 1,000 times more wasteful than bulk chemicals production and oil refining [93, 94]. His approximations are now well accepted. He also recognized that inorganic salts accounted for the bulk of the waste, most of which resulted from neutralization of acidic or basic solutions. Apart from polluting soil and groundwater, salts can lower the pH of atmospheric moisture and may contribute to acid-dew [95] or acid-rain. For cleaner production, their minimization is essential [96].

The use of high-temperature water as a reaction medium can assist in this regard. Reactions necessitating addition of acid or base usually require less agent for high-temperature processes than do those at and below 100 °C. In many instances the requirement is orders of magnitude lower and consequently, neutralization for work-up produces much less salt. For example, a conventional route to 3-methylcyclopent-2-enone from 2,5-hexanedione employed 2–3% aqueous base at reflux and was low-yielding. With microwave heating at 200 °C, only 0.01% base was required, competing reactions were suppressed, salt formation was lowered and the enone was obtained in conversions of over 90% and isolated yields of about 85%. The preparation was readily scaled-up to a continuous process with the CMR [97].

Such impressive outcomes have led to aqueous high-temperature media, particularly with microwave heating, becoming increasingly significant for the development of clean processes. The conditions can be beneficial for a diverse range of reactions and high selectivity is often obtained through seemingly minor variations. Scale-up has been demonstrated for batchwise processes and to continuous operation with the CMR. Obvious additional advantages of the use of water include low-waste work-up, low cost, negligible toxicity and safe handling and disposal.

By working with pressure resistant quartz vessels, Kappe and Kremsner recently extended the temperature of microwave-assisted organic synthesis

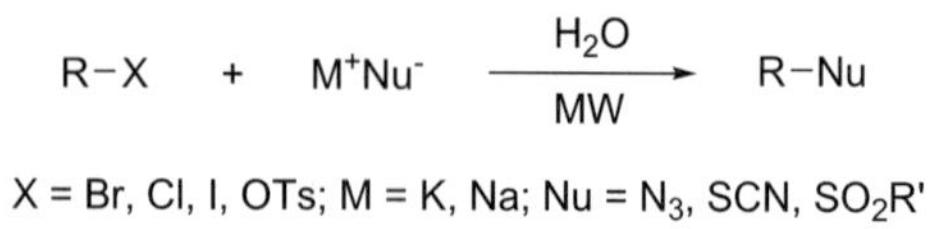

$$R-X \quad + \quad M^+Nu^- \quad \xrightarrow[\text{MW}]{\text{H}_2\text{O}} \quad R-Nu$$

$$X = Br, Cl, I, OTs; \ M = K, Na; \ Nu = N_3, SCN, SO_2R'$$

Scheme 23 Microwave-expedited nucleophilic substitutions in water

in water, to 300 °C [86]. Improved conditions were found for several different known transformations that had been carried out at lower temperatures by earlier workers. Examples included hydrolysis of esters and amides, hydration of alkynes, Diels–Alder cycloadditions, pinacol rearrangements and the Fischer indole synthesis described above, without the addition of an acid catalyst. In most instances, at the higher temperatures, enhanced yield and selectivity were obtained in lower reaction time.

9
Resin-Based Adsorption Processes
for Reactions in High-Temperature Water

Customarily, products of reactions in aqueous media are recovered by extraction with organic solvent. However, this method is not ideal for green and sustainable chemistry. The aqueous phase becomes saturated with the extracting solvent and the solvent with water. This can complicate disposal of waste and offset the environmental benefits of using water as the reaction medium in the first place. To avoid solvent extraction, hydrophobic resins have been demonstrated for concentration and isolation of the products from aqueous media. Organics were retained on the resin and desorbed with ethanol, which is a useful environmentally acceptable solvent as it is readily recyclable and is both renewable and biodegradable. It is produced fermentatively on the industrial scale and can be readily removed and recycled. Advantages of non-extractive processes include convenience, high throughput and low waste owing to ready disposal of the spent water, recyclability of the resin and the solvent used for desorption [97].

10
Differential Heating

A property seemingly unique to microwave chemistry is that the individual phases in multi-phase systems can be heated at different rates owing to differences in the dielectric properties. In some cases, a sizeable temperature difference can be maintained for several minutes. This technique has been applied usefully to produce aryl vinyl ketones batchwise by Hofmann elimination in a two-phase system comprising water and chloroform [77]. Although reactions took place in the aqueous phase, the thermally unstable products simultaneously were extracted and diluted into the cooler organic phase, which could be recycled. Yields were nearly quantitative and twice those obtained by traditional pyrolysis-distillation under vacuum.

11
Advantages of CVMC

Advantages of the CVMC approach for green and sustainable chemistry have been discussed elsewhere [77, 81, 96] and are summarized below for convenience:

11.1
Elevated Temperature

The use of higher temperatures than normal can decrease reaction times substantially. High-boiling solvents, however, are more inconvenient to use than those boiling at or below 110 °C, as they are difficult to remove and to repurify afterwards. In modern CVMC systems, temperatures exceeding the normal boiling point are routinely obtained with low-boiling solvents under pressure. Thus common solvents such as EtOH, EtOAc, MeCN, MeOH and Me_2CO, which boil below 80 °C at ambient pressure, can be used routinely for reactions at temperatures up to 200 °C or so. Not surprisingly, reaction times obtained under conventional conditions typically are lowered by up to three orders of magnitude. Volatile reactants or products such as formaldehyde, aqueous ammonia, methylamine, dimethylamine and iodine can be retained by cooling the contents of vessels before relieving the pressure.

Solvents such as EtOH and EtOAc are particularly attractive as they are obtainable from renewable resources. Although they can be readily purified, if disposal becomes necessary they are relatively environmentally benign. In contrast, several of their higher boiling counterparts have higher toxicity and can be prone to decomposition when heated to near the boiling point.

Although some benefits of high temperature can be gained with traditional autoclaves as an alternative to microwave heating, the energy is usually applied to the reaction mixture conductively, via the vessel. Thus, the rate of temperature increase is usually low, thermal gradients develop and even with stirred batch reactions, not all of the sample will be at the temperature of the applied heat. With microwaves, the whole sample can be bulk-heated and the energy input readily adjusted to match that required.

11.2
Significance of Reaction Vessels for Cleaner Processing

In microwave chemistry, vessels that are microwave-transparent will be no hotter than their contents. Those fabricated from insulating polymeric materials like polytetrafluoroethylene (PTFE) have unique advantages for cleaner processing. First, PTFE is resistant to strong bases or HF, which will attack glass. It is also stable toward halide ions which will corrode stainless steel. Minimizing waste during the cleaning of reaction vessels is another significant aspect that should not be ignored in green chemistry although it often is. The low adhesivity of materials like PTFE has already proved beneficial with domestic appliances, such as non-stick frypans. Detergent and organic solvent usage can be significantly lowered in the laboratory through the use of PTFE microwave vessels.

11.3
Exothermic Reactions

High thermal inertia of glass vessels or stainless-steel bombs not only inhibits rapid heating but also rapid cooling after reaction, which can be essential for products that require heat for their formation, but are thermally unstable. With microwaves, the applied energy can be withdrawn instantaneously at any time. The MBR has facilities for cooling and heating concurrently, if required. This helps to prevent thermal runaway that could result in unwanted by-products or decomposition. The capacity for cooling while heating now has gained prominence with the promotion of commercial systems possessing that function.

11.4
Viscous Reaction Mixtures

Typically, viscous materials transfer energy poorly. With conventional conductively heated vessels, thermal decomposition on the walls can occur at the same time as incomplete reaction towards the center of the sample. Such substantial thermal gradients can afford sub-optimal conversions. That in turn leads to loss of product and renders product isolation difficult. Furthermore, at high temperatures, heat losses with conductive heating increase and the efficiency declines. With microwave systems, these problems are not as pronounced and in some cases, even the heating efficiency can increase with temperature.

11.5
Metal-Catalyzed Reactions

The first reports of the use of microwave heating to accelerate Heck, Suzuki and Stille reactions on solid phase and in solution appeared from the laboratories of Hallberg and Larhed in 1996 [98, 99]. Their efforts and those of subsequent workers adopting their approach, have generated broad know-how for metal-catalyzed processes, which typically now can be completed within minutes by microwave "flash heating" in pressurized systems. These types of reactions have been highly beneficial for high-throughput syntheses in the area of bio-active compound discovery, but constitute extensive research areas on their own and are beyond the present scope. The reader is invited to consult an excellent recent book, several recent reviews in which numerous applications are described [18, 100–102], as well as the chapter by Nilsson, Olofsson and Larhed in this book.

11.6
Energy Savings

Gronnow et al. [103] studied the energy consumed in preparing one mole of a chemical compound by a variety of technologies, including traditional oil bath, supercritical CO_2 and microwave reactors. Reactions included two different Suzuki couplings, a Knoevenagel condensation and a Friedel–Crafts acylation. The most notable result was an 85-fold reduction in energy demand on switching from oil bath to microwave reactor for a Suzuki reaction, indicating that microwave heating can be beneficial.

12
Technology for Transposing Reaction Conditions

With conventional equipment used for the past 150 yearsor so, the vast majority of syntheses (approximately 80%) have been carried out at temperatures below 110 °C (around the boiling point of PhMe or lower) and in times between 1 hour and 4 h.

Thus, in theory at least, microwave heating could be used advantageously to reduce the time required for many traditional literature processes. With emphasis now upon speed and efficiency in synthesis, many old established processes could even be lost if the conditions for performing them cannot be improved. To address that emerging problem, a generic technology was developed [13]. It was aimed at enabling known conditions for *any* thermal reaction to be translated readily and directly into improved protocols, without the need for extensive empirical studies regarding kinetics and/or optimization. Based upon an iterative, interactive approach, the technology treated time, temperature and yield or conversion as equally weighted parameters that could be traded against one another according to the chemist's wishes. From one set of experimental data for a reaction, conditions could be calculated for any time (between 0.1 and 10^5 min) and temperature (between 0 °C and 270 °C) to afford a nominated yield or conversion between 5 and 100%. A satisfactory result usually was obtained within two iterations and a third iteration was rarely required. The technology was shown to be versatile and effective for a diverse range of unrelated reactions.

13
Conclusion

In this chapter, we have outlined roles of microwave chemistry in the establishment of green and sustainable chemistry. Many examples, mostly from the authors' laboratories, have been presented of green microwave processes under solvent-free conditions or with solvents, including high-temperature water. New reactions or improved conditions have been discussed, along with predictive technology for optimization of reactions.

References

1. Szmant HH (1989) Organic Building Blocks for Organic Chemistry. Wiley, New York
2. Blumberg AA (1994) J Chem Educ 71:912
3. Garfield S (2000) Mauve. Faber and Faber, London
4. Carson R (1962) Silent Spring. Houghton Mifflin, New York
5. Strauss CR, Scott JL (2001) Chem Ind (London), p 610
6. McKinnon D (1981) Chem & Eng News 59(24):5
7. Allen B (2000) Green Chem 2:G56
8. Anastas PT, Warner J (1998) Green Chemistry; Theory and Practice. Oxford Science Publications, Oxford
9. Anastas PT, Williamson TC (1998) Green Chemistry: Frontiers in Benign Chemical Syntheses and Processes. Oxford University Press, Oxford
10. Jackson T (1996) Material Concerns: Pollution, Profit and Quality of Life. Routledge, London
11. Gedye R, Smith F, Westaway K, Ali H, Baldisera L, Laberge L, Rousell J (1996) Tetrahedron Lett 27:279
12. Giguere RJ, Bray TL, Duncan SM, Majetich G (1986) Tetrahedron Lett 27:4945
13. Roberts BA, Strauss CR (2005) Acc Chem Res 38:563
14. Trost BM (1991) Science 254:1471
15. Trost BM (1995) Angew Chem, Int Ed Engl 34:259
16. Gedye RN, Smith FE, Westaway KC (1988) Can J Chem 66:17
17. Loupy A (2002) Microwaves in Organic Synthesis. Wiley, Weinheim
18. Kappe CO, Stadler A (2005) Microwaves in Organic and Medicinal Chemistry. Wiley, Weinheim
19. Loupy A, Bram G, Sansoulet J (1992) New J Chem 16:233
20. Gutierrez E, Loupy A, Bram G, Ruiz-Hitzky E (1989) Tetrahedron Lett 30:945
21. Bram G, Loupy A, Majdoub M (1990) Synthetic Commun 20:125
22. Villemin D, Labiad B, Ouhilal Y (1989) Chem Ind (London), p 607
23. Villemin D, Ben Alloum A (1990) Synth Commun 20:925
24. Ben Alloum A, Labiad B, Villemin D (1989) Chem Commun, p 386
25. Latouche R, Texier-Boullet F, Hamelin J (1991) Tetrahedron Lett 32:1179
26. Pilard J-F, Klein B, Texier-Boullet F, Hamelin J (1992) Synlett, p 219
27. Varma RS, Varma M, Chatterjee AK (1993) J Chem Soc Perkin Trans I, p 999
28. Varma RS, Lamture JB, Varma M (1993) Tetrahedron Lett 34:3029
29. Bose AK, Manhas MS, Ghosh M, Shah M, Raju VS, Bari SS, Newaz SN, Banik BK, Chaudhary AG, Baraket KJ (1991) J Org Chem 56:6968

30. Banik BK, Manhas MS, Kaluza Z, Barakat KJ, Bose AK (1992) Tetrahedron Lett 33:3603
31. Alvarez C, Delgado F, García O, Medina S, Márquez C (1991) Synth Commun 21:619
32. Delgado F, Alvarez C, García O, Penieres G, Márquez C (1991) Synth Commun 21:2137
33. Loupy A, Petit A, Hamelin J, Texier-Boullet F, Jacquault P, Mathé D (1998) Synthesis, p 1213
34. Varma RS (1999) Green Chem 1:43
35. Varma RS (2001) Pure Appl Chem 73:193
36. Diaz-Ortiz A, de la Hoz A, Langa F (2000) Green Chem 2:165
37. Fubini B, Otero Arean C (1999) Chem Soc Rev 28:373
38. Ricard M, Villemin D (1984) Tetrahedron Lett 25:1059
39. Varma RS, Kumar D, Liesen PJ (1998) J Chem Soc Perkin Trans I, p 4093
40. Varma RS (2002) Tetrahedron 58:1235
41. Csiba M, Cléophax J, Loupy A, Malthête J, Gero S (1993) Tetrahedron Lett 34:1787
42. Varma RS, Saini RK, Dahiya R (1997) Tetrahedron Lett 38:7823
43. Varma RS, Saini RK (1998) Tetrahedron Lett 39:1481
44. Varma RS, Dahiya R (1997) Tetrahedron Lett 38:2043
45. Laszlo P (1986) Acc Chem Res 19:121
46. Cornelius A, Laszlo P (1994) Synlett, p 155
47. Varma RS, Dahiya R, Saini RK (1997) Tetrahedron Lett 38:7029
48. Varma RS, Saini RK, Dahiya R (1998) J Chem Res (S), p 120
49. Varma RS, Saini RK, Meshram HM (1997) Tetrahedron Lett 38:6525
50. Varma RS, Saini RK (1997) Tetrahedron Lett 38:4337
51. Erb WT, Jones JR, Lu S-Y (1999) J Chem Res (S), p 728
52. Elander N, Jones JR, Lu S-Y, Stone-Elander S (2000) Chem Soc Rev 29:239
53. Fodor-Csorba K, Galli G, Holly S, Gacs-Baitz E (2002) Tetrahedron Lett 43:4337
54. Varma RS, Dahiya R (1998) Tetrahedron 54:6293
55. Tanaka K (2003) Solvent-Free Organic Synthesis. Wiley, Weinheim
56. Vidal T, Petit A, Loupy A, Gedye RN (2000) Tetrahedron 56:5473
57. Perreux L, Loupy A, Volatron F (2002) Tetrahedron 58:2155
58. Perreux L, Loupy A, Delmotte M (2003) Tetrahedron 59:2185
59. Varma RS, Namboodiri VV (2001) Chem Commun, p 643
60. Deetlefs M, Seddon KR (2003) Green Chem 5:181
61. Vo Thanh G, Pegot B, Loupy A (2004) Eur J Org Chem, p 1112
62. Varma RS (2003) In: Rogers R, Seddon KR (eds) Ionic Liquids as Green Solvents Progress and Prospects. ACS Symposium Series 856. Am Chem Soc, Washington, p 82
63. Namboodiri VV, Varma RS (2002) Tetrahedron Lett 43:5381
64. Kim YJ, Varma RS (2005) J Org Chem 70:7882
65. Varma RS (1999) J Heterocycl Chem 36:1565
66. Kappe CO, Kumar D, Varma RS (1999) Synthesis, p 1799
67. Varma RS, Kumar D (1999) Tetrahedron Lett 40:7665
68. Ju Y, Li C-J, Varma RS (2004) QSAR Comb Sci 23:891
69. Varma RS, Dahiya R (1998) J Org Chem 63:8038
70. Varma RS, Kumar D, Liesen PJ (1998) J Chem Soc Perkin Trans I, p 4093
71. Ješelnik M, Varma RS, Polanc S, Kočevar M (2001) Chem Commun, p 1716
72. Loupy A, Pettit A, Bogdal D (2002) In: Loupy A (ed) Microwaves in Organic Synthesis. Wiley, Weinheim, p 147
73. Bose AK, Manhas MS, Banik BK, Robb EW (1994) Res Chem Intermed 20:1

74. Baghurst DR, Mingos DMP (1992) Chem Commun, p 674
75. Gabriel C, Gabriel S, Grant EH, Halstead BSJ, Mingos DMP (1998) Chem Soc Rev 27:213
76. Dabirmanesh Q, Roberts RMG (1993) J Organomet Chem 460:C28
77. Strauss CR, Trainor RW (1995) Aust J Chem 48:1665
78. Cablewski T, Faux AF, Strauss CR (1994) J Org Chem 59:3408
79. Raner K, Strauss C, Constable D, Somlo P (1992) J Microwave Power Electromag Energy 26:195
80. Raner KD, Strauss CR, Trainor RW, Thorn JS (1995) J Org Chem 60:2456
81. Strauss CR (2002) In: Clark J, Macquarrie D (eds) Handbook of Green Chemistry & Technology. Blackwell, London, p 397
82. Bagnell L, Cablewski T, Strauss CR (1999) Chem Commun 283
83. An J, Bagnell L, Cablewski T, Strauss CR, Trainor RW (1997) J Org Chem 62:2505
84. Strauss CR, Trainor RW (1996) In: Takeoka GR, Teranishi R, Williams PJ, Kobayashi A (eds) Biotechnology for Improved Foods and Flavors. ACS Symposium Series 637. Am Chem Soc, Washington, DC, p 272
85. Bagnell L, Cablewski T, Strauss CR, Trainor RW (1996) J Org Chem 61:7355
86. Kremsner JM, Kappe CO (2005) Eur J Org Chem 3672
87. Strauss CR, Trainor RW (1998) Aust J Chem 51:703
88. Ju Y, Varma RS (2004) Green Chem 6:219
89. Ju Y, Varma RS (2005) Org Lett 7:2409
90. Ju Y, Varma RS (2005) Tetrahedron Lett 46:6011
91. Ju Y, Varma RS (2006) J Org Chem 71:135
92. Varma RS, Naicker KP, Kumar D (1999) J Mol Catal A: Chem 149:153
93. Sheldon RA (1994) CHEMTECH 24(3):38
94. Sheldon RA (1992) Chem Ind (London), p 903
95. Okochi H, Kajimoto T, Arai Y, Igawa M (1996) Bull Chem Soc Jpn 69:3355
96. Strauss CR (1999) Aust J Chem 52:83
97. Bagnell L, Bliese M, Cablewski T, Strauss CR, Tsanaktsidis J (1997) Aust J Chem 50:921
98. Larhed M, Lindeberg G, Hallberg A (1996) Tetrahedron Lett 37:8219
99. Larhed M, Hallberg A (1996) J Org Chem 61:9582
100. Kappe CO (2004) Angew Chem, Int Ed 43:6250
101. Lidstrom P, Tierney J, Watthey B, Westman J (2001) Tetrahedron 57:9225
102. Larhed M, Moberg C, Hallberg A (2002) Acc Chem Res 35:717
103. Gronnow MJ, White RJ, Clark JH, Macquarrie DJ (2005) Org Process Res Dev 9:516

Top Curr Chem (2006) 266: 233–278
DOI 10.1007/128_048
© Springer-Verlag Berlin Heidelberg 2006
Published online: 15 July 2006

The Scale-Up of Microwave-Assisted Organic Synthesis

Jennifer M. Kremsner[1] · Alexander Stadler[2] · C. Oliver Kappe[1] (✉)

[1]Christian Doppler Laboratory for Microwave Chemistry (CDLMC) and Institute
of Chemistry, Karl Franzens University Graz, Heinrichstraße 28, 8010 Graz, Austria
oliver.kappe@uni-graz.at

[2]Anton-Paar GmbH, Anton-Paar-Straße 20, 8054 Graz, Austria

Abstract Single-mode microwave reactors have been very successful in the past few years in the field of method development and optimization. Future trends clearly indicate the use of microwave technology for the development of completely new reaction routes for organic synthesis, with increasing demand for large-scale microwave production of chemical substances (> 100 g per run). For microwave-assisted synthesis to become a fully accepted industrial technology in the future, there is a need to develop techniques that can ultimately routinely provide products on a multikilogram scale. Thus, the further development of large reactors, at least in the pilot plant scale to enable multikilogram production of lead compounds, is required. Herein are discussed several already known scale-up processes in microwave-assisted organic chemistry in a ≥ 50 mL batch scale and experiments performed with both continuous flow and stop-flow techniques. Furthermore, the available instrumentation for scale-up at laboratory scale and beyond is presented.

Keywords Batch · Continuous flow · Microwave · Scale-up · Stop-flow

Abbreviations

AcOH	Acetic acid
aq	Aqueous

BTF	Benzotrifluoride (trifluoromethyl benzene)
t-Bu	BuOH, 1-butanol
CF	Continuous flow
CFR	Continuous flow reactor
conv	Conversion
CMDR	Continuous microwave dry-media reactor
Cy	Cyclohexyl
DBU	1,8-Diazabicyclo[5.4.0]undec-7-ene
DI(P)EA	Diisopropyl ethylamine
dmphen	2,9-Dimethyl-[1,10]phenanthroline
GHz	Gigahertz
GLP	Good laboratory practice
GMP	Good manufacturing practice
HIV	Human immunodeficiency virus
HPLC	High-performance liquid chromatography
K10	Montmorillonite K10 clay
KSF	Montmorillonite KSF clay
LC-MS	Liquid chromatography-mass spectroscopy
MAOS	Microwave-assisted organic synthesis
MBR	Microwave batch reactor
MW	Microwave irradiation
NMM	*N*-Methyl morpholine
NMP	*N*-Methyl pyrrolidinone
ppb	Parts per billion
ppm	Parts per million
PTFE-TFM	Modified branched polytetrafluoroethylene
RCM	Ring closing metathesis
SF	Stop-flow
TBAB	Tetrabutylammonium bromide
TBAI	Tetrabutylammonium iodide
TEA	Triethylamine
o-tolyl	*ortho*-Methylphenyl

1
Introduction

1.1
Microwave-Assisted Synthesis – A Brief History

In the past two decades, heating and accelerating chemical reactions by microwave energy have been increasingly popular themes in the scientific community [1, 2]. Microwave energy has found a variety of technical applications in chemical and related industries since the 1950s, in particular in the food-processing, drying, and polymer industries. Other applications range from analytical chemistry (microwave digestion, ashing, extraction) [3], to biochemistry (protein hydrolysis, sterilization) [3], pathology (histoprocessing, tissue fixation) [4] and medical treatments (diathermy) [5]. However, microwave heating was not introduced to organic synthesis until the mid 1980s. The first academic reports on the use of microwave heating to mediate organic chemical reactions were published by the groups of Gedye [6] and Giguere [7] in 1986. These early experiments of microwave-assisted organic synthesis (MAOS) were typically carried out in sealed Teflon or glass vessels in a domestic household microwave oven without any temperature or pressure measurements. Although there were several violent explosions due to the rapid uncontrolled heating of organic solvents in those early days, an increasing number of scientists started to investigate this new technology. In the 1990s the first attempts at solvent-free microwave chemistry (so-called dry-media reactions), which eliminated the danger of explosions, were made [8–13]. Particularly at the beginning of MAOS, the solvent-free approach was very popular since it allowed the safe use of domestic microwave ovens and standard open vessel technology. While a large number of interesting transformations using dry-media reactions have been published [8–13], technical difficulties relating to non-uniform heating, mixing, and precise determination of the reaction temperature remained unsolved, especially when scale-up approaches needed to be considered.

Besides the dry-media attempts, microwave-assisted synthesis in solution has been carried out under open vessel conditions. However, if solvents are heated by microwave irradiation at atmospheric pressure in an open vessel, the boiling point of the solvent limits the reaction temperature. In order to nonetheless achieve high reaction rates, good microwave-absorbing solvents with high boiling points have been frequently used in open-vessel microwave synthesis [14, 15]. However, the use of such solvents (e.g. DMF, NMP, ethylene glycol) presented serious challenges during product isolation and recycling of the solvent. Because of the recent availability of modern microwave reactors with on-line monitoring of both temperature and pressure, MAOS in dedicated sealed vessels using standard solvents – a technique pioneered by Christopher R. Strauss in the mid 1990s [16–18] – has renewed its attractiveness to chemists in recent years. This is clearly evident from a survey of the recently published (since 2001) literature in the area of controlled MAOS. Due to the beneficial combination of rapid heating by microwaves with sealed vessel (autoclave) technology, this approach will most likely be the method of choice for performing MAOS on a laboratory scale in the future. Additionally, recent innovations in microwave reactor technology allow controlled parallel and automated sequential processing under sealed vessel conditions, and the use of continuous flow (CF) reactors or stop-flow (SF) reactors for scale-up purposes, as will be discussed in the following sections.

Regardless of the nature of the observed rate-enhancements ("microwave effects" [19, 20]), which will undoubtedly be an ongoing debate for many years in the academic world, microwave synthesis has now truly matured and has moved from a laboratory curiosity in the late 1980s to an established technique in organic synthesis, popular in both academia and industry. The initial low interest in the technology in the late 1980s and 1990s has been attributed to its lack of controllability and reproducibility, coupled with a general lack of understanding of the basics of microwave dielectric heating. The risks associated with the flammability of organic solvents in a microwave field and the lack of available dedicated microwave reactors allowing for adequate temperature and pressure control were major concerns.

Today's available instrumentation (Sect. 2), "dedicated" microwave reactors, allow for careful control of time, temperature, and pressure profiles, ensuring reproducible protocol development, scale-up, and transfer from laboratory to laboratory and from instrument to instrument. In the so-called multimode instruments (conceptually similar to a domestic oven), the microwaves that enter the cavity are being reflected by the walls and the load over the typically large cavity. In the much smaller single-mode cavities, only one mode is present and the electromagnetic irradiation is directed through an accurately designed rectangular or circular wave guide onto the reaction vessel mounted in a fixed distance from the radiation source, creating a standing wave. The key difference between the two types of reactor systems is that whereas in multimode cavities several reaction vessels can be irradiated simultaneously in multivessel rotors (parallel synthesis), in single-mode systems only one vessel can be irradiated at a time.

Since the year 2000, when the first dedicated single-mode reactors were established among academic laboratories, the number of publications related to MAOS has increased dramatically, reaching an overall number of about 3000 by the end of 2005. Considering this fact, it might be assumed that in a few years most chemists will use microwave energy as the standard procedure to heat chemical reactions on a laboratory scale [1, 2]. Besides the drastic reduction of reaction times, microwave heating is also known to suppress side reactions, increase yields, and improve purity and reproducibility. Therefore, many academic and industrial research groups are already using MAOS as a technology for rapid reaction optimization, efficient synthesis of new chemical entities, and for exploring chemical reactivity.

1.2
The Need for Scale-Up of Microwave-Assisted Transformations

Most examples of microwave-assisted chemistry published until this day have been performed on a scale of less than 1 g, with a typical reaction volume of 1–5 mL. The main applications have been to accelerate and optimize well-known and established synthetic procedures. This is in part a consequence of the availability and popularity of dedicated single-mode microwave reactors (i.e., Biotage Initiator, CEM Discover models) that allow the safe processing of small reaction volumes under sealed vessel conditions by microwave irradiation. Due to limitations in the vessel and microwave cavity size of these single-mode reactors, microwave-assisted synthesis so far has focused predominantly on reaction optimization and method development on the small scale (< 10 mmol). Although these instruments have been very successful in this field, future trends indicate the use of microwave technology for the development of completely new reaction avenues for organic synthesis. Today, the demand for large-scale microwave production (> 100 g per run) of chemical substances increases not only within research and development departments.

To accomplish the task of industrial scale-up, the development of effective flow-through systems is currently under investigation. It seems to be clear that for microwave-assisted synthesis to become a fully accepted industrial technology in the future, there is a need to develop larger scale MAOS techniques that can ultimately routinely provide products on a multikilogram (or even higher) scale. Thus, the further development of large reactors, at least in the pilot plant scale, to enable multikilogram production of lead compounds, is required. As discussed in this chapter and in particular in Sect. 3, several microwave-assisted scale-up processes are already known today.

1.3
Scale-Up Limitations

Microwave-enhanced chemistry is based on the efficient heating of materials by "microwave dielectric heating" effects. Microwave dielectric heating is dependent on the ability of a specific material (i.e., a solvent or reagent) to absorb microwave energy and convert it into heat. When irradiated at microwave frequencies, the dipoles or ions of the sample align in the applied electric field. As the applied field oscillates, the dipole or ion field attempts to realign itself with the alternating electric field and, in the process, energy is lost in the form of heat through molecular friction and dielectric loss. The ability of a specific substance to convert electromagnetic energy into heat at a given frequency and temperature is determined by the so-called loss tangent $\tan \delta$. A reaction medium with a high $\tan \delta$ is required for efficient absorption and, consequently, for rapid heating. Further details on the rather complex microwave dielectric heating phenomena are described in relevant review articles [21–23].

The big challenge for process scale-up involving microwave technology is to establish a reliable and safe process setup where issues like physical properties (i.e., penetration depth), temperature control, and reactor design have to be carefully considered. Using batch reactors, the user is confronted with problems in heating large volumes due to the limited penetration depth of microwave irradiation. Dependent on the dielectric properties of the solvent (loss tangent, $\tan \delta$, see Table 1), microwave penetration into absorbing media (i.e., reaction mixtures) is usually only in the order of a few centimeters (Table 2), which limits the maximum size of a batch reactor.

Table 1 Loss tangents (tan δ) of different solvents (2.45 GHz, 20 °C)[a]

Solvent	tan δ	Solvent	tan δ
Ethylene glycol	1.350	DMF	0.161
Ethanol	0.941	1,2-dichloroethane	0.127
DMSO	0.825	Water	0.123
2-Propanol	0.799	Chlorobenzene	0.101
Formic acid	0.722	Chloroform	0.091
Methanol	0.659	Acetonitrile	0.062
Nitrobenzene	0.589	Ethyl acetate	0.059
1-Butanol	0.571	Acetone	0.054
2-Butanol	0.447	Tetrahydrofuran	0.047
1,2-Dichlorobenzene	0.280	Dichloromethane	0.042
NMP	0.275	Toluene	0.040
Acetic acid	0.174	Hexane	0.020

[a] Data from [24]

Table 2 Microwave penetration depth (245 GHz) in some common materials[a]

Material	Temperature [°C]	Penetration Depth [cm]
Water	25	1.4
Water	95	5.7
Ice	−12	1100
Glass	25	35
Porcelain	25	56
Poly (vinyl chloride)	20	210
Teflon®	25	9200
Quartz glass	25	16 000

[a] Data from [25]

These limitations have been overcome in part by using a parallel scale-up approach, which means that the same experiments are run simultaneously using several (smaller) vessels. However, this methodology again creates two major problems. Firstly, as the reaction volume increases, it becomes more and more difficult to heat up the reaction mixtures. Consequently, more microwave power is needed to reach identical reaction temperatures when performing large scale experiments. Generally, the most common multimode instruments for microwave-assisted synthesis (Biotage Advancer, Milestone MicroSYNTH and UltraCLAVE, CEM MARS-S, Anton Paar Synthos 3000) provide a comparatively high microwave power output (> 1000 W) from standard air-cooled magnetrons, which prove sufficient to effectively heat mixtures of up to 500 mL in comparable rates comparable to those in single-mode experiments [26]. As the capacity of the microwave reactor increases to 5000 W and beyond, more sophisticated oil-based or water-based cooling is required and this introduces extra size, complexity, and cost to microwave reactors [16–18]. Another aspect is that the energy efficiency of the conversion of electricity into microwave power can be relatively low (70% or less), which can make the microwave approach less attractive for very large scale preparations [16–18].

Obviously, heating rates are somewhat lower when processing larger volumes as compared to small volumes, but no adverse effects on the yield or product purity are usually noted. The crucial point for the reaction outcome is generally the overall hold time at the desired reaction temperature. On the other hand, difficulties may arise in attempting the scale-up of small-scale procedures if low absorbing solvents, such as toluene, are employed [27]. In this case, a certain concentration of high microwave-absorbing reagents should be present in the reaction mixture to introduce some heat. If the concentration is low and the used vessel materials are not microwave self-absorbing (thus heating up the reaction mixture by convection and conduction phenomena), almost no temperature increase can be observed.

An additional key point in processing comparatively large volumes under pressure in a microwave field is the safety aspect, as any malfunction or rupture of a large pressurized reaction vessel will have a significant impact. Thus, minimizing the reaction volumes in the microwave contact zone would reduce the risk of this hazard significantly. Furthermore, time becomes an issue when performing parallel reactions, keeping in mind that the individual reaction vessels need to be filled with reagents and loaded into the cavity manually. Thus, flow systems operate more effectively on a larger scale. Accordingly, published examples of MAOS scale-up experiments (> 500 mL) are rare, particularly those involving complex organic reactions. However, plans to manufacture larger microwave systems in batch and CF mode are currently being considered as the demand of industrial laboratories, in particular from the pharmaceutical and fine chemical industries, is increasing.

These are the main reasons for the development of CF reactors, where the reaction mixture is passed through a relatively small microwave-heated flow cell, thus not only minimizing the hazards but also avoiding the above-mentioned penetration depth issues. Other benefits of CF reactors are financial. CF reactors are usually less expensive than commercially available batch reactors and they can be implemented in industrial processes where their operation can be fully automated. Additional benefits of using CF processing have been demonstrated in small-scale microwave protocols involving heterogeneous catalysis, employing a custom-built microflow reactor [28–30].

Filling the flow cell with an immobilized catalyst provides effective product isolation and improved control of the catalyst–reactant contact time. Moreover, the catalyst is only required in comparatively small quantities and can be easily recovered for reuse.

The major drawback of CF processing, however, is the incompatibility with heterogeneous reaction mixtures and highly viscous liquids. Apart from the fact that obtaining a temperature measurement directly from a reaction mixture under flow can be complicated, the adaptation of conditions from small-scale batch reactions to a CF cell may be time-consuming.

A combination of certain advantages from batch and CF processing can be achieved with a so-called stop-flow reactor [27, 31]. Such a system provides a small reaction cavity and a peristaltic pump that pumps the reaction mixture in and out of the reaction vessel (e.g. CEM Voyager reactor). Apparently, considerable time is required for the entire process, including loading the vessel, irradiation of the reaction mixture including the time for reaching the set temperature, cooling down, and finally pumping out the product. But nevertheless, in contrast to batch processing, this can be a fully automated procedure improving the efficiency of microwave-assisted synthesis [27, 31].

In conclusion, there are three different approaches for microwave synthesis on a large scale (> 100 mL volume). While some groups have employed larger batch-type multimode or single-mode reactors ($\leq$ 1000 mL processing volume), others have used CF or SF techniques (multi- and single-mode cavities) to overcome the inherent problems associated with MAOS scale-up.

1.4
The Range of Scale-Up

The scale-up of microwave-mediated transformations can be defined in different ranges, leading to the use of different concepts and varying instrumentation. Depending on the user, the term "scale-up" will have different meanings. In the case of method development, scale-up starts with processing a 50 mL reaction mixture, corresponding to a ten- to 100-fold scale-up of reactions performed in standard single-mode microwave vials with a 0.5–5.0 mL processing volume. In fact, this is a significant amount from a medicinal chemist's point of view, providing multigram product quantities for, e.g., biological screening or pharmacokinetic studies. Employing different vessel types, today's single-mode instruments enable similar scale-up ranges for batch synthesis from 0.2 to 20 mL (Biotage Initiator EXP Series) or from 0.5 to 50 mL (CEM Discover Platform, see Sect. 2), respectively.

A possibility for further scale-up using the above-mentioned instrumentation would be using the "numbering up" approach, e.g., running the same reaction several times in sequence. This approach is aided by existing robotic equipment, which allows unattended vessel transfer in and out of the microwave cavity. Alternatively, those reactions can also be performed by parallel synthesis in corresponding multivessel rotor systems switching to multi-mode instruments ([32] and references cited therein).

From an industrial viewpoint, scale-up means process development and highest possible throughput that virtually excludes the use of batch reactors. In fact, the productivity and not the size of the vessel is important, which clearly indicates that flow systems, regardless of whether applied in SF or CF manner, have distinct benefits over batch process reactors.

Scale-up as defined for this chapter covers batch reactions in closed vessels at the ≥ 50 mL scale, flow systems employing flow cells ≥ 5 mL, and SF containers of ≥ 50 mL volume. Microwave-mediated scale-up reactions under open vessel conditions are not discussed in detail as up to date only a few examples have been published [33–37] and most of the beneficial rate enhancements have been reported under sealed vessel conditions.

2
Different Techniques and Instrumentation

The scale-up of microwave-assisted reactions is of significant interest to many industrial laboratories. Scale-up can be accomplished in different ways, and these methods are presented in more detail in the following section. After an initial discussion of batch synthesis we will present the currently commercially available instrumentation for flow processing, which can be divided into SF or CF techniques.

When going to larger scale with successfully optimized reactions, a major issue is the direct scalability of the method, i.e., the operation at different scales without the need for further optimization steps. In this context, batch synthesis (defined as processing a mixture in one vessel at a time) has a definitive advantage over flow techniques since reaction times are usually identical to small-scale experiments, whereas flow processes require longer reaction times. Batch synthesis in general allows for the use of heterogeneous and highly viscous mixtures or suspensions, whereas the risk of clogging of lines is evident when performing such transformations under CF. Furthermore, precipitation of the products or starting materials may be troublesome in flow reactors. On the other hand, the safety issues are less of a concern for flow systems, as the heating zones are smaller and lower volumes are processed in the microwave irradiation area. Moreover, the energy transmission is higher in smaller cavities and therefore the process is more economical. Furthermore, physical limitations regarding penetration depth and power constraints limit the maximum size of a single batch reaction container.

Today's laboratory-scale batch microwave reactors generally offer a maximum batch size of 1 L reaction volume, in most cases divided into several smaller reaction vessels (multivessel rotor systems). According to the definition given at the beginning, this would not exactly match the "one vessel" philosophy, but the parallel setup allows for the processing of several batches of either the same reaction mixture or of closely related mixtures. In fact, this combination of batch synthesis and parallel processing can furnish either compound libraries on the gram scale or a larger amount of one single compound in a short time. Although the Milestone UltraCLAVE system provides a 3.5 L single vessel cavity, it has turned out to be more effective to heat smaller volumes in parallel rather than one big batch, given that identical microwave output power is applied [27, 37]. This is caused by limitations due to the penetration depth (see above), as microwaves lose their intensity when interacting with dielectric media. On the other hand, this indicates clearly that any attempt to process larger volumes has to be applied using flow-through systems, either in CF or SF mode.

2.1
Batch Processing and Parallel Synthesis

Laboratory scale-up in *single-mode reactors* to produce gram amounts of material can be performed either by the above-mentioned sequential batch processing using various vessel sizes (up to 50 mL) or by employing CF or SF reaction cells (5–50 mL). Conversely, *multimode instruments* allow for parallel synthesis or applications in large batches up to 1 L total volume and even CF and SF approaches utilizing $\geq$ 300 mL cells (see below).

Batch synthesis in single-mode reactors is definitely limited in scale as the size of the utilized microwave cavities is restricted to being monomodal. However, the Biotage Initiator EXP series allows a 100-fold linear scale-up when employing the different available vessel sizes, going from 0.2 mL to 20 mL operation volume (Fig. 1). Repetitive reaction cycles using the au-

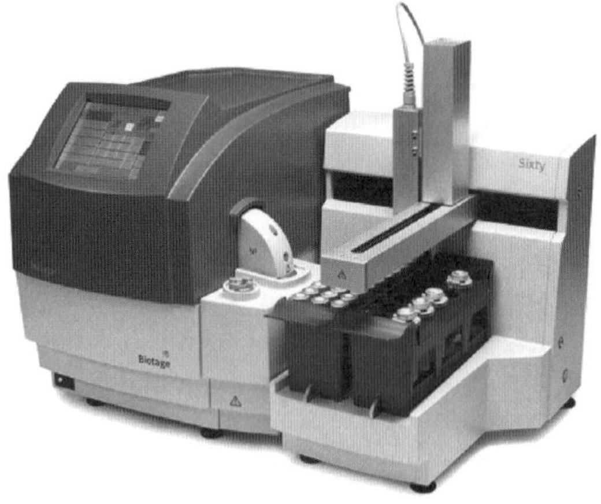
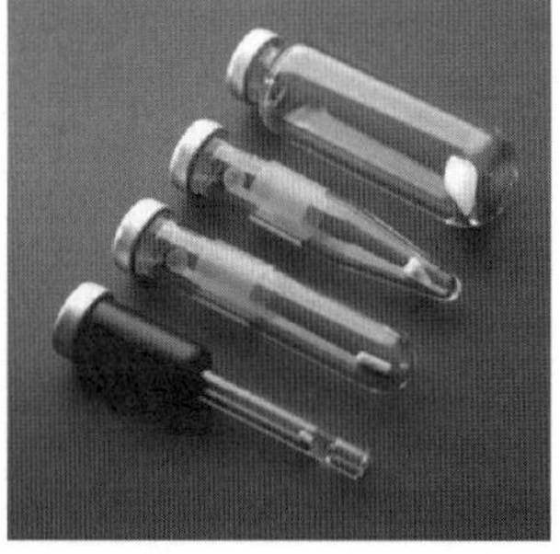

Fig. 1 Biotage Initiator EXP Sixty and its various reaction vessels

tomation upgrade Initiator Sixty (up to 60×5.0 mL or 24×20 mL) allows multigram synthesis by sequential batch reactions [38].

Similarly, the CEM Discover platform (Fig. 2) allows a 100-fold scale-up of small-scale batch synthesis when switching from the standard vial (0.5–5.0 mL operation volume) to the large reaction vessel (50 mL maximum filling volume). Automation is only possible for the small standard vials with the "Explorer" robotic extension (24×5.0 mL) but the large vessel can be utilized for SF processing as well [27, 39].

An interesting contribution to the field of single-mode equipment has been recently presented by Milestone. The MultiSYNTH instrument (Fig. 3) is a kind of "hybrid" reactor as it combines the inherent advantages of single-

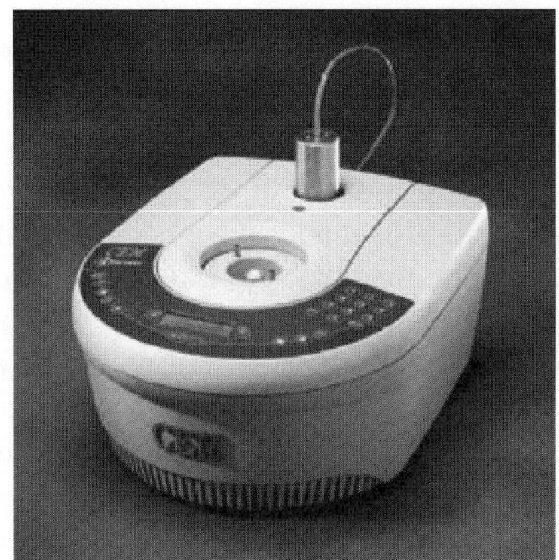

Fig. 2 CEM Discover

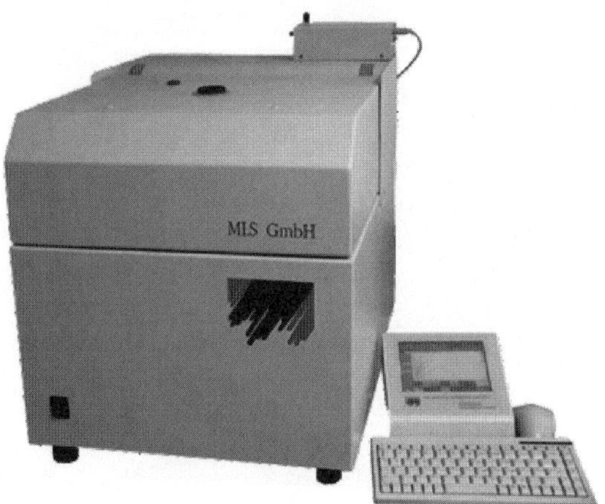

Fig. 3 Milestone MultiSYNTH Hybrid Instrument

mode technology with beneficial features of multimode cavities. A single magnetron delivers 800 W pulsed or unpulsed microwave output power. In the single-mode setup the corresponding reaction vessel (0.25–35 mL operating volume) is located in a position correlated with the highest energy intensity. A unique vessel vibration system ensures homogeneous temperature distribution even in the large vessel. In the multimode setup the cavity can be equipped with a carousel suitable for up to 12 reaction vessels (1–5 mL operation volume). Furthermore simple laboratory glassware can be employed for reflux reactions at atmospheric pressure up to 500 mL volume. Sufficient agitation in each vessel can be accomplished by adding magnetic stir bars. Temperature measurement is achieved by an IR sensor mounted in the side wall, optionally an immersed fiber-optic probe can be utilized. Each vessel is assembled with a pressure jacket comprising a spring-type safety valve for individual pressure control. The multimode carousel comprises glass or quartz vessels in three sizes featuring operation limits of 200 °C and 20 bar. For single-mode applications additional 70 mL TFM-vessels (35 mL operation volume) are available for reactions up to 200 °C and 30 bar. Reactions under reflux can be carried out at a maximum of 250 °C. However, at the time of writing, no applications with this device have been published.

In contrast to single-mode reactors, dedicated multimode instruments allow scale-up to be performed in multivessel rotor systems utilizing various types of sealed vessels. In these systems, reactions can be carried out in batch to synthesize multiple gram quantities (≤ 250 g) of material in typically up to 1 L processing volume. Most of the multimode instruments available for organic synthesis have been derived from closely related sample preparation equipment [39–41]. The MARS Microwave Synthesis System (Fig. 4) is based

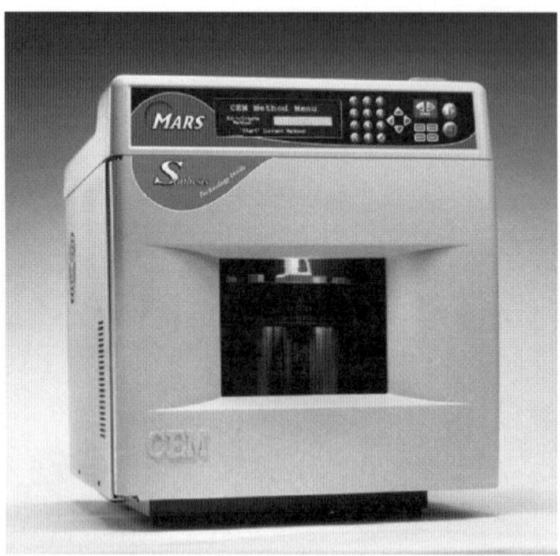

Fig. 4 MARS Synthesis Reactor

on the MARS 5 digestion reactor and offers different sets of rotor systems with several vessel designs and sizes for various synthesis applications [39].

Even reactions at atmospheric pressure can be performed, employing standard laboratory glassware like round bottom flasks of 0.5–3 L [35]. A protective mount in the ceiling of the cavity allows for connection of a reflux condenser or distillation equipment as well as for addition of reagents and sample withdrawal. However, publications presenting large-scale microwave reactions under atmospheric pressure are rare [35] as the vast majority of efficient microwave reactions are carried out under sealed vessel conditions. For this purpose, the HP-500 Plus rotor (14×100 mL vessels), the XP-1500 Plus rotor (12×100 mL vessels) or the Xpress rotor (40×55 or 75 mL vessels) for high-throughput synthesis at elevated pressure can be used with the MARS instrument (Fig. 5).

Temperature measurement, as the most crucial parameter control in MAOS, is conducted in those rotor systems by an immersed fiber optic probe in one reference vessel or by an IR sensor on the surface of the vessels from the bottom of the cavity. Pressure measurement in HP- and XP-rotors is achieved by an electronic sensor in one reference vessel, whereas the MARS Xpress employs "self-regulating" vessels to prevent over-pressure. All the high-pressure vessels have a unique open-architecture design that allows airflow within the cavity to cool the vessels quickly [39]. The general maximum output power of the instrument is 1200 W, but two low-energy levels with unpulsed microwave output power of 300 and 600 W, respectively, are available as well. This feature avoids overheating of the reaction mixture and unit, if just small amounts of reagents are used.

Similar to the CEM equipment, Milestone offers the modular MicroSYNTH platform, which is based on the ETHOS digestion instrument [40]. The diversity of different rotor and vessel systems enables reactions from 3 to 500 mL under open and sealed vessel conditions in batch/parallel manner up to 50 bar of pressure. The START package offers simple laboratory glassware for reactions at atmospheric pressure under reflux conditions (Fig. 6). A protective

Fig. 5 Parallel pressure rotors HP-500, XP-1500m, Xpress (*left* to *right*)

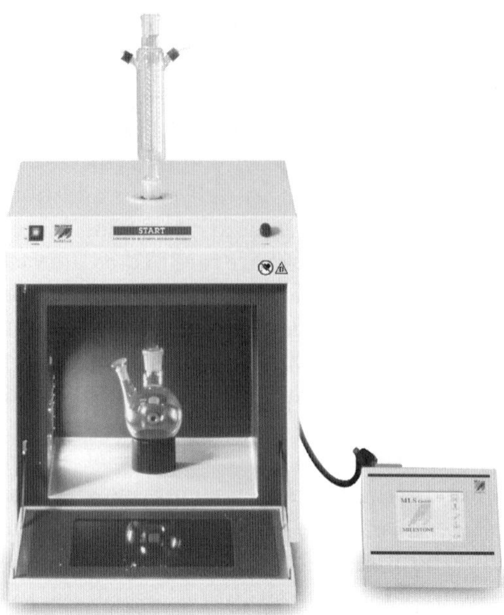

Fig. 6 MicroSYNTH Start package

mount in the ceiling of the cavity enables the connection of reflux condensers or distillation equipment. An additional mount in the sidewall allows for sample withdrawal and flushing of gas to create inert atmospheres. The basic system can be upgraded by several accessories like the research laboratory kit, equipped with the so-called MonoPREP module for single small-scale batch experiments (3–30 mL) in the multimode cavity. In addition, equipment for combinatorial chemistry approaches (CombiCHEM kit, microtiter well-plates up to 96×1 mL), and other special applications is available [40].

Accurate temperature measurement is achieved by the use of a fiber optic probe, immersed in one single reference vessel. Optionally available is an IR sensor for monitoring the outside surface temperature of each vessel, mounted in the sidewall of the cavity. The reaction pressure is measured by a pneumatic sensor connected to one reference vessel. Therefore, the parallel rotors should be filled with identical reaction mixtures to ensure homogeneity. The available parallel rotors employing sealed vessels for elevated pressure have been adapted for synthesis purposes from the original digestion systems (Fig. 7). The PRO 16/24 high-throughput rotor utilizes 16 or 24 reaction containers. Each PTFE-TFM vessel offers 35 mL working volume at 200 °C up to 20 bar. Two variations of parallel high-pressure rotors are available. The MPR-12 comes with 12 segmented 100 mL PTFE-TFM vessels for reactions up to 260 °C or 35 bar. For somewhat extended conditions the HPR-10 serves ten segmented 100 mL PTFE-TFM vessels enabling reac-

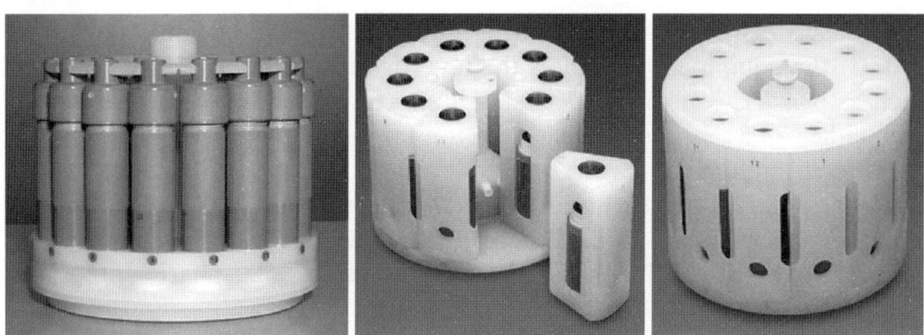

Fig. 7 Parallel pressure rotors Pro24, HPR-10, MPR-12 (*left* to *right*)

tions up to 250 °C at 55 bar. Several other rotor systems are available for the MicroSYNTH platform but a detailed description would extend the scope of this book. For further information the reader is referred to the Milestone website [40].

The most recently released microwave synthesis equipment is the Anton Paar Synthos 3000 (Fig. 8) [41]. This microwave reactor is dedicated for scale-up synthesis in quantities of up to approximately 250 g per run and designed for chemistry under high-pressure and high-temperature conditions. The instrument enables direct scalability up to 1 L total volume of already elaborated and optimized reaction protocols from single-mode cavities without changing the reaction parameters.

Two magnetrons (1400 W continuously delivered output power) allow mimicking of small-scale runs to produce large amounts of the desired compounds within a similar time frame. The homogeneous microwave field guarantees identical conditions at every position of the different rotors, resulting in good reproducibility of experiments, as has been verified for several syn-

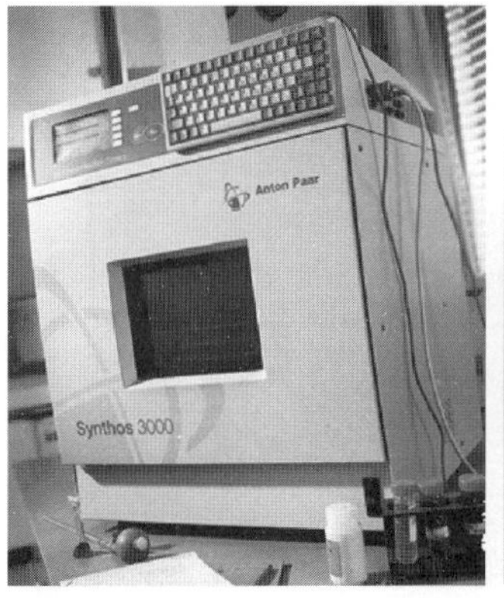

Fig. 8 Synthos 3000 – rotors and vessel types

thetic transformations [26]. Offering advanced operation limits (80 bar at 300 °C) the equipment facilitates the investigation of new reaction avenues, such as near-critical water chemistry [42]. The instrument can be operated with either an 8-, 16- or 48-position rotor, for generation of small compound libraries in multigram scale. The rotors can be equipped with several vessel types for different pressure and temperature conditions. Various accessories allow for special applications like creation of inert/reactive gas atmospheres, reactions in prepressurized vessels [26, 43], as well as solid-phase synthesis or photochemistry.

A dedicated instrument exclusively for one-vessel batch-type reactions is the Biotage Advancer (Fig. 9). This equipment features a multimode cavity for operations with a single 350 or 850 mL Teflon reaction vessel for use at elevated pressure conditions [38]. An operating volume of 50–500 mL at a maximum of 20 bar enables the production of 10–100 g product within one run. Homogeneous heating is ensured by a precise field-tuning mechanism and vigorous magnetic and/or overhead stirring of the reaction mixture. The maximum output power of the Advancer is 1100 W to reach the maximum temperature of 250 °C for 300 mL reaction volume in comparable times to the single-mode experiments. Several connection ports in the chamber head (Fig. 9) enable addition of reagents during irradiation, sample removal for

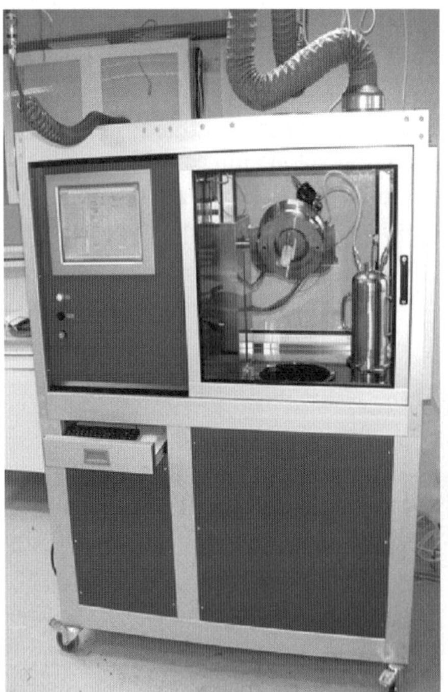

Fig. 9 Biotage Emrys Advancer Scale-Up and its multifunctional chamber head (*right*)

analysis, creation of inert/reactant gas atmosphere, and even in situ monitoring by real-time spectroscopy. Cooling is achieved by an effective certain
gas-expansion mechanism to ensure drastically shortened cooling periods
(e.g. 200 mL ethanol within 1 min from 180 to 40 °C). This instrument (160 ×
85 × 182 cm) is a custom-built, user-specified product, manufactured on re-
quest. Initial examples of Mannich-type reactions [44] or oxidative Heck
couplings [45] to verify the performance and direct scalability by translating
optimized reaction conditions from the Emrys/Initiator system to the larger
scale have already been published.

Another alternative for microwave-assisted batch synthesis is the Milestone UltraCLAVE (Fig. 10). In similarity to many of the other instruments,
this reactor was initially designed for sample preparation applications but has
recently been adapted for organic synthesis. Its 3.5 L stainless steel vessel allows single-batch reactions in a 400 mL to 2.5 L range at maximum operation
limits of 260 °C and 200 bar [40]. The cavity is also capable of accommodating
6 × 120 mL or 40 × 20 parallel rotors employing Teflon reaction containers.
The maximum output power of this device is 1000 W; the vessels are kept
closed by an external pressure of nitrogen. Thus, extremely high pressures of

Fig. 10 Milestone UltraCLAVE

up to 200 bar can be reached. However, at the time of writing no scientific synthesis work with this equipment has been published.

2.2
Flow Reactors for Single-Mode Instruments

Laboratory scale-up utilizing flow techniques in single-mode instruments is currently only possible with the CEM Voyager system. Special flow-through cells for both CF and SF modes are available [39]. The CF system (Fig. 11) offers reaction coils made of glass or Teflon with a maximum flow rate of 20 mL min^{-1} and operation limits of 250 °C or 17 bar. In addition, active flow cells, charged with catalysts or scavengers on solid support, can be applied. The Voyager SF system is operated with a special 80 mL vessel (Fig. 12), utilizing reaction limits of 200 °C or 14 bar and a maximum filling volume of 50 mL for heterogeneous mixtures, slurries, and solid phase reactions.

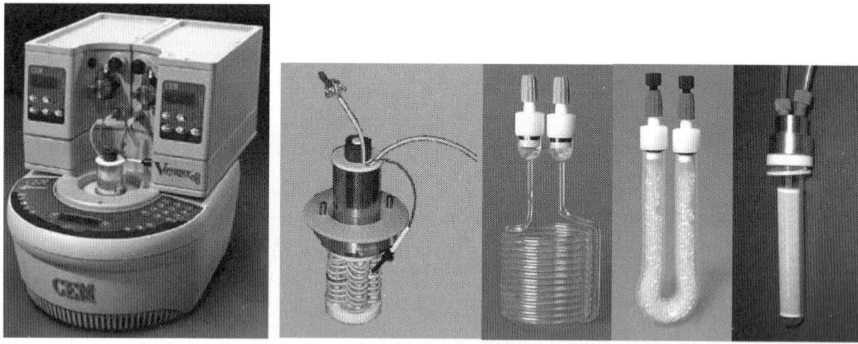

Fig. 11 CEM Voyager CF and its flow cells

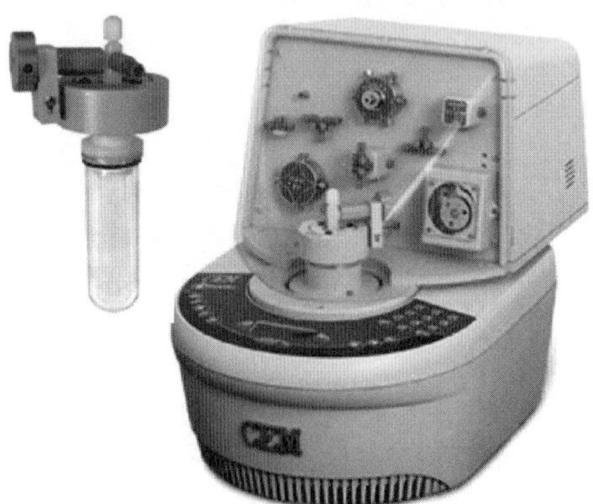

Fig. 12 CEM Voyager SF and its 80 mL reaction vessel

2.3
Scale-Up Beyond Laboratory Scale

The instrumentation presented so far enables scalability in academic and industrial labs from milliliters to approximately 1 L reaction volume or flow rates of several milliliters per minute, respectively. A daily output of some 100 g should be possible with those units. However, industrial users of microwave technology expect solutions for even larger applications as well, dealing with kilogram production per day. Investigations and instrument developments are currently being pursued by most of the instrument vendors. Milestone, for example, already offers two large-scale flow systems (Fig. 13) to close the gap between technical limitations and customer demands [40].

The original CF equipment ETHOS CFR applies quartz or ceramic flow-through cells of various sizes (10–60 mm diameter) inside the regular ETHOS cavity. The reagents are pumped through the microwave field from the bottom to the top of the cavity at maximum operating conditions of 250 °C or 40 bar. The flow rate is dependent on the cell used, but their design generally also allows reactions involving suspensions and inhomogeneous mixtures. Temperature and pressure control via the entire course of the process is achieved by an in-line thermal sensor and an in-line pressure-control valve, respectively. Published examples with this equipment involving methylation reactions have been presented by the group of Shieh [46, 47]. Additionally, available 380 mL MRS-batch tubes, made from fused silica or ceramics, are also applicable in the SF technique at operation limits of 200 °C or 14 bar.

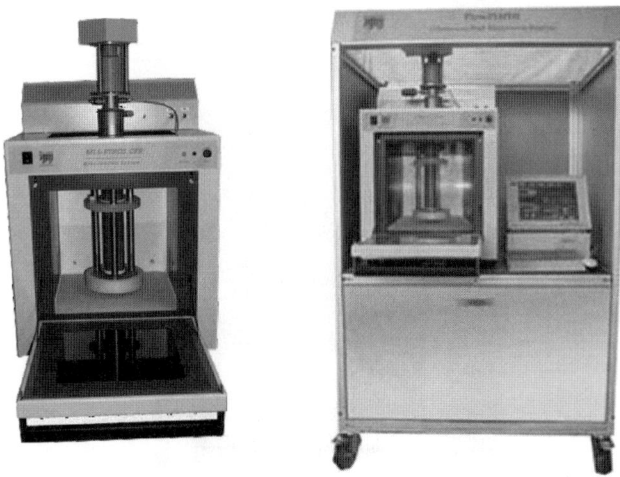

Fig. 13 Milestone scale-up reactors: contFLOW (*left*) and FlowSYNTH station (*right*)

A recent update of Milestone's initial flow equipment resulted in the FlowSYNTH station. The setup is similar to the CF reactor, but operation limits have changed to 200 °C and 30 bar. The most important improvement is an incorporated mechanical stirrer in the 300 mL reaction tube (PTFE in ceramics), providing sufficient agitation of the reaction mixture. The products are pumped out at the top into a heat exchanger for rapid cooling. So far, no experiments have been published with this new device.

The prototype ETHOS pilot 4000 labstation (Fig. 14) is designed for scaleup in the kilogram scale and is built from two regular Milestone cavities. The reaction mixtures can be heated either in CF or batch-type manner. The delivered microwave output power is 2500 W, which can be extended to 5000 W if required [40]. The reaction tubes (quartz or ceramics) are custom built with several diameters and lengths available, covering a broad range of flow rates and pressure conditions. For reaction control, the temperature is monitored over the whole length of the reaction cell, similar to the contFLOW system. As a model reaction to demonstrate the high performance capabilities of this reactor, the esterification of linalool at a 25 kg scale, applying a flow rate of $2.2 \, \mathrm{L\,h^{-1}}$, has been performed [48].

The largest microwave reactor for organic synthetic applications so far is a pilot plant scale prototype installed at Sairem in France, developed and designed in collaboration with BioEurope and De Dietrich. This custom-built $1 \, \mathrm{m^3}$ reactor (Fig. 15) with a powerful 6 kW microwave generator is used for the production of Laurydone [49]. Running in a batch-type recycling process, the equipment accomplished a 40% power reduction compared to the

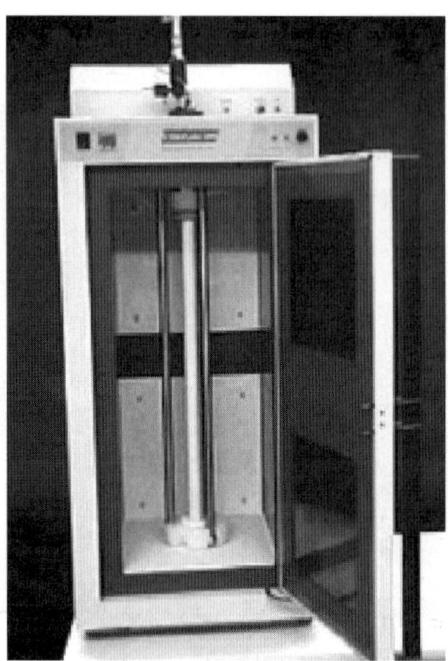

Fig. 14 ETHOSpilot 4000

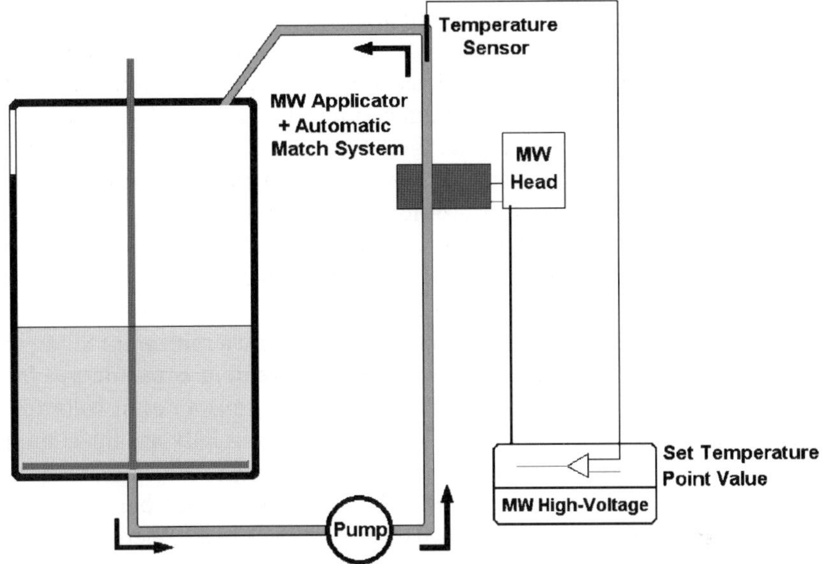

Fig. 15 Recycle batch 1 m^3 prototype microwave reactor

classical thermal approach. Moreover, the overall processing time could be reduced by 80%, which clearly shows the potential of microwave-mediated applications even at production scale.

3
Literature Survey

This survey on microwave-assisted organic chemistry will cover only batch reactions at a $\geq$ 50 mL scale and experiments performed in CF or SF manner.

3.1
Batch and Parallel Processing

Whereas batch synthesis on the small scale is the standard procedure in microwave-assisted synthesis and has been extensively reviewed ([50–52] and references cited therein), protocols in the 50 mL range are rather rare. In this section, scale-up of volumes > 50 mL in sealed vessels will be discussed. An important issue for the process chemist is the potential of direct scalability of microwave reactions, allowing rapid translation of previously optimized small-scale conditions to a larger scale. Several authors have reported independently the feasibility of directly scaling reaction conditions from small-scale single-mode (typically 0.5–5 mL) to larger scale multimode batch microwave reactors (20–500 mL) without re-optimization of the reaction conditions [26, 44, 45, 53–55].

Modern single-mode microwave technology allows the performance of MAOS in batches of very small reaction volumes (< 0.2 mL). A detailed study by Takvorian and Combs highlights the advantage of performing microwave chemistry in very small reaction vessels applicable in, e.g., the Biotage Initiator EXP equipment [56]. The ultralow volume vials utilized in this study (0.2 mL) enabled the authors to run reactions in a very concentrated fashion, minimizing reaction times and utilizing only limited amounts of sometimes expensive scaffolds and reagents (Scheme 1).

With those single-mode reactors that do not require a minimum filling volume (CEM Discover platform; temperature measurement is performed from the bottom and not from the side by an external IR sensor) even volumes as low as 50 μL can be processed [57]. With the commercially available single-mode cavities of today, the largest volumes that can be processed under sealed vessel conditions are ca 50 mL, with different vessel types being available to upscale in a linear fashion from 0.05 to 50 mL. Under open vessel conditions higher volumes (> 1000 mL) have been processed under microwave irradiation conditions, without presenting any technical difficulties as, e.g., described for the synthesis of various ionic liquids on a 2 mol scale [35].

A comprehensive study on the scalability of optimized small-scale microwave protocols in single-mode reactors to large-scale experiments in a multimode instrument has been presented by Kappe and coworkers [26]. As a model reaction, the classical Biginelli reaction in acetic acid/ethanol (3 : 1) as solvent was run in parallel in an eight-vessel rotor system of the Anton Paar Synthos 3000 synthesis platform (Fig. 8) on a 8×80 mmol scale [26]. Here, the temperature in one reference vessel was monitored with the aid of a suitable probe, while the surface temperature of all eight quartz reaction vessels was also monitored (deviation less than 10 °C, see Fig. 16). The yield in all eight vessels after 20 min hold-time at 120 °C was nearly identical, resulting in an overall amount of approximately 130 g of the desired dihydropyrimidine.

To extend this study, numerous other transformations have been carried out on different scales. The results of these scale-up experiments are summarized in Scheme 2 [26]. In all cases, the yields obtained in the optimized small

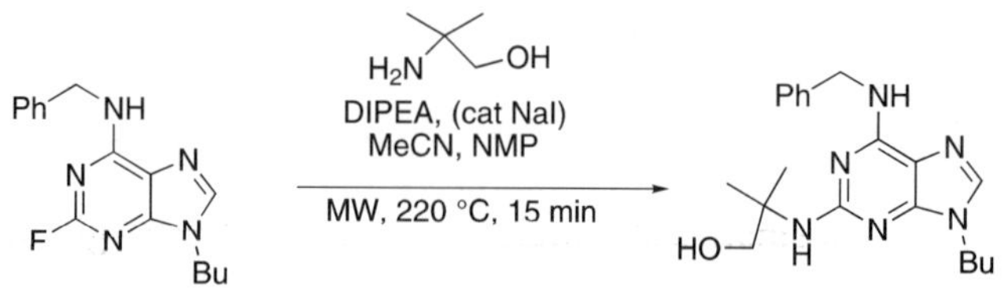

Scheme 1 Amination of purine scaffolds on a 0.2 mL scale

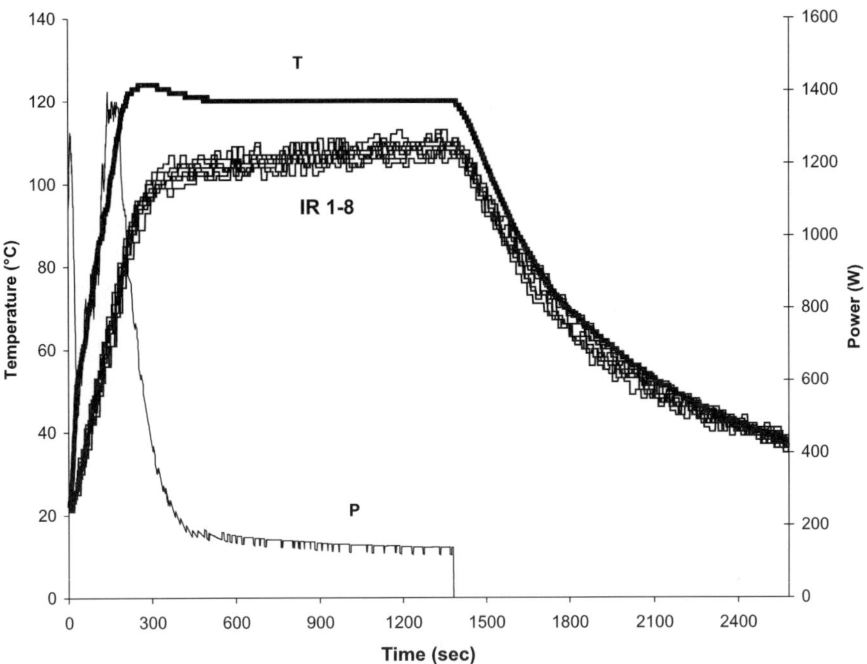

Fig. 16 Temperature and power profiles for a Biginelli condensation (Scheme 2a) under sealed vessel/microwave irradiation conditions. Shown is the temperature measurement in one reference vessel via an internal gas balloon thermometer (T), the surface temperature monitoring of the eight individual vessels by IR thermography (*IR 1–8*), and the magnetron power (*P*, 0–1400 W). Reproduced with permission from [26]

scale single-mode experiments (1–4 mmol) performed on an Emrys Liberator (Biotage) could be reproduced on a larger scale (40–640 mmol) without modifications of the reaction conditions. Dependent on the chemistry and the applied scale, 10–100 g of product could be isolated from a single run. It has to be noted, however, that in many cases the rapid heating and cooling profiles seen in a small-scale single-mode reactor with high power density ("microwave flash heating") cannot be replicated on a larger scale. The heating profile for the Biginelli cyclocondensation shown in Scheme 2a is reproduced in Fig. 16. Despite the somewhat longer heating and cooling period, no appreciable difference in the outcome of the reactions studied was found.

Similar scale-up results were obtained by Luthman and coworkers for Mannich reactions [44] and by the group of Larhed for oxidative Heck couplings [45] (Scheme 3) utilizing a different multimode batch reactor with a single reaction vessel (Emrys Advancer). As expected, yields were comparable on going from a small-scale single-mode reactor to a larger multimode reactor. Here, rapid cooling after the microwave heating step is possible by a patented expansion cooling process.

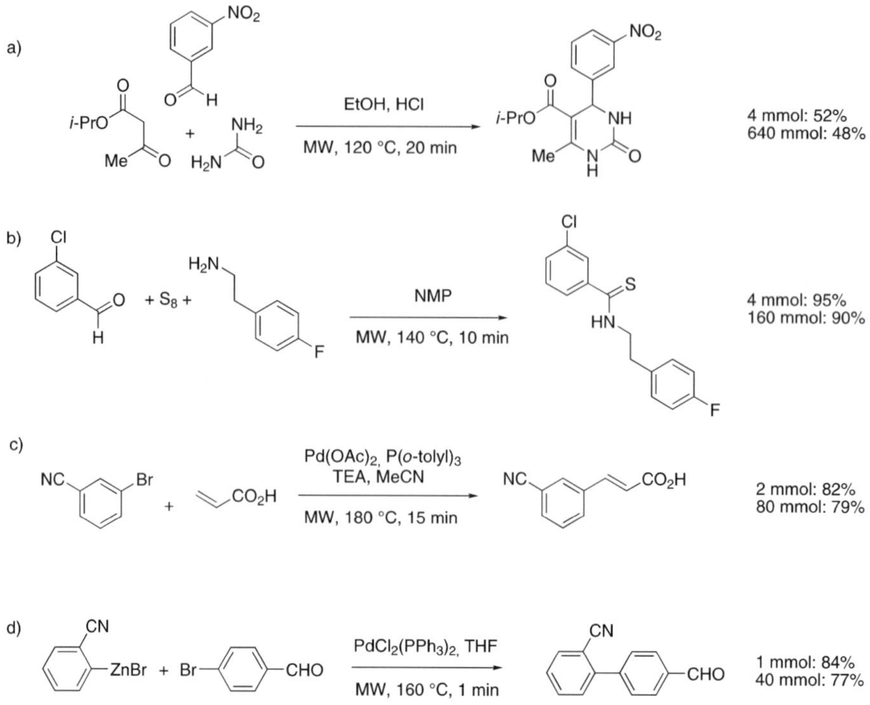

a)

EtOH, HCl

MW, 120 °C, 20 min

4 mmol: 52%
640 mmol: 48%

b)

NMP

MW, 140 °C, 10 min

4 mmol: 95%
160 mmol: 90%

c)

Pd(OAc)$_2$, P(o-tolyl)$_3$
TEA, MeCN

MW, 180 °C, 15 min

2 mmol: 82%
80 mmol: 79%

d)

PdCl$_2$(PPh$_3$)$_2$, THF

MW, 160 °C, 1 min

1 mmol: 84%
40 mmol: 77%

Scheme 2 Direct scalability of microwave synthesis from small scale single mode reactors to large scale multimode batch reactors

Similar work on comparatively large scale microwave synthesis was earlier published by Strauss [58] utilizing a prototype laboratory-scale microwave batch reactor (MBR), and by other authors using a larger single-mode device (Synthewave 1000 reactor by Prolabo) [36, 37]. The MBR consists of a large reaction vessel for 100 mL operating volume equipped with a fiber optic thermometer and a cold finger. To demonstrate the performance of this prototype, several transformations such as oxidations, esterifications, Claisen rearrangement in water, and Willgerodt reactions have been performed on various scales [58]. This MBR prototype was the early predecessor of the above-mentioned Emrys Advancer, therefore reaction details are not presented in this section.

An early investigation on small- and large-scale applications was discussed by the group of Besson [34]. Therein the authors compared two microwave reactors of the French company Prolabo, namely the Synthewave S402 for small-scale and the Synthewave S1000 reactor for larger scale applications, which were commercially available in the 1990s [59]. Two reactions, the conversion of phenyl isocyanate to phenyl-n-butyl-thiocarbamate and the transformation of N-arylimino-4-chloro-5H-1,2,3-dithiazole to 4-n-butoxyquinazoline-

a)

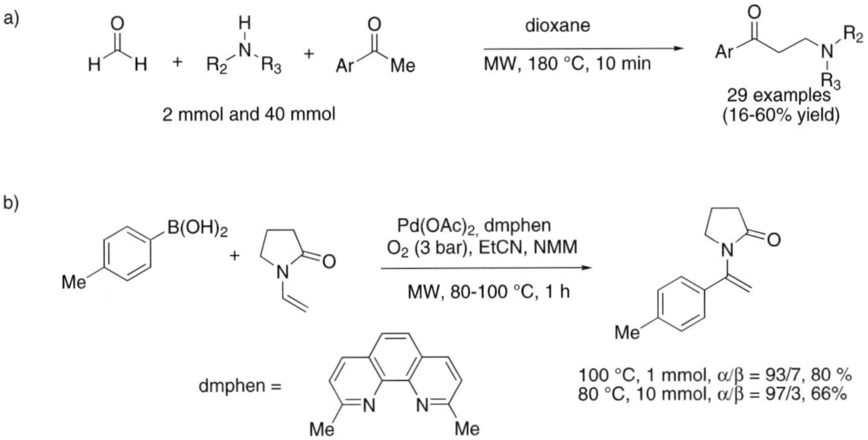

b)

Scheme 3 Direct scalability of microwave synthesis from small scale single mode reactors to large scale multimode batch reactors

2-carbonitrile, were carried out in order to evaluate the differences in reaction time and yield when going from small scale (1 g) up to a 20 g scale (Scheme 4). In the latter example the same yields were obtained but with dramatically reduced reaction times as compared to the small-scale experiments, probably due to the higher power capacity of the large-scale reactor (800 W for the S1000 instead of 300 W for the S402).

In a related study, the scale-up and synthesis of dioxolanes, dithiolanes, and oxathiolanes was performed by Hamelin and coworkers [36]. Employing the Synthewave S1000 apparatus from Prolabo, the authors investigated the synthesis of the protected carbonyls on a 2 mol scale under open vessel conditions employing high-boiling glycols and K10, an acidic clay, as catalyst (Scheme 5). Proving that the reaction conditions (regarding time and temperature) were exactly the same going from 10 mmol to a 2 mol scale,

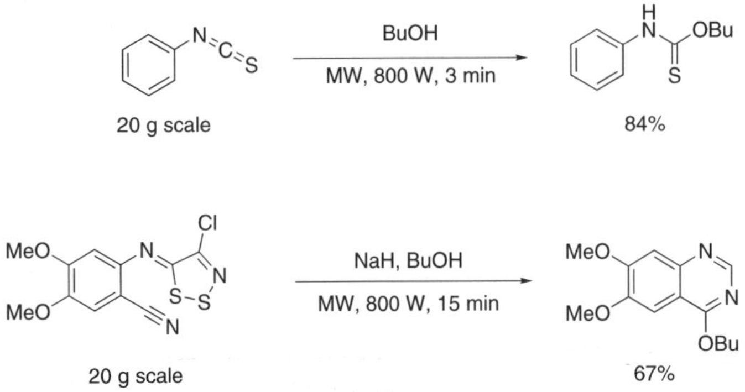

Scheme 4 Scale-up synthesis of thiocarbamates and quinazolines in single-mode reactors

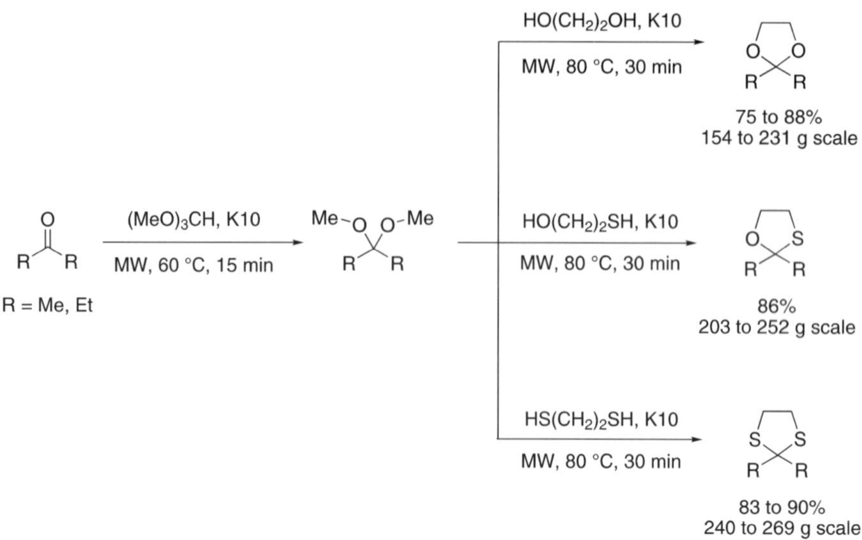

Scheme 5 Scale-up synthesis of dioxolanes, oxathiolanes, and dithiolanes

they observed an easier work-up for the large scale experiments owing to the possibility of removing the formed alcohol by continuous distillation under microwave irradiation in the Synthewave S1000.

In a more recent study, the group of Loupy presented a series of solvent-free reactions scaled up to several hundred grams utilizing the Synthewave 1000 batch reactor [37]. After optimization at a 50 mmol scale employing the Synthewave 402, the alkylation of potassium acetate with n-bromooctane was performed on a 2 mol scale (622 g product) within 5 min, although the heating ramp was somewhat slower than in the small-scale run. An interesting result was obtained for the phenacylation of 1,2,4-triazole (Scheme 6). Whereas thermal heating furnished a mixture of the 1- and 4-alkylated triazoles as well as the quarternary salts, the microwave-assisted phenacylation resulted in the exclusive formation of the 1-alkylated product, regardless of the scale used [37]. As an additional example, the selective dealkylation of 2-ethoxyanisole using KOt-Bu has been carried out solventless, with 20 min irradiation yielding 108 g of the corresponding phenol.

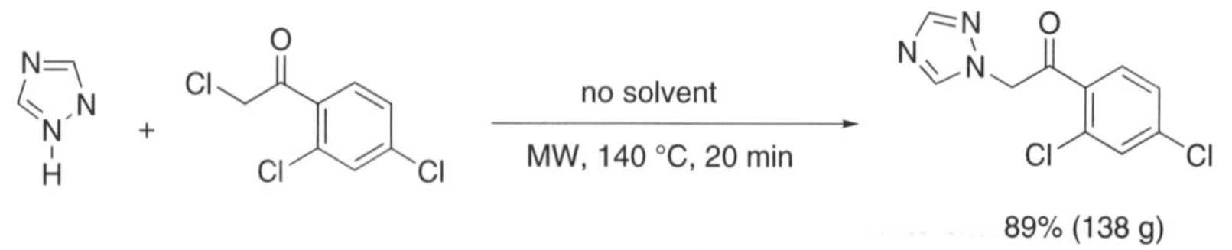

Scheme 6 Phenacylation of 1,2,4-triazole

With these results in hand, several examples in carbohydrate chemistry have been performed including glycosylations, peracetylations, saponifications, and epoxidations of glucose derivatives. Within 2–10 min (depending on the chemistry and the scale) 60–220 g of the desired compounds have been generated, showing the easy access of products in the multigram level by solventless microwave-assisted chemistry [37].

Maes and coworkers recently presented one of the rare direct comparisons of single-mode and multimode instruments in batch and flow processing, respectively [27]. The authors compared the performance of the multimodal Milestone MicroSYNTH and the CEM MARS reactors with the single-mode instruments CEM Discover and its SF extension Voyager SF (Sect. 2). Buchwald–Hartwig aminations in toluene were used as test reactions to investigate initially the direct scalability going from a 10 mL (1 mmol) vessel to an 80 mL (20 mmol) vessel in a Discover apparatus. In contrast to the small-scale experiment, the final temperature could not be reached in the 80 mL vessel due to the poor coupling properties of toluene with microwave irradiation (tan $\delta = 0.04$). Although polar reagents are present in the reaction mixture in a higher concentration compared to the small-scale experiment, the poor coupling characteristics of toluene are obviously responsible for the failure of the scale-up experiment [27].

A 20-fold scale-up has been performed using a more polar solvent (BTF), going from a 10 mL to an 80 mL vessel in the Discover, 3×20-fold (60 mmol) by employing the Voyager SF, and 6×20-fold (120 mmol) by employing parallel rotors in the MARS and microSYNTH reactors. Similar results regarding yield and product purity were obtained with each platform, demonstrating that the success of the reactions is neither dependent on the equipment used nor on the scale applied (Scheme 7).

To investigate these findings further the authors determined heating rates of the employed multimode instruments and the Discover unit, once again using toluene as solvent. After 10 min irradiation at a constant maximum power output for each microwave reactor, different final temperatures were measured (Fig. 17). Furthermore, it could be shown that the observed differences in temperature are not only related to the different heating efficiencies of the instruments but also to the specific vessel material [27]. Usually the vessel material itself is not completely microwave-transparent and therefore it is at least partially responsible for heating of the irradiated solvent via conventional thermal conduction [42].

Not surprisingly, the final temperature of the solvent relies on the volume used, especially if experiments are performed at constant power. In such experiments, a decrease of the final temperature was observed with increased volume. Obviously, it is not possible to directly compare single-mode experiments with multimode experiments at an identical output power. Due to the significantly higher power density, the heat transfer of single-mode reactors is substantially higher.

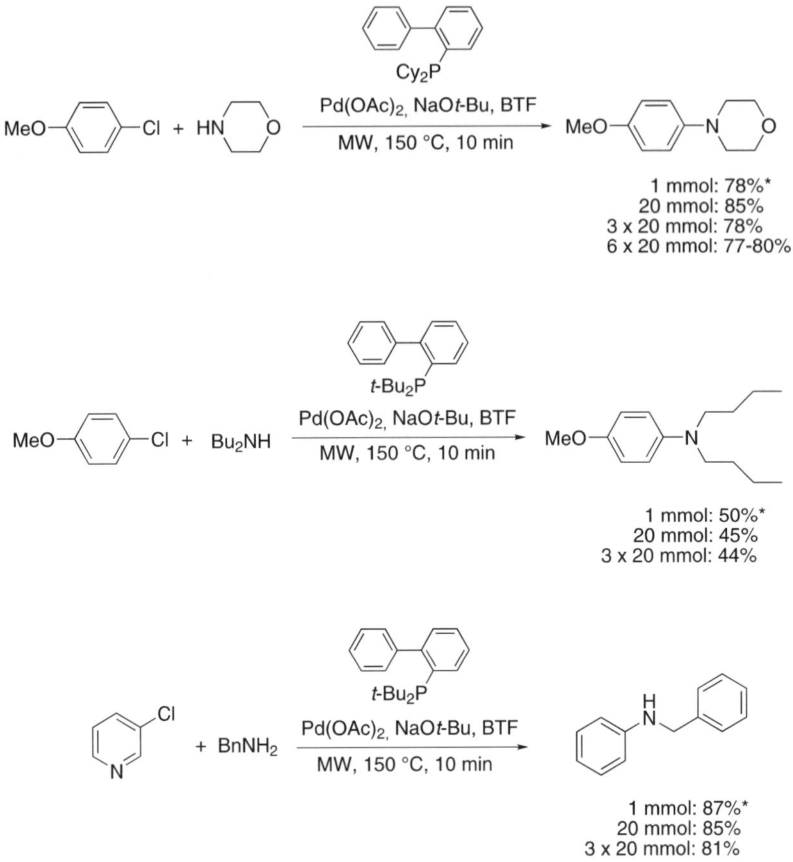

Cy2P
Pd(OAc)2, NaOt-Bu, BTF
MW, 150 °C, 10 min
MeO
Cl
HN
O
N
1 mmol: 78%*
20 mmol: 85%
3 x 20 mmol: 78%
6 x 20 mmol: 77-80%
t-Bu2P
Bu2NH
1 mmol: 50%*
20 mmol: 45%
3 x 20 mmol: 44%
BnNH2
H
N
1 mmol: 87%*
20 mmol: 85%
3 x 20 mmol: 81%

BTF = benzotrifluoride

* performed in toluene

Scheme 7 Scale-up of Pd-catalyzed Buchwald–Hartwig aminations utilizing different microwave instruments

In case of the Buchwald–Hartwig reactions, all those previously mentioned problems could be minimized using alternative solvents like trifluoromethylbenzene (BTF), which is a far better microwave absorber than toluene. Utilizing such modified reaction mixtures the authors found comparable yields in single-mode and in multimode reactors for small- as well as for large-scale experiments (Scheme 7). Thus, the above-mentioned aminations have been performed efficiently on a multigram scale producing a daily output of 261 g of a morpholine-containing compound [27].

Baxendale and Ley [60] employed the Smith Synthesizer as well as the Emrys Optimizer from Biotage [59] for conducting neat KO*t*-Bu mediated trimerizations of various liquid nitriles to give aryl- and alkyl-substituted 4-

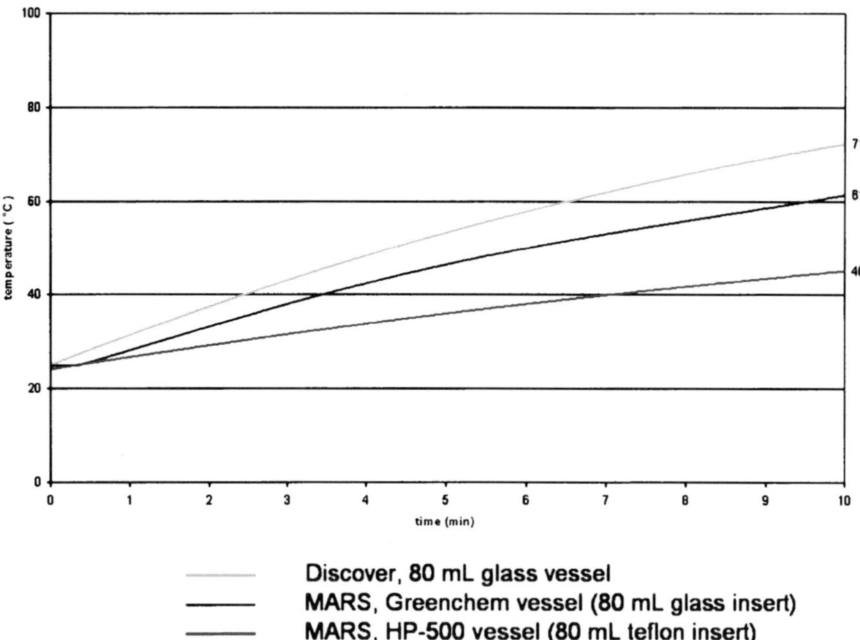

Fig. 17 Heating profile of 20 mL of toluene in several commercially available microwave systems, irradiated in a power/time experiment at a constant power of 300 W for 10 min. Reproduced with permission from [27]

aminopyrimidines (Scheme 8). A set of 23 different nitriles was reacted on 1, 5, and 20 g scales. For the large-scale attempts, a 3 × 25 min heating circle was required whereas three successive 15 min runs were found to be enough for high overall yields and excellent purities in the 1 g and 5 g scale reactions.

Bose and coworkers have performed known chemical processes in various kitchen microwave devices to explore microwave chemistry at a larger scale [33]. Representative examples for multigram-scale synthesis without optimization of reaction conditions have been presented, such as the rapid preparation of 500–800 g of acetylsalicylic acid (Aspirin). Another valuable pharmaceutical compound, Tylenol (acetaminophen, paracetamol), could be

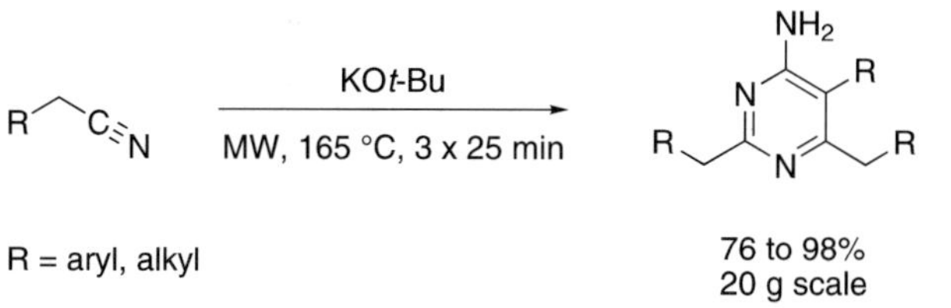

Scheme 8 Gram-scale generation of a 23-membered aminopyrimidine library

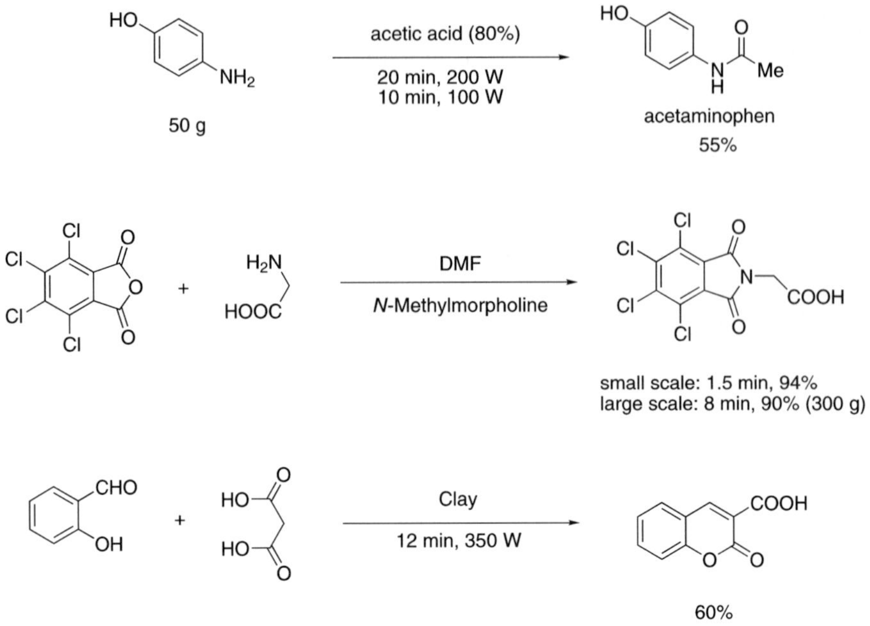

Scheme 9 Model reactions for microwave-scale-up in open vessels (no temperature control)

produced in good yield within 4 min irradiation in a simple domestic microwave oven (Scheme 9). Up-scaling of this process (25-fold) required sequential irradiation at different power outputs to achieve an almost similar yield. The other examples presented in Scheme 9 also required somewhat prolonged reaction times in the scale-up attempt compared to the initial optimized protocols [33].

In the context of a medical chemistry project, Hazuda and coworkers proved the necessity for large-scale production of an HIV integrase inhibitor in order to be able to carry out several clinical tests [61]. The required precursor presented in Scheme 10 was prepared by a team from Merck in

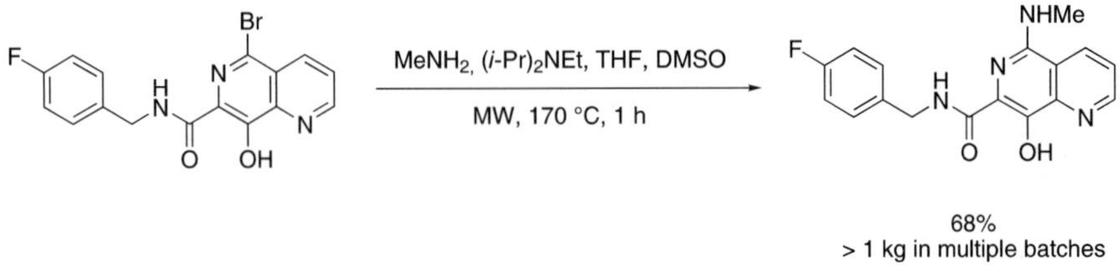

Scheme 10 Microwave-mediated process for the synthesis of an HIV integrase inhibitor precursor

multiple batches employing a 300 mL single-mode batch reactor (Emrys Advancer) [62–64]. The total required amount of the corresponding bromide was generated within 35 h total processing time under microwave conditions in several batches, whereas the process performed at the same scale under thermal heating would have required 30 days [64].

3.2
Scale-Up Using Continuous Flow Methods

Mainly because of safety concerns and issues related to the penetration depth of microwaves into absorbing materials such as organic solvents, the preferable option for processing volumes of > 1 L under sealed vessel microwave conditions is a CF technique, although here the number of published examples using dedicated microwave reactors is still limited [46, 47, 65–68]. In such systems the reaction mixture is passed through a microwave-transparent reaction container that is placed in the cavity of a single- or multimode microwave reactor. The previously optimized reaction time under batch microwave conditions now needs to be related to a "residence time" (i.e., the time the sample stays in the microwave-heated coil) at a specific flow rate. While the early pioneering work in this area stems from the group of Strauss [65], others have made notable contributions to this field in the past, often utilizing custom-built microwave reactors or modified domestic microwave units [69–73].

Already in 1990, the group of Wang pointed out the inherent hazard of violent explosions due to the high pressure/high temperature conditions in closed vessels under microwave irradiation [74]. Organic transformations like esterifications, racemizations, hydrolysis, cyclization, and substitution reactions have been reported on a scale > 20 g using a CF reaction process in a modified kitchen-type microwave oven [74].

An efficient application of microwave-mediated synthesis by a CF technique has been presented by Esveld and coworkers [72]. In their work, the formation of waxy esters using montmorillonite clay as a catalyst under solventless microwave conditions has been intensively studied. To avoid side reactions and to aim for the highest reaction rate, microwave heating was used in order to keep the temperature constant for a certain period of time. Equimolar composition, microwave heating, and montmorillonite clay catalysts have been combined to achieve the scale-up of these esterification reactions by a continuous process without the use of any organic solvent. The employed equipment, a prototype continuous microwave dry-media reactor (CMDR), consists of a previously described multimode tunnel microwave cavity ($150 \times 40 \times 35$ cm), delivering an output power of 4.4 kW from a diagonal slotted waveguide (Fig. 18) [71]. The solid mixtures are transported on a Teflon-coated glass fiber web conveyor (1.5 m) up to 17 cm min^{-1}. With this apparatus – on a scale of 12 kg h^{-1} – it has been shown that dry-media and

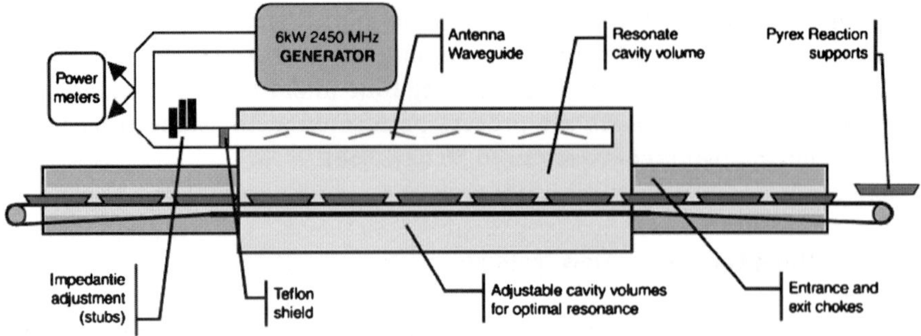

Fig. 18 Continuous microwave dry-media reactor (CMDR) for kg-scale dry-media reactions. Reproduced with permission from [71]

solvent-free reactions can be scaled to pilot processes, without altering the main reaction conditions. The continuous production capacity of the CMDR has been fully exploited, resulting in a day production of about 100 kg of corresponding waxy esters in high purity (Scheme 11).

A more recently published example of organic microwave synthesis under CF conditions is the 1,3-dipolar cycloaddition chemistry in the CEM CF Voyager system (Fig. 11). Savin and coworkers presented the cycloaddition of dimethyl acetylene dicarboxylate with benzyl azide in toluene, which was first carefully optimized with respect to solvent, temperature, and time under batch conditions. The best protocol was then translated to a CF procedure where a 0.33 M solution of both building blocks was pumped through a Kevlar-enforced Teflon coil (10 mL total capacity) heated in the single-mode reactor at 110 °C (10 min residence time) [66]. This method provided a 91% conversion to the desired triazole product (Scheme 12).

In the context of elaborating a degradation method for hazardous and toxic halogenated aromatic hydrocarbons, Varma and coworkers reported on

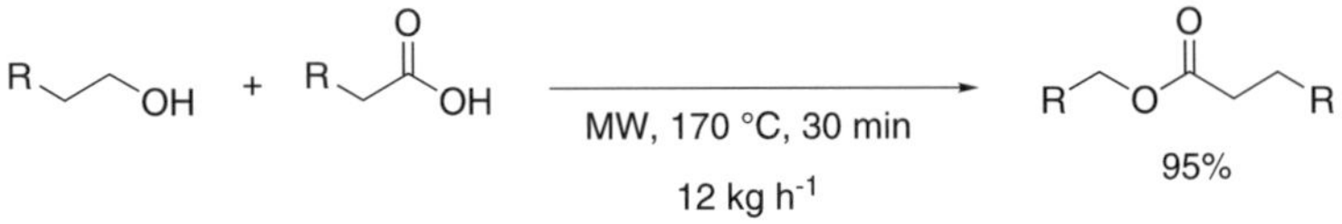

Scheme 11 Waxy ester production in pilot scale

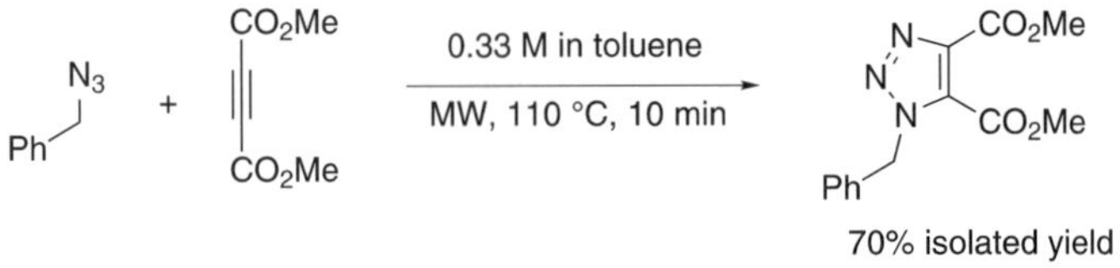

Scheme 12 1,3-Dipolar cycloaddition reactions under CF conditions

the hydrodechlorination of chlorinated benzenes using a microwave-assisted CF process that involves "active flow cells" (Scheme 13) [75]. Here a 15 mL quartz U-tube filled with spherical particles of 0.5% Pd/Al_2O_3 catalyst was fitted into the CEM Voyager CF reactor. Chlorobenzene was fed into the reactor at an adjustable flow rate along with a controlled flow of hydrogen from a cylinder. Comparison studies between the microwave and the thermally heated process indicated that the microwave method was both more efficient and selective than the thermal process, which gave formation of undesired cyclohexane byproduct under standard, thermal conditions.

Current single-mode CF microwave reactors allow only processing of comparatively small volumes. Much larger volumes can be processed in CF reactors that are housed inside a multimode microwave system. In a publication from 2001, Shieh and coworkers described the methylation of phenols, indoles, and benzimidazoles with dimethyl carbonate under CF microwave conditions, using a Milestone ETHOS-CFR reactor [46]. In a typical procedure, a solution containing the substrate dimethyl carbonate, 1,8-diazabicyclo[5.4.0]undec-7-ene (DBU), tetrabutylammonium iodide (TBAI), and a solvent was circulated by a pump through the microwave reactor, which was preheated to 160 °C and 20 bar by microwave irradiation (Scheme 14). Under these conditions, the methylation rate for phenols was accelerated from hours to minutes, representing a nearly 2000-fold rate increase. Similar results were also achieved for benzylations employing dibenzyl carbonate [47], and the same authors also reported the usefulness of this general method for the esterification of carboxylic acid [68]. Benzoic acid, for ex-

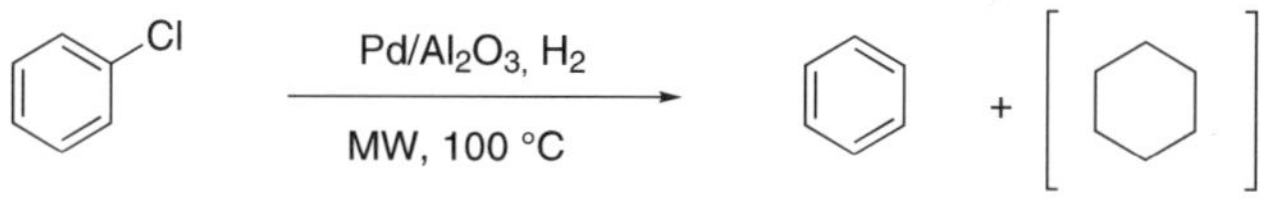

Scheme 13 Hydrodechlorination of chlorobenzene using active flow cells in a CF microwave process

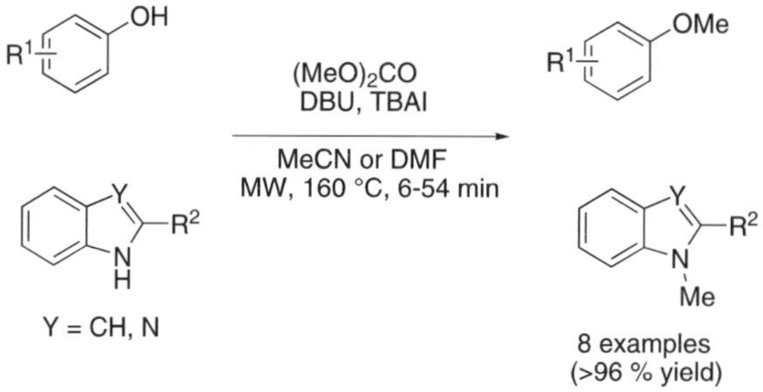

Scheme 14 Methylation of phenols, indoles, and benzimidazoles in a multimode CF microwave reactor

ample was converted within 20 min (microwave residence time) to its methyl ester on a 100 g scale, utilizing the dimethyl carbonate, DBU-mediated CF protocol [68].

The CF technique has attracted several research groups to construct unique and unusual prototypes of small-size flow-through reactors for single-mode instruments as well as for domestic ovens. Although these kinds of applicators do not really meet our definition of scale-up we will nevertheless mention the most interesting CF approaches in this context.

In order to investigate the possibility of efficiently performing organic transformations employing immobilized catalysts under microwave irradiation, the group of Bagley [76] developed a special CF microwave reactor for use in the CEM Discover unit. The reactor consists of a standard 10 mL glass tube, fitted with a custom-built steel head, and filled with sand ($\sim$ 12 g) between two drilled frits in order to create a lattice of microchannels charged with solvent ($\sim$ 5 mL). The tube is sealed using PTFE washers and connected to a regular HPLC flow system with back-pressure regulation (Fig. 19). In order to investigate the aptitude of this reactor for chemical transformations, a simple hydrolysis of thiazole and a Fischer indole synthesis have been performed on the gram scale under CF conditions (Scheme 15). Furthermore, in order to compare the efficiency of this setup with batch processing, a Bohlmann–Rahtz reaction was carried out furnishing pyridines by cyclodehydration of the corresponding aminodienones (Scheme 16) employing a Teflon heating coil. Applying conditions that gave almost quantitative conversion to the pyridine, the processing rates using the glass tube reactor

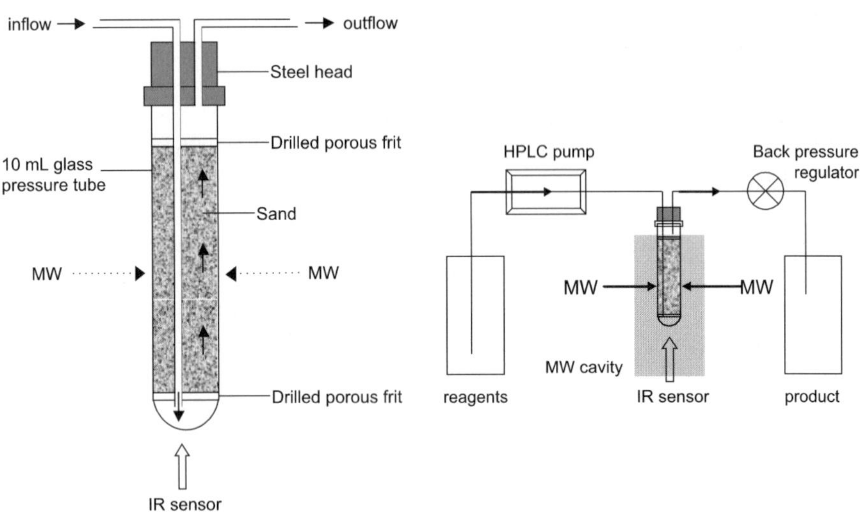

Fig. 19 Schematic diagram of the flow cell/CF reactor. Reproduced with permission from [76]

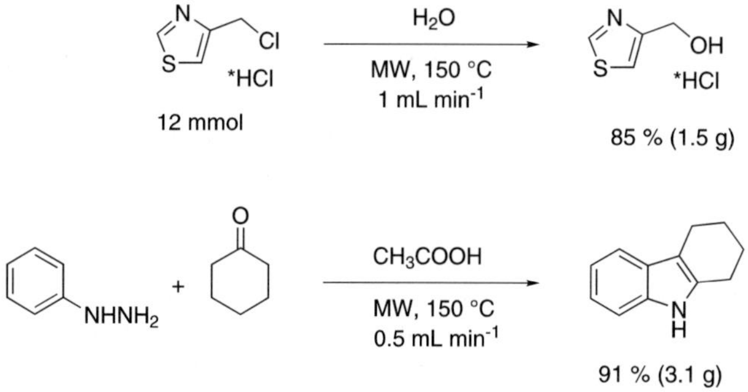

Scheme 15 Model reactions for optimization of the sand-filled CF reactor

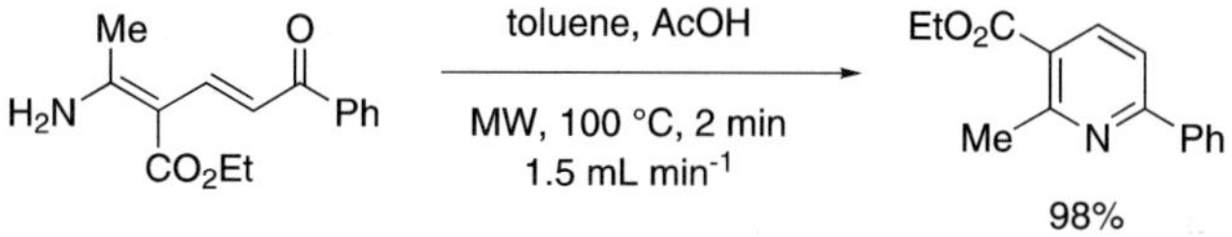

Scheme 16 Bohlmann–Rahtz synthesis under CF conditions

were considerably higher, and CF reactions run at the same flow rate also used less magnetron energy in a glass tube than in the heating coil. This demonstrates clearly that a glass tube CF reactor offers (i) improved heating efficiency, (ii) the potential for operation on a large scale, (iii) successful transfer from batch to CF processing, and (iv) improved performance over commercial Teflon heating coils. In principle, replacing the sand by an immobilized catalyst would allow for transformations involving heterogeneous catalysts under CF conditions.

In a related study, the group of Kappe reported on Biginelli reactions and Dimroth rearrangements comparing batch and CF techniques (Scheme 17) [77]. Similar to the above-mentioned setup, a corresponding CF coil for the CEM Voyager unit was charged with 2 mm-sized glass beads in order to create microchannels, which result in increased residence time of the reaction mixture in the microwave heating zone (Fig. 20). The reaction mixture was introduced into the flow cell at the bottom of the vial via a Teflon tube using standard HPLC pumps, and the reaction pressure was controlled by a back-pressure regulator connected to the end of the outlet tubing. Monitoring of the reaction temperature was achieved at the bottom of the flow cell by the IR sensor incorporated into the instrument. The described setup was initially evaluated with the well-known Biginelli reaction going from a milligram to a $25\,\mathrm{g\,h^{-1}}$ scale, and furthermore extended to the microwave-assisted rearrangement of thiazines (Scheme 17). The Biginelli reaction was carried out with equimolar ratios of the building blocks in a concentration of 1.3 M at an adjusted flow-rate of $2\,\mathrm{mL\,min^{-1}}$, resulting in an

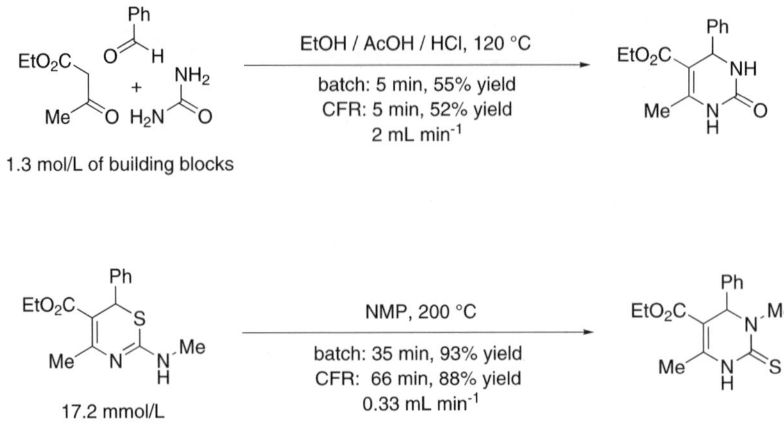

Scheme 17 Biginelli reaction and Dimroth rearrangement via microwave-assisted batch and CF processing

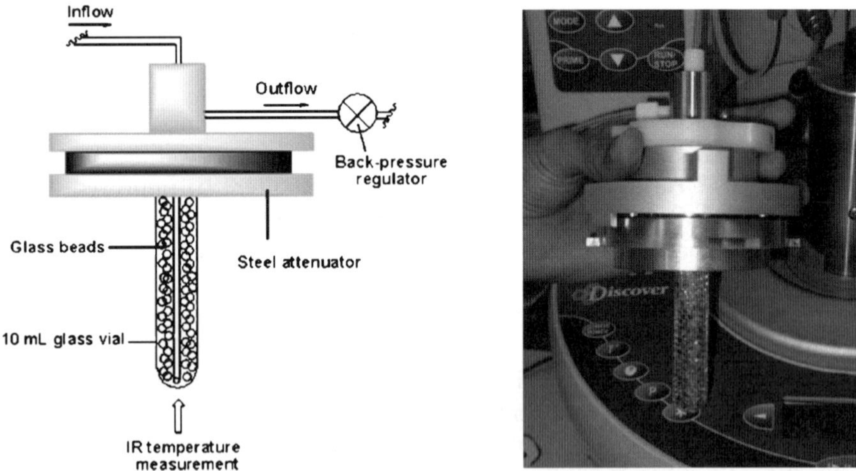

Fig. 20 Flow cell for performing CF microwave synthesis. Reproduced with permission from [77]

identical residence time compared with the batch attempt. In contrast, the Dimroth rearrangement was performed with a flow-rate leading to an almost doubled residence time of the substrate within the microwave heating zone compared to the batch experiments. The results were nicely comparable with the corresponding batch yields.

Wilson and coworkers described a custom-made flow-reactor for the Biotage Emrys Synthesizer single-mode batch reactor (Fig. 21) that was fitted with a glass-coiled flow cell [67]. The flow cell was inserted into the cavity from the bottom of the instrument and the system was operated either under

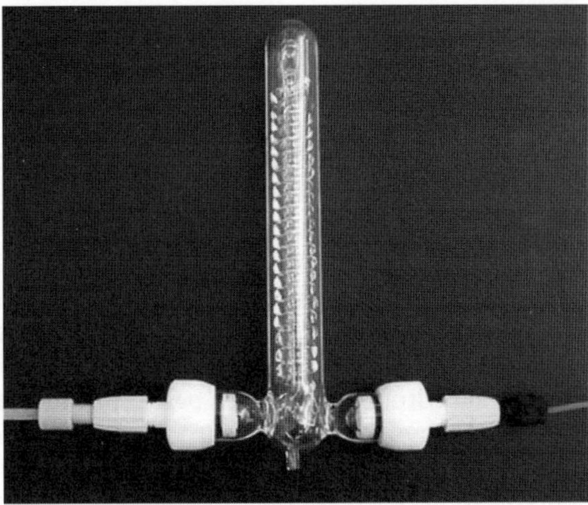

Fig. 21 Flow reactor applied with the Emrys Synthesizer. Reproduced with permission from [67]

open- or closed-loop mode. The temperature was monitored and controlled through the internal IR sensor (located in the cavity) and the instrument software.

The different synthetic transformations investigated with this system included nucleophilic aromatic substitutions, esterifications and Suzuki reactions and are shown in Scheme 18. In all cases, product yields were similar to those using conventional heating and were easily scalable to multigram quantities [67]. Within 5 h more than 9 g of the resulting phenethylamine could

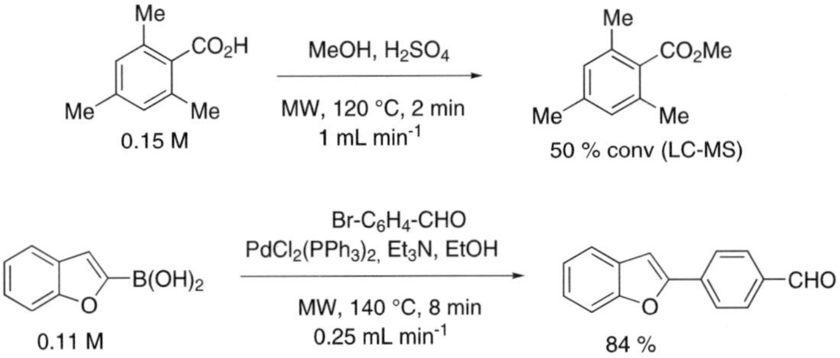

Scheme 18 Various chemistries under CF conditions

be prepared by nucleophilic substitution of the corresponding fluorinated nitrobenzene. However, clogging of the lines and over-pressurization have been observed, and those phenomena are significant limitations of this processing technique.

In a related work, Organ and coworkers portrayed a microcapillary flow-through reactor attached to an Emrys Synthesizer [78, 79]. The microreactor (Fig. 22) can be used in CF as well as in SF mechanisms and has been extended to a simultaneous parallel flow device [78]. Standard glass capillaries with different diameters (200–1200 μm) coated with Pd on the inner surface have been employed. Several classic organic transformations, such as Wittig olefinations, Suzuki–Miyaura couplings, ring closing metathesis (RCM) and nucleophilic aromatic substitutions have been performed in microscale to optimize this tool. Flow rates of 2–40 μL, equal to an average residence time of 4 min, were applied in order to investigate the optimum microwave power and reaction concentration. Furthermore, internal coating of the capillaries with thin films of Pd metal showed tremendous rate accelerations as the metal films themselves are capable of inducing Pd-catalyzed reactions with no ex-

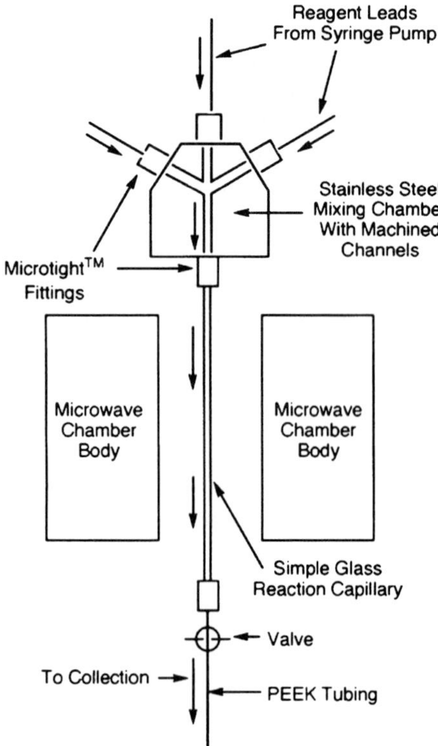

Fig. 22 Basic CF reactor design (consists of a stainless steel holding/mixing chamber with three inlet ports that merge into one outlet). Reproduced with permission from [79]

ogenous catalyst added. Although very thin capillaries have been used, clogging was not an issue as this newly developed flow-reactor consists of short and straight reaction cells. Other systems like the above-mentioned flow cells are coiled or contain spirals and are therefore more prone to clogging [66, 67]. With this data in hand, a sophisticated modification of the microreactor was developed, allowing for sequential synthesis in parallel mode to generate a 2×4 library of biaryls, as demonstrated for Suzuki couplings [78]. With this setup, parallel preparation of drug candidates on the milligram scale can be performed with all reactions heated simultaneously in a microwave reactor with the reaction mixture flowing through the individual capillaries.

Earlier, the group of Laporterie reported on another prototype CF microwave reactor [80]. Solvent-free Friedel–Crafts reactions have been successfully carried out employing only catalytic amounts of the $FeCl_3$ catalyst (Scheme 19). At a flow rate of 20–22 mL min^{-1} the corresponding substrates have been circulated in a molar scale ($2:1$ ratio) in the apparatus. Thus, 150–250 g products could be isolated. Excess substrates have been recovered by evaporation and recycled in the process.

A very interesting approach to process intensification was recently presented by Jachuck and coworkers. The authors described the development and performance of an isothermal CF reactor to be used in a domestic microwave oven [81]. The small ($270\,\mu L$) CF reactor consists of two sections, a microwave transparent PTFE part for the reaction side and an alumina part for heat transfer (Fig. 23). The heat generated due to the activation by microwave irradiation was rapidly absorbed by the heat transfer liquid (H_2O) pumped through the alumina part. Inlet and outlet temperatures of both the reaction mixture and the heat transfer liquid were monitored using a PICO temperature recorder.

The beneficial effect of isothermal conditions on chemical reactions under microwave irradiation was investigated by performing the simple oxidation

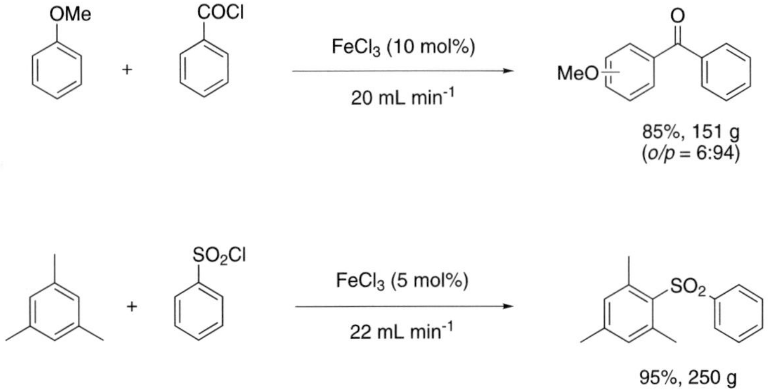

Scheme 19 Friedel–Crafts reactions in a prototype CF reactor

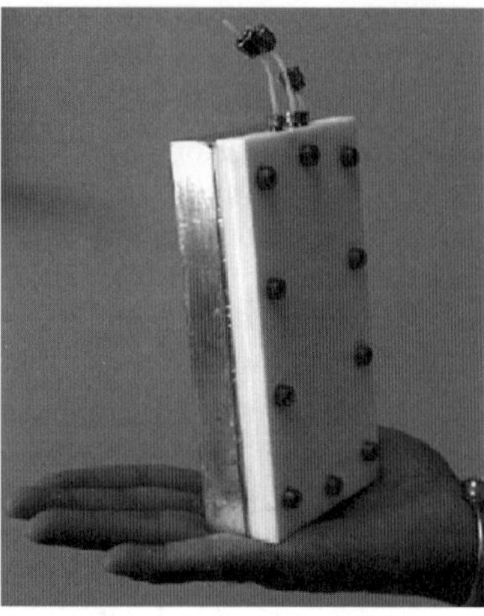

Fig. 23 Isothermal CF reactor. Reproduced with permission from [81]

Scheme 20 Oxidation of benzyl alcohol employing the novel isothermal CF reactor

of benzyl alcohol (Scheme 20). A range of residence times corresponding to different flow rates at varying microwave densities was applied to find the optimum conditions. Best results were obtained employing a residence time of 17 s, equal to a flow rate of $1\,mL\,min^{-1}$. Further development of this reactor might have some potential commercial value as dozens of inexpensive domestic microwave ovens can be used in parallel to produce chemical feedstock by isothermal microwave conditions.

Another rather unusual application of CF microwave processing within a CEM Discover unit was described by the group of Haswell [28, 29]. Microwave energy was used to deliver heat locally to a heterogeneous Pd-supported catalyst (Pd/Al$_2$O$_3$, catalyst channel: $1.5 \times 0.08 \times 15$ mm) situated within a microreactor device (Fig. 24). A 10–15 nm gold film patch, located on the outside surface of the base of a glass microreactor, was found to efficiently assist in the heating of the catalyst, allowing Suzuki cross-coupling reactions to proceed very effectively (Scheme 21). However, under these conditions the catalyst surface temperature is hard to estimate as it is simultaneously cooled by the reagent flow and heated by microwave absorption into, mainly, the catalyst.

In order to investigate those temperature sensing difficulties the same group have recently presented an electrical conductivity method for in situ temperature monitoring within the capillary flow reactor under microwave ir-

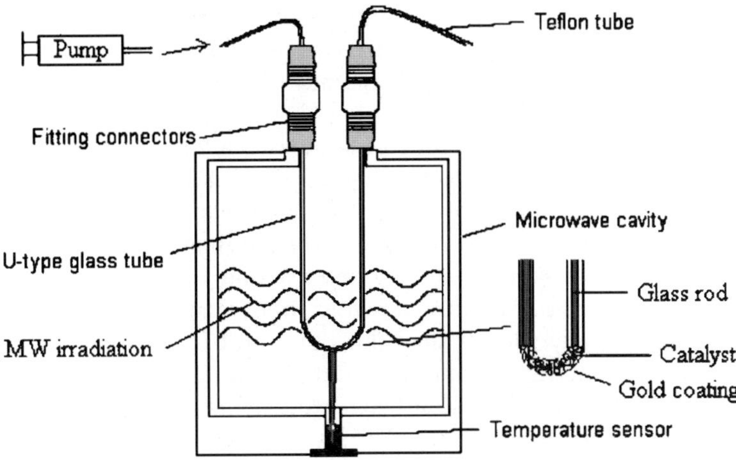

Fig. 24 Schematic diagram of the setup for MW-assisted coupling reactions. Reproduced with permission of [28, 29]

Scheme 21 Suzuki reactions in a microreactor environment

radiation [30]. With two electrodes positioned at both the inlet and the outlet of the capillary tube, the exact temperature of each section can be measured as well as the average temperature of the main U-shaped capillary. To demonstrate the suitability of the proposed methodology, the alkylation reaction of 2-pyridone with benzyl bromide was employed observing similar temperature values for the average conductivity measurements and for the IR sensor. However, the presence of a localized hot zone within the capillary was confirmed by showing that the outlet temperatures were significantly higher than the IR values.

3.3
Scale-Up Using Stop-Flow Methods

Two serious problems with CF reactors is the clogging of lines and the difficulties in processing heterogeneous mixtures. Since many organic transformations involve some form of insoluble reagent or catalyst, both single- and multimode so-called SF microwave reactors have been developed where peristaltic pumps – capable of pumping slurries and even solid reagents – are used to fill a batch reaction vessel (80–380 mL) with the reaction mixture.

After microwave processing in batch, the product mixture is pumped out of the system, which is then ready to receive the next batch of the reaction mixture. As this technique has been just recently developed only few published reports are available [27, 31].

In a recent publication, the group of Leadbeater examined the scale-up of Suzuki and Heck reactions in water using ultralow Pd catalyst concentrations (Scheme 22) [31]. This proof-of-concept study points out the advantages of the CEM Voyager SF system by combining the advantages of a batch reactor with those of a CF reactor. Since only one reaction vessel is used, the time taken to cool the reaction mixture down to room temperature at the end of the run is significantly shorter than those reported for parallel batch reactors employed in a multimode apparatus (20–30 min). Furthermore, applying conditions directly from small-scale batch experiments and transferring them to the Voyager SF system was possible with only minor modifications of the reaction conditions. Thus, reactions were scaled up from 1 mmol to ten cycles of 10 mmol giving comparable yields. Similar results were also found by Maes and coworkers when comparing single-mode SF with multimode parallel technique, as recently reported (Scheme 7 above) [27].

Critically evaluating the currently available instrumentation for microwave scale-up in batch and CF, one may argue that for processing volumes of < 1000 mL a batch process may be preferable. By carrying out sequential runs in batch mode, kilogram quantities of product can easily be obtained. When larger quantities of a specific product need to be prepared on a regular basis, it may be worthwhile evaluating a CF protocol. Large-scale CF microwave reactors (flow rate 20 L h^{-1}) are currently under development [82, 83]. However, at the present time there are no documented published examples of

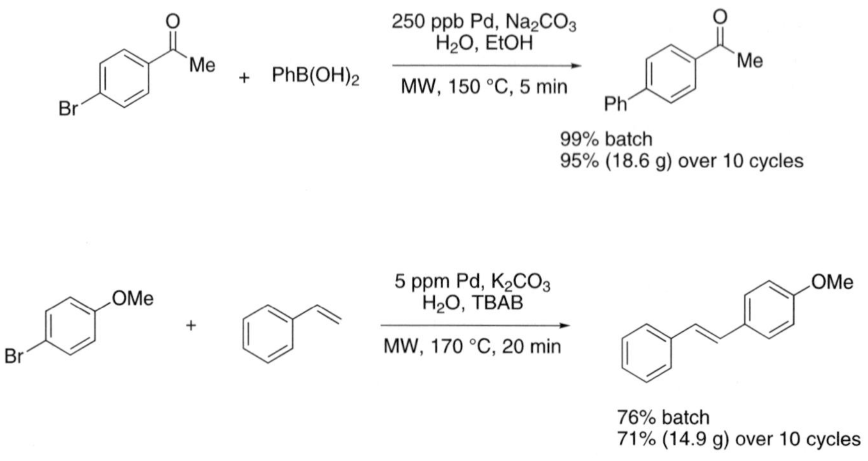

Scheme 22 Microwave-promoted Suzuki and Heck couplings for comparison of batch and stop flow techniques

the use of microwave technology for organic synthesis on a production scale level (> 1000 kg), which is a clear limitation of this otherwise successful technology [84].

4
Conclusion

After two decades of applying microwave irradiation to small-scale organic synthesis, the transformation to intermediate multigram scale has been conducted successfully. Various systems ranging from single-mode to larger multimode instruments enable the efficient multigram synthesis of materials without time-consuming re-optimization steps. Continuing efforts in scaling-up microwave-assisted reactions leading to multikilograms of products, and further to the semi-plant scale are necessary to extend the success story of MAOS. Industrial companies need plant reactors that can perform developed reactions with none or minimal re-optimization of their synthetic methods [64]. Therefore, ongoing investigations and comparisons of steady batch platforms with CF reactors are valuable investments for future instrument development. Due to the physical limitations of batch reactors it is likely that future approaches to the process and production scale will be performed in flow systems. As mentioned above, the penetration depth of microwaves at the established frequency of 2.45 GHz limits the maximum size of reactors. Unless there is a switch to different wavelengths, allowing increased penetration of reaction mixtures, reactors similar to the presented $1\,m^3$ prototype of Sairem [49] will be at the very end of the scale-up range. In daily routine, this reactor can produce kilogram amounts of compounds but the demand for much larger units is obvious. At the time of writing no time frame can be suggested when appropriate instrumentation for process scale-up will be available. Technicians are undoubtedly working hard together with chemical engineers and process scientist to investigate reasonable equipment, and viable solutions will probably be presented within the next couple of years.

Besides technical and physical limitations another important issue, especially for pharmaceutical laboratories, is the guarantee of "good manufacturing practice" (GMP) quality. Whenever new or unusual instrumentation is used in the process of preparing or producing drugs and lead compounds, it has to meet these stringent regulations in order to ensure the clean and reproducible synthesis of pure and contamination-free products. Particularly when using large parallel rotors for batch synthesis, there are still concerns whether the reaction outcome will be identical for all rotor positions. For the laboratory equipment, so far only experimental data can provide corresponding information [26, 27, 31, 82] and it has been shown that the multimode instruments available nowadays provide very homogeneous microwave fields [26, 82]. Thus, rotors with up to 48 parallel positions furnish identical results at all positions, with variations only within the standard deviation, nicely meeting the requirements of "good laboratory practice" (GLP). It is very likely that this homogeneity and reproducibility will also be achieved in future process-scale reactors, but the agreement with GMP rules has to be verified by physical and technical testing and only certified equipment will be allowed to be used in drug manufacturing processes.

However, within 20 years microwave technology has conquered significant areas of preparative organic and medicinal chemistry. Thus, with all the knowledge and experience accessed in these two decades there is no doubt that the principle of MAOS will shortly reach the next level.

References

1. Leadbeater NE (2004) Chem World 1:38
2. Adam D (2003) Nature 421:571
3. Kingston HM, Haswell SJ (eds) (1997) Microwave-enhanced chemistry. fundamentals, sample preparation and applications. American Chemical Society, Washington
4. Giberson RT, Demaree RS (eds) (2001) Microwave techniques and protocols. Humana, Totowa, NJ
5. Prentice WE (2002) Therapeutic modalities for physical therapists. McGraw-Hill, New York
6. Gedye R, Smith F, Westaway K, Ali H, Baldisera L, Laberge L, Rousell J (1986) Tetrahedron Lett 27:279
7. Giguere RJ, Bray TL, Duncan SM, Majetich G (1986) Tetrahedron Lett 27:4945
8. Loupy A, Petit A, Hamelin J, Texier-Boullet F, Jacquault P, Mathé D (1998) Synthesis, p 1213
9. Varma RS (1999) Green Chem 1:43
10. Kidawi M (2001) Pure Appl Chem 73:147
11. Varma RS (2001) Pure Appl Chem 73:193
12. Varma RS (2002) Tetrahedron 58:1235
13. Varma RS (2002) Advances in green chemistry: chemical syntheses using microwave irradiation. Kavitha, Bangalore
14. Bose AK, Banik BK, Lavlinskaia N, Jayaraman M, Manhas MS (1997) Chemtech 27:18
15. Bose AK, Manhas MS, Ganguly SN, Sharma AH, Banik BK (2002) Synthesis, p 1578
16. Strauss CR, Trainor RW (1995) Aust J Chem 48:1665
17. Strauss CR (1999) Aust J Chem 52:83
18. Strauss CR (2002) Microwave-assisted organic chemistry in pressurized reactors. In: Loupy A (ed) Microwaves in organic synthesis. Wiley, Weinheim, p 35
19. Perreux L, Loupy A (2001) Tetrahedron 57:9199
20. De la Hoz A, Diaz-Ortiz A, Moreno A (2005) Chem Sov Rev 34:164
21. Baghurst DR, Mingos DMP (1991) Chem Soc Rev 20:1
22. Gabriel C, Gabriel S, Grant EH, Halstead BS, Mingos DMP (1998) Chem Soc Rev 27:213
23. Mingos DMP (2005) Theoretical aspects of microwave dielectric heating. In: Lidström P, Tierney JP (eds) Microwave-assisted organic synthesis. Blackwell, Oxford, p 1

24. Hayes BL (2002) Microwave synthesis: chemistry at the speed of light. CEM, Matthews, NC
25. Bogdal D (2005) Microwave-assisted organic synthesis. one hundred reaction procedures. Elsevier, Oxford, p 11
26. Stadler A, Yousefi BH, Dallinger D, Walla P, Van der Eycken E, Kaval N, Kappe CO (2003) Org Process Res Dev 7:707
27. Loones KTJ, Maes BUW, Rombouts G, Hostyn S, Diels G (2005) Tetrahedron 61:10338
28. He P, Haswell SJ, Fletcher PD (2004) Lab Chip 4:38
29. He P, Haswell SJ, Fletcher PD (2004) Appl Cat A, p 111
30. He P, Haswell SJ, Fletcher PDI (2005) Sen Act B 105:516
31. Arvela RK, Leadbeater NE, Collins MJ Jr (2005) Tetrahedron 61:9349
32. Kappe CO, Stadler A (2005) Microwave processing techniques. In: Microwaves in organic and medicinal chemistry. Wiley, Weinheim, p 57
33. Bose AK, Manhas MS, Ganguly SN, Sharma A, Huarotte M, Rumthao S, Jayaraman M, Banik BK (2001) Article E0047. In: Kappe CO, Merino P, Marzinzik A, Wennemers H, Wirth T, Vanden Eynde JJ, Lin SL (eds) Fifth international electronic conference on synthetic organic chemistry. CD-ROM edition, ISBN 3-906980-06-5, MDPI, Basel
34. Besson T, Dozias MJ, Guillard J, Jacquault G, Legoy MD, Rees CW (1998) Tetrahedron 54:6475
35. Deetlefs M, Seddon KR (2003) Green Chem 5:181
36. Perio B, Dozias MJ, Hamelin J (1998) Org Process Res Dev 2:428
37. Cléophax J, Liagre M, Loupy A, Petit A (2000) Org Process Res Dev 4:498
38. Biotage (2006) http://www.biotage.com
39. CEM (2006) http://www.cemsynthesis.com (US website for CEM synthesis equipment); www.cem.com (main website), www.cem.de (German website)
40. Milestone (2006) http://www.milestonesci.com (US website), www.milestonesrl.com (Italian website), www.mls-mikrowellen.de (German website)
41. Anton Paar (2006) http://www.anton-paar.com
42. Kremsner JM, Kappe CO (2005) Eur J Org Chem 3672
43. Kaval N, Dehaen W, Kappe CO, Van der Eycken E (2004) Org Biomol Chem 2:154
44. Lehmann F, Pilotti Å, Luthman K (2003) Mol Diversity 7:145
45. Andappan MMS, Nilsson P, von Schenck H, Larhed M (2004) J Org Chem 69:5212
46. Shieh WC, Dell S, Repiè O (2001) Org Lett 3:4279
47. Shieh WC, Lozanov M, Repiè O (2003) Tetrahedron Lett 44:6943
48. Nüchter M, Ondruschka B, Bonrath W, Gum A (2004) Green Chem 6:128
49. Howarth P, Lockwood M (2004) The Chemical Engineer 756:29, for further information see http://www.sairem.com
50. Kappe CO (2004) Angew Chem Int Ed 43:6250
51. Kappe CO, Stadler A (2005) Microwaves in organic and medicinal chemistry, Wiley, Weinheim
52. Lidström P, Tierney JP (eds) (2005) Microwave-assisted organic synthesis. Blackwell, Oxford
53. Stadler A, Pichler S, Horeis A, Kappe CO (2002) Tetrahedron 58:3177
54. Iqbal M, Vyse N, Dauvergne J, Evans P (2002) Tetrahedron Lett 43:7859
55. Shackelford SA, Anderson MB, Christie LC, Goetzen T, Guzman MC, Hananel MA, Kornreich WD, Li H, Pathak VP, Rabinovich AK, Rajapakse RJ, Truesdale LK, Tsank SM, Vazir HN (2003) J Org Chem 68:267
56. Takvorian AG, Combs AP (2004) J Comb Chem 6:171
57. Silva AMG, Tomé AC, Neves MGPMS, Cavaleiro JAS, Kappe CO (2005) Tetrahedron Lett 46:4723

58. Raner KD, Strauss CR, Trainor RW, Thorn JS (1995) J Org Chem 60:2456
59. Kappe CO, Stadler A (2005) Equipment review. In: Microwaves in organic and medicinal chemistry. Wiley, Weinheim, p 29
60. Baxendale IR, Ley SV (2005) J Comb Chem 7:483
61. Hazuda DJ et al. (2004) Science 305:528
62. Merck & Co Inc (Anthony NJ et al.) (2002) World Patent WO-00230930
63. Merck & Co Inc (Anthony NJ et al.) (2002) World Patent WO-00230931
64. Wolkenberg SE, Shipe WD, Lindsley CW, Guare JP, Pawluczyk JM (2005) Curr Opin Drug Disc Dev 8:701
65. Cablewski T, Faux AF, Strauss CR (1994) J Org Chem 59:3408
66. Savin KA, Robertson M, Gernert D, Green S, Hembre EJ, Bishop J (2003) Mol Diversity 7:171
67. Wilson NS, Sarko CR, Roth G (2004) Org Process Res Dev 8:535
68. Shieh WC, Dell S, Repiè O (2002) Tetrahedron Lett 43:5607
69. Kazba K, Chapados BR, Gestwicki JE, McGrath JL (2000) J Org Chem 65:1210
70. Khadlikar BM, Madyar VR (2001) Org Process Res Dev 5:452
71. Esveld E, Chemat F, van Haveren J (2000) Chem Eng Technol 23:279
72. Esveld E, Chemat F, van Haveren J (2000) Chem Eng Technol 23:429
73. Pipus G, Plazl I, Koloini T (2000) Chem Eng J 76:239
74. Chen ST, Chiou SH, Wang KT (1990) J Chem Soc Chem Commun, p 807
75. Pillai UR, Sahle-Demessie E, Varma RS (2004) Green Chem 6:295
76. Bagley MC, Lenkins RL, Lubinu MC, Mason C, Wood R (2005) J Org Chem 70:7003
77. Glasnov T, Vugts DJ, Koningstein MM, Desai B, Fabian WMF, Orru RVA, Kappe CO (2006) QSAR Comb Sci 25:509. DOI 10.1002/qsar.200540210
78. Comer E, Organ MG (2005) Chem Eur J 11:7223
79. Comer E, Organ MG (2005) J Am Chem Soc 127:8160
80. Marquie J, Salmoriea G, Poux M, Laporterie A, Dubac J, Roques N (2001) Ind Eng Chem Res 40:4485
81. Jachuck RJJ, Selvaraj DK, Varma RS (2006) Green Chem 8:29
82. Nüchter M, Ondruschka B (2003) Mol Diversity 7:253
83. Bierbaum R, Nüchter M, Ondruschka B (2004) Chem Ing Techn 76:961
84. Hajek M (2002) Microwave catalysis in organic synthesis. In: Loupy A (ed) Microwaves in organic synthesis. Wiley, Weinheim, p 345

Author Index Volumes 251–266

The volume numbers are printed in italics

Escudé C, Sun J-S (2005) DNA Major Groove Binders: Triple Helix-Forming Oligonucleotides, Triple Helix-Specific DNA Ligands and Cleaving Agents. *253*: 109–148
Van der Eycken E, see Appukkuttan P (2006) *266*: 1–47

Fages F, Vögtle F, Žinić M (2005) Systematic Design of Amide- and Urea-Type Gelators with Tailored Properties. *256*: 77–131
Fages F, see Žinić M (2005) *256*: 39–76
Fechter EJ, see Dervan PB (2005) *253*: 1–31
Fensterbank L, see Albert M (2006) *264*: 1–62
Fernández JM, see Moonen NNP (2005) *262*: 99–132
Fernando C, see Szathmáry E (2005) *259*: 167–211
Ferrer B, see Balzani V (2005) *262*: 1–27
De Feyter S, De Schryver F (2005) Two-Dimensional Dye Assemblies on Surfaces Studied by Scanning Tunneling Microscopy. *258*: 205–255
Flood AH, see Moonen NNP (2005) *262*: 99–132
Fujiwara S-i, Kambe N (2005) Thio-, Seleno-, and Telluro-Carboxylic Acid Esters. *251*: 87–140

Gansäuer A, see Daasbjerg K (2006) *263*: 39–70
Garcia-Garibay MA, see Karlen SD (2005) *262*: 179–227
Gelinck GH, see Grozema FC (2005) *257*: 135–164
George SJ, see Ajayaghosh A (2005) *258*: 83–118
Gerenkamp M, see Daasbjerg K (2006) *263*: 39–70
Giaffreda SL, see Braga D (2005) *254*: 71–94
Grepioni F, see Braga D (2005) *254*: 71–94
Grimme S, see Daasbjerg K (2006) *263*: 39–70
Grozema FC, Siebbeles LDA, Gelinck GH, Warman JM (2005) The Opto-Electronic Properties of Isolated Phenylenevinylene Molecular Wires. *257*: 135–164
Guiseppi-Elie A, Lingerfelt L (2005) Impedimetric Detection of DNA Hybridization: Towards Near-Patient DNA Diagnostics. *260*: 161–186
Di Giusto DA, King GC (2005) Special-Purpose Modifications and Immobilized Functional Nucleic Acids for Biomolecular Interactions. *261*: 131–168

Hansen SG, Skrydstrup T (2006) Modification of Amino Acids, Peptides, and Carbohydrates through Radical Chemistry. *264*: 135–162
Heise C, Bier FF (2005) Immobilization of DNA on Microarrays. *261*: 1–25
Heitz V, see Collin J-P (2005) *262*: 29–62
Higson SPJ, see Davis F (2005) *255*: 97–124
Hirst AR, Smith DK (2005) Dendritic Gelators. *256*: 237–273
Holzwarth AR, see Balaban TS (2005) *258*: 1–38
Houk RJT, Tobey SL, Anslyn EV (2005) Abiotic Guanidinium Receptors for Anion Molecular Recognition and Sensing. *255*: 199–229
Huc I, see Brizard A (2005) *256*: 167–218

Ihmels H, Otto D (2005) Intercalation of Organic Dye Molecules into Double-Stranded DNA – General Principles and Recent Developments. *258*: 161–204
Indelli MT, see Chiorboli C (2005) *257*: 63–102
Inoue Y, see Borovkov VV (2006) *265*: 89–146
Ishii A, Nakayama J (2005) Carbodithioic Acid Esters. *251*: 181–225

Subject Index

Printing: Krips bv, Meppel
Binding: Stürtz, Würzburg